Medieval Warfare: Technology, Military Revolutions, and Strategy

This volume explores the topics of military revolutions, strategy, and technology both separately and as they relate to each other. It makes important contributions to understanding European warfare in the Early, High, and especially the Late Middle Ages, as well as the military transition to the Early Modern Period.

Readers will find detailed analysis of how technological and non-technological developments interacted to effect major changes in how wars were fought across the period. The evolution and capabilities of the English longbow and of early gunpowder artillery are examined in depth. Changes in the tools of war naturally affected plans to employ those tools to achieve political ends – military strategy – but strategy was never dictated by technology. That point is illustrated by examinations of English efforts to conquer Wales; the Anglo-Burgundian alliance of the late Hundred Years War; and the economic factors shaping medieval conquests in general.

The nine studies in the volume have all been published previously, but a new introduction shows how they fit together, particularly explaining how they collectively rebut common critiques of Rogers's controversial thesis that European warfare was reshaped by the Infantry and Artillery Revolutions during the era of the Hundred Years War. Two of the chapters have been substantially expanded, so that the versions printed here should be the ones consulted and cited in the future by scholars of medieval warfare and military revolutions.

Clifford J. Rogers is Professor of History and co-Director of the Digital History Center at West Point. He is the author, editor, or co-editor of thirteen books, and twenty volumes of the *Journal of Medieval Military History*. His work has been honored with a Society for Military History Distinguished Book Award, Moncado Prize, and digital military history award; the Royal Historical Society Alexander Prize; three Army Historical Foundation Distinguished Writing Awards; two Verbruggen book prizes and the Bachrach Medal from De Re Militari; and the USMA Dean's Award for Career Teaching Excellence.

Varioum Collected Studies

Also in the Variorum Collected Studies Series

CLIFFORD J. ROGERS
Medieval Warfare: Technology, Military Revolutions, and Strategy (CS1125)

LIZ JAMES
Mosaics, Empresses and Other Things in Byzantium
Art and Culture 330–1453 (CS1124)

DAVID NEAL GREENWOOD
Studies in Late Antiquity (CS1123)

ARTHUR HOLDER
Biblical Exegesis and Mystical Theology in the Venerable Bede (CS1122)

WOLFHART HEINRICHS edited by Hinrich Biesterfeldt and Alma Giese
Wolfhart Heinrichs' Essays and Articles on Arabic Literature
Authors, Semitic Studies, and Islamic Jurisprudence (CS1121)

WOLFHART HEINRICHS edited by Hinrich Biesterfeldt and Alma Giese
Wolfhart Heinrichs' Essays and Articles on Arabic Literature
General issues, Terms (CS1120)

SONJA BRENTJES
The Sciences in Islamicate Societies in Context
Patronage, Education, Narratives (CS1119)

MICHAEL LECKER
Studies on the Life of Muhammad and the Dawn of Islam
Idol Worshippers, Christians and Jews in Pre- and Early Islam (CS1118)

DONALD F. DUCLOW
Engaging Eriugena, Eckhart and Cusanus (CS1117)

VICTOR MALLIA-MILANES
The Winged Lion and the Eight-Pointed Cross
Venice, Hospitaller Malta, and the Mediterranean in Early Modern Times (CS1116)

SVETLANA KUJUMDZIEVA
Studies on Eastern Orthodox Church Chant (CS1115)

SONJA BRENTJES
Historiography of the History of Science in Islamicate Societies
Practices, Concepts, Questions (CS1114)

www.routledge.com/Variorum-Collected-Studies/book-series/VARIORUM

Medieval Warfare: Technology, Military Revolutions, and Strategy

Clifford J. Rogers

Routledge
Taylor & Francis Group

LONDON AND NEW YORK

First published 2025
by Routledge
4 Park Square, Milton Park, Abingdon, Oxon OX14 4RN

and by Routledge
605 Third Avenue, New York, NY 10158

Routledge is an imprint of the Taylor & Francis Group, an informa business

British Library Cataloguing-in-Publication Data
A catalogue record for this book is available from the British Library

ISBN: 978-1-032-50851-1 (hbk)
ISBN: 978-1-032-50852-8 (pbk)
ISBN: 978-1-003-39997-1 (ebk)

DOI: 10.4324/9781003399971

Typeset in Times New Roman
by Apex CoVantage, LLC

VARIORUM COLLECTED STUDIES SERIES CS1125

CONTENTS

INTRODUCTION

The first scholarly article I ever wrote, which was published in 1993, argued for two military revolutions during the period of the Hundred Years War: an Infantry Revolution in the fourteenth century, and an Artillery Revolution in the fifteenth, both of which it described as part of a larger pattern of "punctuated equilibrium evolution" in military affairs. The essay was well received at the time, and has been cited in hundreds of other works since then.[1] One reason for its popularity was that it tried to make sense of a long period of military history (including how military changes affected larger historical developments) in relatively few pages.[2] That same fact, however, meant that the ideas it presented were not as fully developed, and the evidence offered in support of its claims was not as extensive, as would have been possible if it had been a monograph rather than an article. I think that was at least one reason why a significant minority of the later works that referred to it did so in order to dispute its facts, its analysis, or its conclusions. There were two main lines of attack: the first argued that my proposed "revolutions" were only periods of rapid evolution, and the second claimed that my analysis suffered from "technological determinism."

The first six chapters of this book all derive, at least in part, from my reflection on these criticisms. Chapter one, "The Idea of Military Revolutions in Eighteenth- and Nineteenth-Century Texts," puts my arguments for an Infantry and an Artillery Revolution into the context of an intellectual tradition that was already deep and broad long before Michael Roberts put forward his famous thesis regarding an Early Modern military revolution in 1956. Of course, coming from a long historiographical tradition is far from being a guarantee of correctness. And, as the

---

1   "The Military Revolutions of the Hundred Years' War," *The Journal of Military History* 57 (1993): 241–278. Reprinted with revisions in C. J. Rogers, ed. *The Military Revolution Debate* (Boulder: Westview, 1995), and reprinted in Paul E. J. Hammer, ed. *Warfare in Early Modern Europe, 1450–1660* (London: Ashgate, 2007). The article received the Society for Military History Moncado Prize in 1994, and as of 2023 has been cited in some 350 other works, according to Google Scholar.

2   Even in an article devoted to attacking my version of the Infantry Revolution thesis, John Stone admits it "is seductive in its parsimony" and "tempting." John Stone, "Technology, Society, and the Infantry Revolution of the Fourteenth Century," *The Journal of Military History* 68 (2004), 368.

DOI: 10.4324/9781003399971-1

chapter demonstrates, there has also been a long history of *disputing* putative military revolutions, both retrospective and prospective ones. In 1906, for example, historian W. C. H. Wood scoffed at "people who are all agog with every newspaper nonsense-tale about the something or other that is always going to 'completely revolutionize' the art of war." Strategic principles, he asserted, were timeless, while "tactics have always been undergoing the same slow process of evolution."[3]

In line with Wood's views, the most common criticism of the Infantry Revolution part of my article has been that it exaggerated both the magnitude and the rapidity of the changes it described. One line of thought held that the thesis of an Infantry Revolution in the fourteenth century rested on an underestimation of the role of footmen in the preceding centuries.[4] The more extreme version of this position asserted not only that infantry was *important* during the High Middle Ages (which actually I had explicitly said myself),[5] but that it was the "dominant" arm, tactically as well as numerically, across the whole period.[6] The age of cavalry dominance before the fourteenth century, several historians argued, was just a "myth."[7] Although this idea came to be applied much more broadly, its origin was the work of Bernard S. Bachrach on the Carolingian period.[8] I have not yet published a comprehensive defense of the traditional idea that in High Medieval warfare the cavalry was the "decisive," "dominant," "predominant," "preponderant," or "principal" arm,[9] but as a first step in that direction I reexamined the (scanty) evidence for the tactical role of horsemen in the Carolingian period. As explained in the second chapter of this volume, my finding was that Bachrach was wrong to say that "there is no reason to believe" that mounted

---

3 In this volume, p. 27.

4 E.g. Nicholas Hooper and Mathew Bennett. *Cambridge Illustrated Atlas [of] Warfare. The Middle Ages, 768–1487* (Cambridge: Cambridge U.P., 1996), 156–7; Stephen Morillo, Jeremy Black, and Paul Lococo, *War in World History* (Boston, McGraw Hill, 2009), 230–1 ("radically overdrawn and misleading" revolution thesis; actually "no major change in tactical fundamentals" around 1300, only a "slow shift to more effective infantry.") Alexander Querengässer, *Before the Military Revolution. European Warfare and the Rise of the Early Modern State, 1300–1490* (Oxford: Oxbow, 2021), 151–3.

5 Rogers, "Military Revolutions," in Rogers, ed., *The Military Revolution Debate*, 57: "infantry could be very important"; 58: "In the thirteenth century, infantry played an important role on the battlefield."

6 Bernard S. Bachrach and David S. Bachrach, *Warfare in Medieval Europe, c. 400–c. 1453* (New York: Routledge, 2017), 276.

7 For example in James G. Patterson, "The Myth of the Mounted Knight," in *Misconceptions About the Middle Ages,* ed. Stephen Harris and Bryon L. Grigsby (New York: Routledge, 2010), p. 91, and Matthew Bennett, "The Myth of the Military Supremacy of Knightly Cavalry," in Matthew Strickland, ed., *Armies, Chivalry and Warfare* (Stamford: Paul Watkins, 1998), 315–16.

8 Including Bernard S. Bachrach, "Charlemagne's Cavalry: Myth and Reality," *Military Affairs* 47 (1983): 181–87, and idem, "Charles Martel, Mounted Shock Combat, the Stirrup, and Feudalism," *Studies in Medieval and Renaissance History* 7 (1970): 49–75.

9 Karl Marx and Friedrich Engels, *Collected Works,* vol. 18 (Moscow: Progress Publishers, 1982), 104, and Charles Oman, *A History of the Art of War in the Middle Ages* (London: Methuen, 1898), 355; R. A. Brown, *Origins of Modern Europe* (Woodbridge: Boydell, 1972, repr. 1996), 116; Oman, *History of the Art of War,* 16, 268, 355; Geoffrey Parker, "The Military Revolution," in Rogers, ed., *The Military Revolution Debate,* 44.

forces were the most effective part of Charlemagne's army.[10] Despite the wide acceptance of Bachrach's views in recent years, the limited evidence available actually makes it clear that by 782 (just fifty years after Charles Martel's army won the battle of Poitiers or Tours in 732 fighting on foot) cavalry was at least *perceived* by the Frankish elite to be the most important arm of their forces, so much so that they were willing to send forces composed entirely of cavalrymen, without any infantry, to conduct major operations. They seem to have considered a cavalry charge as their "accustomed way" of fighting, and they clearly believed that a well-executed mounted attack could be expected to rout a numerically superior force of infantry. In the ninth century, our sources show even more clearly that the Franks thought of fighting on horseback as their normal and preferred mode of combat, though they were willing and able to dismount and fight on foot when the situation demanded it. Cavalry, in short, had become the tactically dominant arm of the Franks, who were the dominant military power of western Europe. And, although this is not demonstrated within the chapter, cavalry remained the principal arm of most European armies for the following four centuries.[11]

When the ascendancy of the cavalry was finally overturned in the fourteenth century, the battle that was most important in catalyzing change was the Flemish victory of Courtrai in 1302. That engagement was won by close-order heavy infantry wielding long spears and spiked clubs called *goedendags*, fighting on very favorable terrain.[12] But the most dramatic victory won by footmen, the one that firmly established the revolution, was Crécy in 1346, where Edward III of England defeated one of the largest armies the French monarchy ever assembled during the Middle Ages. Indeed, despite being outnumbered by around 2:1, Edward did not just defeat a French royal army, but crushed it, with only minimal English losses suffered in return. Here too a phalanx of armored footmen (in this case dismounted armored cavalrymen) played an important role. But the real shock to European opinion came from the frightful effectiveness of the English archers, who launched long, heavy arrows from powerful longbows, first routing the mercenary crossbowmen serving the French crown, then decimating the charging French horsemen with volley after volley. These arrows not only wreaked havoc on the men-at-arms' warhorses, they killed or wounded well-armored knights and esquires by the dozen and the hundred, "their armor, of whatever sort, rarely

---

10 Bachrach, "Charlemagne's Cavalry," 184.

11 See my articles in Jeremy Black, ed. *Cavalry Warfare from Ancient Times to Today* (Rome: Fucina de Marte 18, 2024), 59–82, and forthcoming in the *Journal of Military History*.

12 Since I made Courtrai's importance quite clear in my 1993 article (as Stone recognizes on p. 372), and since I nowhere suggested that the success of the Flemings was due to the technology they employed, I cannot understand why Stone thinks that "in early formulations of his thesis, Rogers evidently viewed technological innovation as a prime mover in the Infantry Revolution," or that I "performed something of a *volte face*" or made a point "contrary to [my] original views," when in a slightly more recent article I said specifically that the Infantry Revolution was "*not* based primarily on technological innovations." Stone, "Infantry Revolution," 370, 371.

preventing it."[13] By the time the day was done, 1,542 noblemen lay dead in front of the Black Prince's position, in addition to commoners and to those killed on other parts of the field or borne away during the fighting. In sharp contrast, the English knights and esquires slain numbered only three, and the dead of all ranks fewer than fifty.[14] Because this result was so lopsided and so unexpected, the battle has been seen by many historians as emblematic of the Infantry Revolution, which in turn makes it seem that the longbow was even more important an element of the broader Infantry Revolution than it actually was.

Although I certainly do not think that the development of the longbow was the *driver* of the Infantry Revolution – that would conflict with simple chronology, since the cause must come before the effect – I do think the weapon was a distinct innovation of the fourteenth century: a far more powerful bow than the shorter selfbows used in Europe during the previous century. In design and in effectiveness it represents a sharp discontinuity with the past. Such discontinuities are at the core of the nature of "revolutions," and although the longbow was not the prime mover of the Infantry Revolution, the rise of infantry tactics would likely not have been rapid or widespread enough to qualify as a military revolution without the impetus of the victories won by the English using it. The defeat at Courtrai did not cause the French to change their tactics, but after Crécy even their men-at-arms switched from mostly fighting on horseback to mostly doing battle on foot. So my understanding of the "revolutionary" nature of the Infantry Revolution rests in some part (though certainly not entirely) on the longbow being a new and highly effective weapon. Both of those characteristics of the longbow, however, have been challenged.

Readers who are familiar with the broad outlines of the Hundred Years War but have not been keeping up with the recent scholarship may find that surprising. It might seem that the undisputed and unquestionably discontinuous string of English battlefield victories against the French, the Scots (who, it should be remembered, had repeatedly defeated the English during the reign of Edward's father), and everyone else who fought against them from 1332 to 1424, would be *prima facie* evidence that the longbow was something new and was highly effective. Nonetheless, based in part on a flawed interpretation of the results of methodologically flawed tests,[15] Kelly DeVries claimed that longbow arrows

---

13 Adam Murimuth, *Continuatio chronicarum*, ed. E. M. Thompson (London: Rolls Series, 1889), 247: "sagittis densissime intervenientibus, quae non minus hominum quam equorum viscera sunt scrutatae, armatura qualimcumque raro prohibente." For Crécy as a whole, see Andrew Ayton and Sir Philip Preston, *The Battle of Crécy, 1346* (Woodbridge: Boydell, 2005); for the effectiveness of the longbow more broadly, see Clifford J. Rogers, "The Efficacy of the Medieval Longbow: A Reply to Kelly DeVries," *War in History* 5 (1998): 233–42.

14 *The Battle of Crécy: A Casebook*, ed. Michael Livingston and Kelly DeVries (Liverpool: Liverpool U.P., 2016), 58, 60, 213, 225; Clifford J. Rogers, *Soldiers' Lives Through History: The Middle Ages* (New York: Greenwood, 2007), 215, 228–30.

15 Kelly DeVries, "Catapults are Not Atom Bombs: Towards a Redefinition of 'Effectiveness' in Premodern Military Technology," *War in History* 4 (1997), 462, says that Peter Jones's "findings show that unless an arrow discharged from a longbow hits an armoured target at a 90° angle, which

were rarely capable of killing armored enemy soldiers, or indeed of penetrating armor at all except at short range, and then usually only minimally.[16] DeVries has also disputed the novelty of the longbow, joining a small chorus of historians who claim (in the words of Matthew Strickland, who has made the most developed argument in support of this thesis) that archers throughout the Middle Ages used "essentially the same weapon" employed in 1346, and that the bows of Crécy were no more powerful than those used at Bourgthéroulde (1124) or the Standard (1138).[17] DeVries and John Stone, finally, have also accused me of "technological determinism" and "removing the human factor" from warfare because of my analysis of the effects of the longbow and of the broader Infantry Revolution.[18]

Thus, to sustain my position I had to provide fuller evidence both that the longbow was indeed a lethal weapon, even against armored combatants, and that earlier European bows were very much less so. The former task was relatively easy, especially with respect to the fourteenth century, since there is abundant testimony from contemporary sources of the effectiveness of the English longbowmen, and

---

would rarely have occurred, the amount of penetration into either chain or plate armour would have been minimal." But there are two problems there. First, Jones's tests were methodologically flawed, because the arrows were shot from a longbow of only 70 pounds draw weight at a draw length of only 28", imparting only around 45–55 joules to the projectiles – less than half what a median *Mary Rose* bow could do. Robert Hardy, *Longbow: A Social and Military History*, 3rd ed. (Goldthwaite: Bois d'Arc Press, 1992), 231; Peter N. Jones, "The Metallography and Relative Effectiveness of Arrowheads and Armor During the Middle Ages," *Materials Characterization* 29 (1992), 115–16. Second, Jones actually found that up to 45° *from the normal* (i.e. across a 90° arc of incoming trajectories) the arrowhead in his tests still fully penetrated the plate. The penetration was minimal for a hit at *60° from the normal*, i.e. at very oblique angle (30° from the plane of the armor's face, since the arrowhead then broke after only the very tip was through the plate. In Hardy, *Longbow*, 235, photographs and captions, and 236. Moreover, far from finding that "penetration into . . . chain . . . armour would have been minimal," Jones did not test mail armor, but did say "it is easy to see why mail armour was *less effective* than plate." Ibid., 236. For confirmation from subsequent testing that longbow arrows, unlike arrows from ordinary shortbows, could be lethal through mail and padding, see in this volume, p. 56.

16 DeVries suggests that at Agincourt (and, apparently, "in all the battles since the beginning of the fourteenth century") longbowmen did not do "any more damage than the killing of a few horses and the wounding of even fewer men," and more specifically "did not kill many men." DeVries, "Catapults," 462; idem, *Medieval Military Technology* (Peterborough, ONT: Broadview, 1992), 38. In the second edition of that book, co-authored with Robert Douglas [Kay] Smith, it is said that "it is unlikely that arrows were able to penetrate armor, except at . . . less than 50 yards." (Toronto: University of Toronto Press, 2012), 40. Minimally: see the previous note in this introduction.

17 See pages 44 and 54.

18 Stone, in "Infantry Revolution," says that I "endow technology . . . with deterministic qualities," (368) and that my treatment of the longbow and the Infantry Revolution makes me "vulnerable to the same forms of criticism that the work of [Lynn] White and [Geoffrey] Parker has attracted" (368), after earlier in the article associating Parker (367, 366, 364) and White (365) with "technological determinism." DeVries, "Catapults," 455, includes me in a group of historians who, he says, show "an almost constant willingness to describe premodern military technology in . . . deterministic terms," which "has also too often and too easily removed the individual soldiers and their leaders from the military historical equation, replacing them with a technological, deterministic explanation."

since the revisionist position would make the outcome of the battles of Crécy, Poitiers, Agincourt, and Verneuil (among others) simply inexplicable. I made that case in an article published in 1998.[19] But was the sharp change in what the English archers could accomplish actually the result of their employing a new, more powerful weapon, or was it, rather, just a consequence of much larger-scale recruitment and massed tactical employment of bowmen, with no significant technological change involved? The latter would be a logical conclusion if it were true that, as Matthew Strickland and others have argued, the idea of a transition from the shortbow to the longbow was merely a "myth."[20] So in "The Development of the Longbow in Late-Medieval England and 'Technological Determinism,'" the third chapter of this volume, I mustered archaeological, documentary, iconographic, and chronicle evidence to demonstrate that in fact the traditional view was essentially correct. Short bows and other "ordinary bows" (standing no more than shoulder high and allowing for the use only of relatively short arrows, launched with relatively little power) were definitely in use throughout the High Middle Ages. Indeed, they were practically the only kinds of selfbows employed in the realms of the European "core" in those three centuries. The transition to the longbow in England, I found, was even more sudden than previously believed. The development of the 6'-plus longbow, capable of sustaining a 30–32" draw to the ear, is often associated with the Welsh Wars of Edward I (1277–83). In fact, however, there is no unambiguous evidence that the longbow was in use by the English until the reign of Edward II (1307–1327). Even presuming a rapid phasing-out of the ordinary bow for military use thereafter, it is therefore likely that the first battles in which most of the English archers employed proper longbows were also the first major victories over the Scots in a generation, and the first in a century-long string of almost uninterrupted tactical successes: Dupplin Moor in 1332, and Halidon Hill in 1333.

The version of that article published in the *Journal of Medieval History* had to be trimmed rather heavily to meet the journal's word-count maximum. Here it is presented in substantially expanded form, with in particular a much fuller treatment of the iconographic evidence for change. For this version, in addition to restoring cuts made to the original draft, I have provided more archaeological and documentary evidence of High-Medieval European selfbows of less-than-longbow stature, and also noted some new experimental results that confirm what the narrative sources tell us: that the long, heavy arrows of fourteenth-century longbows could achieve far greater results against mail-armored men than could ordinary selfbows with ordinary arrows (or even strong Eastern composite bows of the sort encountered by Western knights during the crusades).

The same article also addresses the accusation of technological determinism with respect to my analysis of the Infantry Revolution. I emphasize that

---

19 Rogers, "Efficacy."
20 Matthew Strickland and Robert Hardy, *The Great Warbow: A History of the Military Archer* (Stroud: Sutton Publishing, 2005), ch. 2.

acknowledging that technological change can be *important* – an assertion that no one seriously disputes – is far from the same as claiming that new weapons by themselves "determine" outcomes of battles or wars, or the course of socio-political developments.[21] That is even true in the case of military revolutions that (unlike the Infantry Revolution or the military transformation resulting from the French Revolution) *were* technologically driven, such as the fifteenth-century Artillery Revolution and the sixteenth-century Artillery Fortress Revolution, the subjects of the fourth chapter in this volume. In addition to again addressing concerns of "technological determinism," that study digs deeper into the question of what makes a technological-military change qualify as "revolutionary": essentially, the magnitude and rapidity of the change it involves. An important point made is that the presence of evolution does not disallow the possibility of revolution. Both cannon and anti-cannon defenses certainly did change over time in evolutionary ways. But, as with the gradual addition of more and more weight to one side of a balance scale, evolution can lead to sudden revolution when a tipping point is reached – in these two cases, a tipping point that shifts the operational and strategic-level advantage from the defense to the offense, or vice versa. No revolution "spring[s] fully formed from the head of Zeus," but that does not mean there are no revolutions.[22]

Perhaps the best illustration of how evolution can produce revolution is laid out in detail in the fifth chapter, which traces the changes in the design and employment of gunpowder artillery across its first 175 years of use in Europe, from 1326 to the end of the Middle Ages. It has been a commonplace since the eighteenth century that the introduction of firearms "changed the whole art of war" and was part of the "general revolution . . . in human affairs" that ushered in modernity.[23] But the early cannon were not really superior, overall, to existing weapons, and actually for at least a half-century they had only a marginal impact on the conduct of warfare, even though they were rapidly evolving in design and increasing in efficacy. "Gunpowder Artillery in Europe, 1326–1500: Innovation and Impact" shows that during the first decade of the fifteenth century the trajectory of growing power for individual cannon witnessed a sharp inflection, from a linear to a geometric progression.[24] Even the following leap in the strength of artillery, however, did not immediately bring about a revolution in *results*. That came after another decade or two of gun development, combined with changes in firing methods that allowed for more shots per hour, and a steady increase in the number of big bombards in use. These evolutionary improvements interacted with the growing punch of individual guns in a *multiplicative* way, so that from 1405 to 1425, the

---

21 Cf. DeVries, who says "it has been thought that this weapon [the longbow] alone determined many victories for England." "Catapults," 461.

22 Cf. DeVries, "Catapults," 466.

23 See the following, ch. 1, p 16, quoting David Hume.

24 The version published in this volume is mostly the same as the one previously published in a Festschrift for Joe Guilmartin, but with a dozen or so minor improvements and additions made.

average daily destruction that could be inflicted on a besieged fortress in a major siege increased probably somewhere between twenty-five-fold and fifty-fold.[25] This rapidity and magnitude of change does much to explain the equally swift and dramatic transformation of the practical conduct of war that began to be observable once the tipping point was reached in the 1420s. Fortresses that formerly would have been almost impossible to capture except by starvation now could be battered into submission in a matter of weeks, and campaigns of conquest that demanded the success of multiple major sieges in a single campaigning season became feasible once again (though of course not guaranteed to succeed.)

Obviously, not every development in warfare is part of a mutually reinforcing network of changes that leads to a non-linear quantum leap in effectiveness. Much of the time, warfare in general, and weapons and tactics in particular, evolve in more steady and gradual ways. If that were not true, then there would be no point in singling out a few revolutions as salient exceptions to the norm. The sudden rise in the importance of footmen in the fourteenth century may have been revolutionary, but the subsequent changes in how and with what equipment infantry fought during the following three centuries was, in my opinion, essentially evolutionary.[26] Big changes certainly did occur. First there was the largely separate development of pike squares and arquebuses; then their combination in refined pike-and-shot formations; then (as some have considered revolutionary) the move to small units arrayed in multiple thinner lines, with intervals between formations; and finally (also pitched as a revolution by at least one scholar) the move to true "linear tactics" when the invention of the socket bayonet for the flintlock musket made it possible for a single soldier to assume the functions both of pike and of shot. These processes of change, along with simultaneous and interrelated developments in cavalry, field artillery, and combined-arms tactics, are explained in the sixth chapter, "Tactics and the Face of Battle, 1350–1750." This revised version of an essay originally published in 2010 is significantly expanded, especially by the addition of more supporting or illuminating quotations in the footnotes.

***

The second article I ever published, three decades ago, argued that, contrary to the prevailing historiography, Edward III of England consistently employed a battle-seeking strategy in his wars with Scotland and France. Unlike the first

---

25　This calculation follows the same logic and uses the same information as the one in this volume, p. 182, but for a narrower time-period; 20 to 50 is based on 3 times as much powder per shot * powder 1.5 times as powerful * 2.5 times as many guns * 2 to 4 times as many shots per day, and allowing a small increase for the larger proportion of the power of the powder captured by longer barrels.

26　The same goes for the development of gunpowder artillery for four centuries after the mid-fifteenth century, until the advent of smokeless powder, rifled barrels, high explosive shells, and hydrostatic recoil systems, which in combination (again with a multiplicative interaction) again revolutionized warfare.

article, it provoked very little disagreement. The general acceptance of my argument, which I developed more fully in my first monograph, *War Cruel and Sharp: English Strategy under Edward III, 1327–1360*, gave me some reputation as an expert on medieval strategy, and in consequence I was asked by one of my mentors, Williamson Murray, to contribute to two volumes he co-edited for Cambridge University Press. Each was a collection of case studies intended to cast historical light on a subject of concern to the contemporary defense analysis community. The first was commissioned in 2011, when the U.S. military was attempting a "surge" in Afghanistan, where after a decade of conflict strategic success was proving elusive.[27] Some in the Pentagon seemed to be questioning whether it was even possible to craft a strategic plan that could be carried out largely as designed and achieve success – not just in Afghanistan, but in general. The editors therefore wanted to examine "successful strategies," which came to be the title of the book. To argue, as I did in *War Cruel and Sharp*, that Edward III was a master strategist who sagely employed military force to achieve his political objectives, I had needed to assemble data-points from many sources and, to some extent, to read backwards from actions to intent. There were no extant council minutes or strategic memoranda laying out prospectively how he proposed to win his wars. I could not examine another successful (or unsuccessful) medieval commander using the same approach for this project, because the emphasis was on prior planning.

Aside from plans prepared for crusades, there are not many written strategic plans surviving from the Middle Ages, but the earliest one I know of, Giraldus Cambrensis's analysis of how to conquer Wales, seemed as if it might provide a suitable topic. It was quite detailed, and had the added benefit of resonating with current affairs. Wales and the Welsh posed some of the same problems for the English crown that Afghanistan and the Afghan resistance posed to U.S.-led coalition forces. The terrain was difficult; the adversaries were tough, motivated, and unwilling to fight conventional battles that would play to their enemies' strengths. A concern was that Giraldus wrote his text in 1194, and when King John invaded Wales in 1209, intending to "entirely divest" Prince Llywelyn of his dominion, he neither followed the plan nor succeeded. But I had the impression that Edward I, who *did* succeed in conquering Wales in 1276–1277, had largely adhered to what Giraldus advised. Further research revealed that to be correct, but understated: in fact, Edward followed the plan very closely indeed, so closely that I had no qualms about explaining his success as a consequence of his employment of an exceptionally wise and thorough prospective strategic plan.

The second of the two volumes of strategy-oriented case studies I contributed to in the 2010s originated with dissatisfaction at how the United States had struggled with effective coalition warfare in recent conflicts, despite the nation's long history of working very successfully within alliance frameworks. To some in 2015, America's recent military past seemed characterized by hubris and insufficient

---

27 Only very minor changes or corrections have been made in the version published here.

understanding of what it took to be a good partner to allied nations. The volume's editors, however, were confident that "in the future the United States will need alliance and coalition partners to achieve its strategic goals," and that a better understanding of "the crucial importance that alliances and coalitions have played in the conduct of strategy in peace and war over the centuries" could help the soldiers and policy-makers responsible for overseeing America's military and diplomatic policies and practices do better at their jobs.[28] I agreed to contribute an essay on the creation, sustainment, and eventual collapse of what was probably the most significant of all the many, shifting alliances formed during the Hundred Years War: the Anglo-Burgundian partnership of 1419–1435.[29] The creation of that partnership created the very real possibility that England and France would be lastingly united in a dual monarchy; the end of the alliance restored the possibility (eventually realized) that France would win the war outright.

Neither the foundation of the Anglo-Burgundian coalition, nor its coherence after the early death of Henry V, nor its final collapse can be said to have largely been shaped by great historical tides, or by the dictates of *Realpolitik*. Instead, the alliance was born of the confluence of two unpredictable contingencies: first, the extraordinary military successes of Henry V at Agincourt (1415) and during his subsequent operations in Normandy; second, the murder in 1419 of John the Fearless, duke of Burgundy, by the entourage of the dauphin (the future Charles VII). The success of the alliance in weathering a series of subsequent crises can only be properly understood in terms of the personalities, individual priorities, family relationships, and decisions of three particular individuals: Philip the Good of Burgundy; John of Bedford, regent of France for his nephew the young Henry VI; and Anne of Burgundy, who was sister to the first and wife to the second. Larger economic and political developments surely did play a role in Burgundy's eventual decision to switch sides, which he did with the Treaty of Arras in 1435. But, as the chapter shows, this outcome too is best explained in terms of the same factors that had sustained the partnership in earlier days. The loss of the English martial predominance that Henry V had established, and that Bedford managed for some years to keep up, put pressure on the Anglo-Burgundian alliance from both directions. But equally or more important was the gradual undoing of the personal relationship between Philip and John, which was essentially a matter of family dynamics, not geopolitics. The bad behavior of Henry VI's younger uncle, Duke Humphrey of Gloucester (whose pursuit of his wife's inheritance rights in the Low Countries challenged Philip's core policies there), pitted the regent's brother against his brother-in-law, and left both of them unhappy with Duke John. That

---

28 *Grand Strategy and Military Alliances*, ed. Peter R. Mansoor and Williamson Murray (Cambridge: Cambridge U.P., 2016), 2–3.

29 The only other contender being the Franco-Scottish alliance that began before and ended after the war. However, Scotland had nowhere near the resources of the Burgundian territories of the fifteenth centuries. The version of the article included as a chapter in this volume is practically unchanged from the original publication.

crisis was soothed over in part because of the influence of Anne; her death in 1432, however, removed a very important tie that until then had bound the Lancastrian and Burgundian dynasties, and John and Philip personally, together.

The last chapter of this volume also deals with strategy, but moves the focus back from the particular to the general, examining some key patterns in the big picture of what caused a particular type of war – the war of conquest – to be more favored, and more successful, in some places and times within the medieval era than in others. Drawing inspiration from the discipline of economics (one of my undergraduate majors), it considers the factors that could systematically affect the fiscal and economic costs and benefits that had to be calculated by rulers who sought to seize, occupy, and retain the territory of others. The strategic decision to launch an aggressive war is very different depending on whether or not the context allows war to be waged profitably in the first instance – that is, if an offensive campaign can bring in more wealth in plunder and land than it costs in logistical expenses, wages, and other rewards distributed to the army.

By casting a broad net, I was able to find a significant number of cases in which the sources do tell us, or at least allow us to estimate, how much a particular expedition or siege cost, and how much it brought in. One key observation arising from this data, and some mathematical thinking about it, is that the *inhabitants* of a besieged town were, where slave-taking was possible, generally more economically valuable than their *material possessions* were. It was much easier to make war pay for war, therefore, in times and circumstances when mass enslavement was permissible. Another important finding is that sieges of castles cost nearly as much as sieges of towns, yet yielded far less wealth for the besiegers when successful. Thus, the process of encastellation that Europe experienced starting in the tenth and eleventh centuries made conquest both more expensive (gross) and less rewarding (net). It is far from a new observation that the spread of castles caused a great shift in the strategic balance between offense and defense – one that probably deserves to be considered another military revolution itself, despite the rather long time it took for the phenomenon to fully play out. But the analysis in this chapter should enhance understanding of the mechanisms behind that shift, up at the strategic level rather than down at the tactical and operational level, where the impact of encastellation is more obvious.

***

It has been suggested that those who include military revolutions in their analysis of how warfare has changed over time "appear to leave no real space for the influence of human agency" and "too easily remov[e] the individual soldiers and their leaders from the military historical equation."[30] The combination of the two blocks of essays in this book should make it clear, however, that there is no

---

30 Stone, "Infantry Revolution," 366; DeVries, "Catapults," 455.

inherent conflict between, on the one hand, recognizing that there can be sudden and dramatic shifts in patterns of warfare, including shifts resulting from technological innovation, and, on the other hand, fully appreciating that particular historical events are driven and shaped by individual human beings interacting with each other, with their historical environments, and with what Clausewitz calls the "play of chance." As Karl Marx observed, "Men make their own history, but they do not make it just as they please."[31] Changes in the big picture of military affairs, whether they take the form of sharp discontinuities or gradual evolutions, change the circumstances in which individual actors make their decisions and roll their dice. When the Artillery Revolution caused a big shift in the balance between offense and defense, it did not "determine" the success of any given siege, campaign, war, or grand strategy, but it did shift the odds and change the boundaries of the practically feasible. The longbow was obviously not an "invincible" weapon,[32] and its presence on the field of Crécy in 1346 no more guaranteed Edward III's success there than its absence prevented the success of Edward I's war of 1276–1277. Nonetheless, recognizing the effectiveness of the longbow and its difference from earlier selfbows is as *necessary* to providing a good explanation of the outcome of Crécy as emphasizing good strategy is to explaining the English conquest of Wales. A good historian can neither abstract the individual from his context of place and time, nor ignore the fact that history is shaped as much by the agency of individual human beings, and by contingent outcomes of human choices, as it is by the flow of the "deep currents" and impersonal forces.

31 Quoted in Alex Roland, "Is Military Technology Deterministic," *Vulcan* 7 (2019), n. 5.

32 Despite both me ("Efficacy," 234) and Roland ("Deterministic," 22) pointing out that attacking the "invincibility thesis" is going after a strawman, because no serious historian claims the longbow was "invincible," DeVries continues to insist that it is "far from" being a strawman, because the "tenacious myth" of the invincibility of the longbow "can easily be found in the work of Robert Hardy (1993)," for example. But he does not cite a page, and I cannot find any such claim in Hardy's work. Kelly DeVries, "Catapults Are Still Not Atom Bombs: Effectiveness and Determinism in Premodern Military Technology," *Vulcan* 7 (2019), 42 n. 2.

Part 1

TECHNOLOGY AND MILITARY REVOLUTIONS

THE IDEA OF MILITARY REVOLUTIONS IN EIGHTEENTH AND NINETEENTH CENTURY TEXTS

Historiographical surveys of work on military revolutions, or revolutions in military affairs, often begin with Michael Roberts's seminal 1956 essay on the period from 1560–1660.[1] But although his military revolution (or a variation on it) is nowadays the one invariably meant when one speaks of "the" military revolution, Roberts himself clearly thought that he was describing only "a" military revolution, and took it as a given that there were others of equal or comparable significance. This is clear from the very first sentence of his study, which reads, "It is a historical commonplace that major revolutions in military techniques have usually been attended with widely ramifying consequences."[2]

In fact, by the time Roberts was writing, the idea of military revolutions – in the sense of revolutions in the art or conduct of war, often with great significance for the general course of human history – had been growing increasingly familiar for about 200 years. Although the concept even predates the French Revolution, the latter without doubt contributed to its growing popularity, both because the drama of the Revolution predisposed men to think in those terms, and also because it, and Napoleon, had set before the world's view an example of a military revolution that was so clear and so unambiguously of vast import.

The nineteenth century was, moreover, the time when the Industrial Revolution seemed to be changing so many aspects of human life so rapidly that, again, many observers came naturally to look for, and find, revolutions in all sorts of areas.

---

1 I now regret that my own first article dealing with military revolutions, for example, opened with the statement, "The concept of the 'military revolution' first entered the historical literature with Michael Roberts's famous inaugural lecture" on the subject. C. J. Rogers, "The Military Revolutions of the Hundred Years War," (1993) reprinted in idem, ed., *The Military Revolution Debate* (Boulder, 1995), at p. 55.

2 As other examples of military revolutions "recognized as major turning-points in the history of mankind" he notes "the coming of the mounted warrior, and of the sword, in the middle of the second millennium BC; the triumph of the heavy cavalryman, consolidated by the adoption of the stirrup, in the sixth century of the Christian era; [and] the scientific revolution in our own day." Michael Roberts, "The Military Revolution, 1560–1660," (Belfast, 1956), reprinted in Rogers, ed., *Military Revolution Debate*, at p. 13.

DOI: 10.4324/9781003399971-3

In addition to revolutions in military tactics, affairs, methods, and systems, they described or anticipated revolutions in military harmony, military sanitation, military music, military surgery, military discipline, war correspondence, and military education.[3] In the 1820s, some writers saw revolutions in the art of war as imminent; in the 1830s as under way; in the 1840s, 1850s, 1860s, 1870s, 1880s, and 1890s, they proclaimed both that war had recently been revolutionized and that it was about to be. Meanwhile they identified more and more military revolutions as having taken place over the course of history, from ancient Egypt through the "complete revolution in the art of war" effected by Frederick the Great, who "gave a new code of military rules to the nations of Europe."[4] Indeed, claims that the art of war was about to be "revolutionized" became so commonplace that they inspired skepticism of the validity of the very concept, and even mockery – which shows how entrenched the idea had become, for mockery implies an expectation of the audience's familiarity with its target.

In this essay, I will explore eighteenth- and nineteenth-century ideas about military revolutions. My topic, I must emphasize, is not the military revolutions themselves, but rather the beliefs and conceptions about them that were expressed in books, journals, and newspaper articles.[5] The oldest meaning of the word "revolution" is, as the etymology suggests, a turning around, for example the revolution of the earth or of a wheel. In the sense of a sudden and thoroughgoing change in a situation, it goes back at least to the fourteenth century.[6] This sense was originally most commonly applied to the overthrow of a government, e.g. in the Glorious Revolution of 1688. Hence, even in the nineteenth century, a "military revolution" most often meant the seizure of political power by the army. But already by the middle of the eighteenth century, the idea that especially important changes in military methods or affairs might take the form of "revolutions" was beginning to emerge. Some of its early proponents were very widely read and extremely influential. David Hume, in 1759, noted that "gunpowder changed the whole art of war," and included this transformation as one of the major strands in the "general revolution . . . in human affairs" that created the modern world.[7] In 1761, the military writer Campbell Dalrymple apparently took that as a given, and went on to suggest that recent tactical innovations that emphasized bayonet attacks and downplayed the effectiveness of fire "in time may produce another

---

3 This essay is based on a collection of nearly 300 mentions of military revolutions (by various names, but always including some form of "revolution"; not including mere "revolutions in tactics") in books, newspapers, and journals published between 1759 and 1900, only a fraction of which can be cited here.

4 *The English Review* 10 (1787), p. 423.

5 Many of the "revolutions" noted in this essay are assessed in W. Murray and M. Knox, eds., *The Dynamics of Military Revolution, 1300–2050* (Cambridge, 2001). It is striking, however, how little that volume has to say about the extent to which the revolutions it discusses were *perceived* as such by contemporaries.

6 *Oxford English Dictionary*, s.v.

7 *History of England*, vol. 1 (London, 1759), p. 7.

military revolution, and send us back to the arms in use before the invention of gunpowder."[8] In 1776, Adam Smith's hugely influential *Wealth of Nations* confirmed the idea that the introduction of firearms had been "a great revolution in the art of war." As an economist, Smith emphasized the cost of gunpowder weapons; in his view, their significance was truly epochal, because, "In ancient times, the opulent and civilized found it difficult to defend themselves against the poor and barbarous nations. In modem times, the poor and barbarous find it difficult to defend themselves against the opulent and civilized."[9] Other commentators soon noted other ways in which this military revolution had broad effects on the course of history. The great French *philosophe* Condorcet accepted Smith's points, and added that because of the gunpowder revolution "military expeditions are more expensive; riches can balance force; even the most warlike nations feel the necessity of preparing themselves [for war] by acquiring wealth, through commerce and the arts, thereby assuring to themselves the means of fighting," and that "Great conquests . . . have become nearly impossible."[10] John Gifford in 1793 claimed more specifically that the introduction of *portable* firearms, during the reign of Charles VI of France, "gave a fatal blow to chivalry, and effected a total alteration in the art of war." The opposition between firearms and knightly dominance of war was far from new, but Gifford raised the point to a higher level of insight in observing that

> the bravest warrior could no longer rely on his personal prowess, or the excellence of his arms, as a means of defence against an adversary, who, though destitute of courage, might, with success, attack him at a distance. A tranquil intrepidity, accustomed to give and to receive death without design as without fear, was now substituted in the room of that active valour which had hitherto been deemed the chief support of hostile armies. Battles became more bloody . . . By this new mode of fighting every man was rendered fit for the purposes of war. Armies were more numerous, and nations exhausted their resources in augmenting their military forces.[11]

Hence, it came to be broadly accepted that gunpowder had not only effected a "revolution in the art of war," but also, in Gibbon's words, a revolution in "the history of mankind."[12] By 1808, even the readers of the *Ladies' Repository* knew that gunpowder had "entirely revolutionized the art of war," with broad consequences.[13]

---

8 *A Military Essay* (London, 1761), p. 56.

9 *An Inquiry into the Nature and Causes of the Wealth of Nations* (London, 1776), p. 2:313.

10 N. de Condorcet, *Esquisse d'un tableau historique des progrès de l'esprit humain* (Paris, 1794), pp. 179–80.

11 *History of France* (London, 1793), p. 2:262.

12 *History of the Decline and Fall of the Roman Empire*, new ed., vol. 10 (London, 1791), p. 18.

13 O. M. Spencer, "What We Owe to Germany," *The Ladies' Repository* 18 (1808), p. 526.

The examples in the last paragraph were taken from both before and after the most sweeping and dramatic revolution in European history – the French Revolution. Writers of the early nineteenth century did recognize that warfare was among the many aspects of the human experience that had been transformed by this great discontinuity in history. Carnot was said to have caused a "complete revolution in tactics" and "nearly as great a revolution in the art of war, as had been effected in politics,"[14] and Dumoriez to have "made the war of the Revolution along with the revolution of war."[15] It was also claimed that "Napoleon's system of fighting had occasioned a revolution in the art of war," and indeed caused "the same kind of revolution in war, that Copernicus did in astronomy."[16] Clausewitz recognized that the Revolution had caused "an enormous transformation in war," and indeed the "revolution in the theory of war" he hoped to effect was to a substantial extent based on his analysis of how it had done so, and how Napoleon had directed the military energies created by the Revolution so effectively.[17]

And yet, in early- to mid-nineteenth-century texts, references to the introduction of gunpowder as the prototypical military revolution continued to be far more common than references to the transformation wrought by the French Revolution and Napoleon. Moreover, both before and after the French Revolution, a whole range of other military revolutions were offered up to the reading public as well. The idea that there had been several military revolutions over the course of history was apparently something of a commonplace even by 1774, when Alexander Gerrard, illustrating the point that different people responded to the same stimulus differently, supposed that a scholar presented with news of a recent battle might react by "recollect[ing] some battle rendered famous by classical description, or trac[ing] the revolutions in the art of war."[18] Among such historical revolutions, William Eden in 1779 drew attention to the importance of the change from feudal to paid military service, which he considered perhaps the greatest change in any of the human arts experienced in the late Middle Ages. "The revolution made in the system of warfare," he wrote "induced another in that of military establishments: the art of war from an occasional occupation became a trade."[19] In a way, this revolution was the reversal of another noted by George Ellis in 1796, who – anticipating Heinrich Brunner by nearly a century – argued in passing that Charles Martel's encounter with Muslim cavalry at the battle of Tours/Poitiers in 732 "effected a great change in the manners of Europe, by producing a complete revolution in the art of war": a shift in emphasis from infantry soldiers to cavalrymen, which in turn

---

14 T. B. Johnson, *An Impartial History of Europe* (London, 1812), p. 514; *The French Revolution, Sketches of its History* (London, 1847), p. 144.

15 P. Barras, *Memorien* (Stuttgart, 1896), p. 3:53.

16 E. Odleben, *A Circumstantial Narrative of the Campaign in Saxony, in the Year 1813* (London, 1820), p. 326. J. Bentham, *Works*, vol. 9 (London, 1843), p. 426.

17 *Vom Kriege*, ed. G. von Marées (Berlin, 1883), pp. 255, xiii, 570; note also *United Service Magazine* (1844, pt. 3), p. 366.

18 *An Essay on Genius* (London, 1774), pp. 131–2.

19 A. T. Malkin, *Historical Parallels* (London, 1831), p. 299.

led more or less to the rise of feudalism.[20] The later reversal of that shift – the rise of the infantry, first pikemen, then handgunners, in the Late Middle Ages – was also seen as a revolution by many nineteenth-century writers.[21]

Considering the origins of the twentieth-century historiography on "the" military revolution, it is interesting to see that the "revolutionary" nature of sixteenth- and seventeenth-century changes in the art of war was recognized early on, and often referred to. In 1824, in his *Essai sur l'histoire générale de l'art militaire, de son origine, de ses progres et de ses revolutions*, Col. Marie-Henri Carrion-Nisas described the introduction of gunpowder as "the dominant revolution of the military art" – "the great military revolution" – but considered the Nassau cousins and Gustavus Adolphus to be the "restorers of the military art" whose contributions initiated the modern era in warfare.[22] The following year, in the context of a discussion of Niebuhr's history of Ancient Rome, a writer in *The Quarterly Review* undertook briefly to "illustrate the revolutions which the art of war has undergone under similar circumstances in different ages and countries, to bring ancient and modern history together, and to make them each reflect light upon the other." This author noted how in Servian Rome, in ancient Greece, and in late-medieval Europe alike, the rising prosperity of the free citizenry (reflected in better armor and more time for military exercise), combined with the development of massed infantry formations, led to a rise in both the military importance and the political standing of the commons at the expense of the mounted nobility. Thus, the pikemen of Switzerland and Spain became for long "the terror of Europe, till the improved use of fire-arms, about the middle of the seventeenth century, produced another revolution in the art of war, to which ancient history can afford no parallel."[23] The next year, Sholto and Reuben Percy similarly commented on how the rise of the pike in "the last years of the middle ages indicated the commencement of the military revolution in the general employment of pikemen and musketeers" in the sixteenth century.[24] In 1830, a British writer suggested that contrary to popular opinion gunpowder had *not* caused a "total revolution in the warfare of the middle ages," and indeed that that the rise of modern tactics and "the first modern revolution" in military science "was in no degree owing either to the discovery of gunpowder, or the invention of fire-arms," since the true "revolution in warfare" came only in the sixteenth-century Wars of Italy, with the effective combination of arquebus

20 Preface to G. Legrand, *Fabliaux or tales...* (London, 1796), xi. Interestingly, P. Villari, *I primi due secoli della storia di Firenze* (Florence, 1893), p. 310, saw basically this same *rivoluzione nell'arte della guerra* (replacement of traditional reliance on infantry with the dominance of heavy cavalry) as occurring only after the battle of Montaperti in 1261.

21 M.-H. Carrion-Nisas, *Essai sur l'histoire générale de l'art militaire . . .* (Paris, 1824), pp. 1:418, 422; K. Hagen, *Reden und Vortäge* (Bern, 1861), p. 147; C. Oman, *Art of War in the Middle Ages* (London, 1885), pp. 132–4.

22 Carrion-Nisas, *Essai*, 1:xxxviij-ix, 422, 436, 529; 2:1, 59.

23 *Quarterly Review* 32 (1825), pp. 77–81.

24 *Percy Anecdotes*, vol. 7 (London, 1826), p. 15; H. Hallam, *View of the State of Europe in the Middle Ages* (London, 1856), p. 185.

and pike, after which infantry again "became the nerve [i.e. sinew] of armies."[25] In 1833, the entry for "Gustavo" in the *Dizionario universale della linga Italiana* connected Gustavus Adolphus to "una rivoluzione nell'arte della Guerra."[26] This was seconded by Goodrich's *History of All Nations* in 1851, which described him as initiating a new era in the art of war, and the next year by an article in the *London Quarterly Review*, which ascribed to Gustavus, along with Maurice of Nassau and Henri IV of France, a "complete revolution in the art of war."[27] In 1893, D. Campbell wrote that Maurice "revolutionize[d] the military science of his time as completely as Napoleon did the work for his contemporaries."[28]

Michael Roberts's conception of the Military Revolution was a broad one, extending in many directions, and addressing strategic conceptions on a new scale; the proliferation of military treatises and the founding of military academies; a "revolution in drill," which was as significant for its impact on discipline off the battlefield as on; and increased use of "propaganda, psychological warfare, and terrorism as military weapons," among other topics.[29] Nineteenth-century authors' idea of historical military revolutions tended to be more focused on the adoption of, and tactical adaptation to, new weapons. In addition to gunpowder, cannon, and pikes and arquebuses, past revolutions in the art of war were attributed to: Egyptian horse-drawn chariots; the introduction of bronze weapons; catapults; Macedonian *sarissas*; Roman roads; Roman swords; Greek fire; English longbows; Vauban's invention of the bayonet; Gribeauval's artillery system; and to "flying" (fast-moving, horse-drawn) artillery.[30] In 1860, a military correspondent of the *Times*, with reference to Gustavus among others, sweepingly claimed, "We find invariably in the history of warfare that the invention or adoption of a new or improved weapon goes together with, or is followed by, a complete revolution in the art of warfare."[31]

That did not mean, of course, that *only* new weapons caused military revolutions. Frederick the Great was sometimes cited as having effected a revolution in

---

25 *United Service Magazine* (1830, pt. 1), pp. 452–3; note also A. C. N. Gallenga, *Storie de Piemonte* (Botta, 1856), 2:150.

26 Ed. C. A. Vanzon (Livorno, 1833).

27 Goodrich, p. 1032; *LQR*, Am. ed., p. 247. In 1878, it was argued that sixteenth-century Holland (and Venice) had "effected a revolution in the art of war" by using their wealth to break the tight link between a nation's population and its military strength, a link that would later be restored by the French Revolution. *The Nineteenth Century* (1878, vol. 3), p. 403.

28 *The Puritan in Holland . . .* , 4th ed. (New York, 1893), p. 2:253. See also A. Young, *History of the Netherlands* (Boston, 1884), p. 400.

29 "Military Revolution," esp. pp. 15, 29.

30 J.A. Wylie, *History of the Scottish Nation* (London, 1886), p. 1:477; J. Dümichen, *Geschichte des alten Aegypten* (Berlin, 1879), p. 219; E.A. Freeman, *History of Sicily* (Oxford, 1894), p. 4:64; *Dublin Review* (July–Oct., 1870), p. 485; *Spirit of the XIX Century* (Vol. 2, Iss. 7), p. 409; J. R. Green, *History of the English People* (London, 1885), 1:421; W. E. H. Lecky, *History of the Rise and Influence of the Spirit of Rationalism in Europe* (London, 1866), p. 2:232; *The Academy* 16 (1879), p. 469; B. F. Perry, *Reminiscences* (Philadelphia, 1883), p. 250.

31 5 Jan. 1860, p. 7.

the art of war, mainly by means of his own generalship, and so too were Epaminondas of Thebes, Hamilcar Barca, Gonsalvo de Cordova, and (at sea) Horatio Nelson.[32] The creation of the manipular legion was deemed by Theodor Mommsen "a complete revolution . . . in the military system" of the Romans.[33] Others described the restructuring of the Macedonian army by Philip II, of the Roman army by Marius, of Italian armies by the rise of the condottieri, and of the French army by Charles VII in similar terms.[34] As already noted, military revolutions were also ascribed both jointly and severally to Napoleon and the French Revolution, and these too were not usually tied to advances in weaponry. Nonetheless, the nineteenth century's view of past military transformations was in general strongly linked to technological developments: in 1884, for example, one writer claimed that "the weapon has been the cause and key-note of every tactical change."[35] This should not be surprising, for the same technological focus imbued the views of the writers of the time with regard to military change in their own present and their future.

Less than a decade after Waterloo, Jacob Perkins patented a device which some thought would effect "a revolution in the art of war": an "infernal machine" with the potential "to render all valour useless, and to reduce the science of war to the employment, more or less intelligent, of some moving volcanoes, which will exterminate entire masses in the course of a few hours."[36] The device in question, reportedly able to fire 250 balls per minute with great destructive effect, was a sort of steam-powered machine gun. Despite successful tests with Wellington in attendance, Perkins's steam gun was never used in war, nor were its successors in the 1840s or 1860s, for which comparable claims were made. Still, serious thinkers recognized the potential of steam power early on, particularly for military transport. As early as 1826, the *Militairische Blätter* anticipated that steam would bring about a "total revolution in military science, analogous to the revolution formerly worked by gunpowder."[37] In 1834 a committee of the Tennessee Constitutional Convention predicted that steam-powered locomotion "cannot but tend to produce a revolution in military operations and military science, whereby, nations will, in the future, be attacked with the aid of steam, and consequently,

---

32 *The English Review* 10 (1787), p. 423; C. Anthon, *Manual of Grecian Antiquities* (New York, 1852), p. 257; G. B. Malleson, *Ambushes and Surprises* (London, 1885), p. 3; H. C. Chatfield-Taylor, *Land of the Castanet* (Chicago, 1896), p. 180; *The Academy* 41 (1892), p. 126.

33 *History of Rome*, tr. W. P. Dickson (London, 1862), p. 1:453.

34 C. Rehdantz, intro. to *First Philippic*, tr. T. Gwatkin (London, 1883), p. xxix; H. B. Stuart, *History of Infantry* (London, 1861), p. 14; *North British Review*, 1848, p. 88; E. de la Barre Duparcq, *Elements of Military Art and History* (New York, 1863), p. 109.

35 Cpt. J. Chester, "Dynamite and the Art of War," *The United Service* 10 (1884), pp. 350, 349.

36 *The Gentleman's Magazine* 96, pt. 1 (1826), p. 79; *Vermont Journal*, 20 Feb. 1826, p. 3.

37 *Militairische Blätter* 1 (1826), p. 373. General Lamarque said much the same in the French Chamber of Deputies after a British use of rail to move troops in 1831. F. C. H. Clarke, *Staff Duties . . .* (London, 1884), p. 247. The *Militairische Blätter* noted that for logistical reasons the railroad should strongly favor the strategic defender.

with a celerity of movement which must subject the assailed nation to inevitable destruction, unless she also employs steam power … to oppose to the enemy's rapid approach, disposable force[s] and supplies thrown with increased rapidity" from central locations to frontiers.[38] In 1838, the *Daily National Intelligencer* noted that the development of the trans-Atlantic steamship had greatly strengthened Britain's hold on Canada, "brought Quebec nearer to London for all practical purposes than Marseilles is to Havre," and thereby "wrought a revolution in the art of war" – one to which, however, the United States had an answer, in "the prodigious facility our railroads are giving us of a CONCENTRATION."[39] A British writer in 1843 predicted that as an engine of war the railroad would "prove next in importance to the invention of gunpowder" (which had become the basic standard of comparison for "revolutionary" military changes), contributing to the beginning of a new "era in the history of mankind."[40] Another Briton writing two years later, though agreeing that "steam has revolutionized the art of war," took a pessimistic view of its implications: "it has rendered comparatively ineffective the greater navy for purposes of defense, while it has increased the offensive capabilities of the lesser navy by four-fold … you can never hope again, by means of your navy alone, to prevent an invasion."[41] This point was soon picked up by Benjamin Disraeli himself, who justified plans for a British naval buildup in terms of the need to respond to scientific developments that had occasioned a "revolution in the art of war, which had deprived us of one of our natural sources of defense."[42] Also in 1845, a New York newspaper concurred that the introduction of steamships was by itself enough to "produce a perfect revolution in military tactics," adding that in combination with other developments steam ensured that "the next war between any two powerful nations, will be quite different from any thing which has heretofore been seen in the world."[43] By 1852, Commodore Robert Stockton could confidently presume the U.S. Senate had "often been told" that steam would revolutionize naval war.[44] In 1855, Prussian general Marieluise von Prittwitz noted that much had already been written on the impact railroads would have on war, and declared that, despite the skeptics, it was now clear beyond doubt that "the entire contemporary art of war must undergo a complete revolution," because railroads had rendered worthless all current calculations of tactics and strategy.[45] By 1861 it was widely accepted that "true strength in war may, indeed, be said to consist in a proper development of [steam's] powers. Armies may be organized, fleets may be equipped, fortifications may be built; but unless

---

38 *Journal of the Convention of the State of Tennessee* (Nashville, 1834), p. 345; see also *Spirit of the XIX Century* (1843), vol. 2, iss. 7, p. 409.
39 11 August 1838, col. F.
40 *The Foreign and Colonial Quarterly Review* 1 (1843), p. 525.
41 *The Living Age* 8 (Oct.–Dec. 1845), pp. 540–1.
42 *Annual Register* (1854), p. 9.
43 *The New York Herald*, 27 Dec. 1845, col. B.
44 R. F. Stockton, *A Sketch of the Life of Robert F. Stockton* (New York, 1856), p. 114.
45 *Andeutungen über die künstige Fortschritte* … (Berlin, 1855), p. 190.

the whole system – whether of attack or defense – is based upon principles in harmony with the extraordinary acceleration of movement that steam has produced, it will be but labour thrown away."[46]

At the same time as steamships and railroads were coming into widespread use, two other classes of weapons with potentially revolutionary effects were being developed: long-range rifled artillery pieces, and improved infantry rifles. The 1850 *Annual of Scientific Discovery* concluded that the new Krupp guns, if "brought into practical use," would "work a revolution in war."[47] An article in the 1859 *Edinburgh Review* opened by stating, "During the last few years such progress has been made in the manufacture of all descriptions of fire-arms … as in reality to amount to a complete revolution in the whole art of war."[48] Another piece in that year's *Bentley's Miscellany* held that Minié-ball muskets and rifled artillery (dubbed "armes de précision" by Napoleon III) had "brought about almost a revolution in military tactics and strategy," quoted a newspaper correspondent who described the effect of rifled cannon at Solferino as "like the work of enchantment," and concluded that "improvements in small arms may reasonably be expected to bring about as great a revolution in military tactics and operations as will inevitably be produced by the introduction of rifled guns."[49] Some observers were already prepared to dispense with the use of qualifiers and the future tense: 1859 and 1860 articles in the *New York Times* and the *Times* of London praised the French for their thoroughgoing use of new technology – including railroads, balloons, and the telegraph as well as Minié balls and rifled artillery – and declared that together these amounted to a "complete revolution in military matters."[50] By the end of the American Civil War (1861–5) and the Seven Weeks' War (1866), it was widely accepted that the military revolution brought by "the introduction of the steam engine, the locomotive, the electric telegraph, rifled ordnance, iron-clad ships, and other inventions of this scientific age," was an accomplished fact, rather than an imminent possibility.[51]

One of the most interesting and thoughtful commentaries on these developments was the already-mentioned 1860 essay in the *Times*. After summarizing the history of war in terms of a series of military revolutions – the development of the Macedonian phalanx, the Roman legion, the rise of heavy cavalry, the development of gunpowder, Gustavus' tactical reforms, and the replacement of the pike with the bayonet – the author focused in on the impact of "the great perfection of the modern rifle and its general introduction as an arm of the infantry," which he saw as foreshadowing "as great a revolution in all military matters as any which has ever taken place." The author was wise enough not to make such sweeping

---

46 *United Service Magazine* (1861, pt. 1), p. 43.
47 *Annual of Scientific Discovery* (London, 1850), p. 45.
48 *Edinburgh Review,* 102 (1855), p. 212.
49 Vol. 46, pp. 189–92; see also J. D. Dougall, *The Rifle Simplified* (Glasgow, 1859), p. 10.
50 *The Times,* 5 Jan. 1860, p. 7; *New York Times,* 19 Jul. 1859, p. 4.
51 J. W. Draper, *History of the American Civil War* (New York, 1867), p. 1:17.

claims on the basis of the need to change tactical arrangements or the clear-cut advantage armies with the new weapons would enjoy over those without. Rather, he presumed that all armies would soon have them, and the result would be "nothing more or less than a complete subversion of the very principle on which the military organization was based hitherto." "The old musket," he continued, "was the arm of the masses, and the individual soldier did not exist as such, but was only looked upon as part of a whole; the general adoption of [the modern rifle] gives back the full value of individual excellence and skill to the single soldier. The great aim in the old system was, therefore, to make an unconscious passive machine of the soldier, while in the future everything must tend to develop the individual soldier into an intelligent being." This, he predicted, would prove to the advantages of democratic societies whose citizens were taught self-reliance.[52]

Another noteworthy essay, entitled "Revolutionary Agents in the Art of War," appeared in the *United Service Magazine* the following year. It opened with the assertion that there had historically been two "great revolutions in the art of war," one occasioned by gunpowder, the other by steam, and that the world stood "at the commencement of a third, which promises to be more extensive than either of its predecessors." The third revolution, "springing, as it were, from an alliance of the other two" (since steam-powered machinery was crucial for the manufacture of the new weapons), included more powerful explosive projectiles, rifled cannon, "and the adoption of a fire arm, compared to which the old musket is little better than a pocket pistol." These statements were by this time not unusual, but the article is remarkable both for its impressive foundation of historical detail and for its conclusion regarding "the wonderfully accelerated rate with which change follows upon change in the present day":

> There is no pause. As the world is now constituted a perpetual struggle is going on between races, between nations, between individuals. Where there is rivalry or competition there can be no loitering or standing still; and if we would keep our position we must move onwards at such a pace as will prevent our being overtaken by others. An unceasing progress! The changes of to-day, which seem so perfectly to meet our wants, will have to be superseded by the necessities of tomorrow; and before we have completed any one new plan of action, we must hold ourselves in readiness to forsake it for another.[53]

With such radical ideas of progress in the air, it is no wonder that commentators of the period began to see military revolutions practically around every corner. In 1845, the *New York Herald* described a new design of steam-powered gun, similar to the Perkins gun, which "will satisfy every one that we are on the eve of a greater

52 *The Times*, 5 Jan. 1860, p. 7.
53 *United Service Magazine* (1861, pt. 1), p. 38.

revolution, in military and naval affairs, than even the most sanguine mind could ever imagine,"[54] and Marshal Marmont predicted that Congreve rockets could replace muskets as the basic arm of the infantry and would "effect a revolution in the art of war, and moreover ensure success and glory to the genius who first grasps their importance."[55] The next year it was the invention of guncotton that the *Herald* suggested would to be able to "blow a whole city in the air in a few minutes," and thereby bring "a revolution in the art of war."[56] Over the following years, revolutions in the art of war were predicted to develop from a whole long list of inventions, including infantry rifles firing exploding shells (1857);[57] a new design of guns invented by Napoleon III (1859);[58] torpedo-boats (1864);[59] the "Pollyferi," a Gatling-style gun (reported in 1868 under the headline "A Regiment of Soldiers, a Mile and a Half Distant, to Be Killed in Four Minutes");[60] the "14-mile Iowa gun," a cannon employing a projectile with a series of sequentially firing propulsive charges (1869);[61] "Earth torpedoes" (land-mines) (1883);[62] rockets with dynamite warheads (1884);[63] dynamite, melinite, and lyddite shells (1886, 1887, 1898);[64] submarine torpedo-boats powered by "the decomposition of caustic soda," (1886);[65] anesthetic bullets (1888);[66] shells propelled by liquid gas (1890);[67] gas-operated automatic rifles (1891);[68] a high explosive called "terrorite" (1892);[69] pneumatic rifles powered by compressed air (1893);[70] bullet-proof cloth (1894);[71] and acid-projecting rifles (1897).[72] A carbine firing phosphorus-filled incendiary bullets, it was thought, would produce "una completa rivoluzione nell'arte della guerra maritima" if its principles were scaled up for use in naval guns (1860).[73] The mere arrangement of a meeting between Thomas Edison and

54 *The New York Herald*, 27 Dec. 1845, col B.
55 *De l'esprit des institutions militaires* (Paris, 1845), pp. 83–5.
56 *New York Herald*, 10 Nov. 1846, p. 1.
57 *The Times*, 25 Mar 1857, p. 12
58 *The New Monthly Magazine*, 116 (1859), p. 240.
59 *Daily Richmond Examiner*, 20 Aug. 1864, col. B.
60 *Daily National Intelligencer*, 30 Dec. 1868, col. F.
61 *Idaho Tri-Weekly Statesman*, 28 Oct. 1869, p. 1.
62 *The Times*, 19 Sep. 1883, p. 3; *Rocky Mountain News*, 14 Oct. 1883, p. 2.
63 Chester, "Dynamite."
64 *Daily Evening Bulletin* (San Francisco), 11 Aug. 1886, col. G; *Public Opinion* 3 (1887), p. 138; *Atchison Daily Globe*, 28 Oct. 1898.
65 *Milwaukee Daily Journal*, 24 Nov. 1886
66 *Boston Daily Advertiser*, 1 Feb. 1888, p. 4.
67 *Rocky Mountain News*, 9 July 1890.
68 *Galveston Daily News*, 23 Sep. 1891, p. 4.
69 *Bangor Daily Whig & Courier*, 16 Feb. 1892, col. A.
70 *Rocky Mountain News*, 19 Feb. 1893, p. 4.
71 *Bismarck Daily Tribune*, 21 Apr. 1894, p. 2.
72 *Owyhee Avalanche*, 24 Dec. 1897.
73 *Il Paese: giornale politico-letterario* I (1859–60), 32.

Alfred Krupp in 1889 was described as likely to presage "an entire revolution in the art of war."[74]

More sober commentators limited their claims regarding "revolutionary" potential or results to developments that had moved off the drawing boards and into the hands of actual soldiers, or even to those that had been tested in war. Even though breech-loading rifles, which allowed for a substantially increased rate of fire and could be shot effectively from a prone position, had seen some use in the American Civil War, Europe was shocked by how effective the Dreyse needle-gun proved when employed in the war of 1866. Austria's shockingly rapid defeat was widely attributed to this "revolutionary" firearm.[75] But the needle-gun could not explain the equally stunning German victory in 1870–71, because the French had by then adopted an even better breech-loader, the Chassepot. So in this case, the dominant interpretation was that Helmuth von Moltke and the Prussian General Staff had transformed the conduct of war in much the same way that Napoleon had done at the start of the century:

> General Moltke has revolutionised the art of war. With that depth of insight which belongs to genius alone, he has clearly seen that in these days of railroads and of electric telegraphs, war can be carried on with a rapidity of action and a precision in execution which were utterly impossible in the days of Wellington and of Napoleon. He has also perceived, and in the most startling manner demonstrated, the practicability of one general moving, feeding, and directing with unerring precision, three or four armies, each as numerous and as powerful as the largest which under the old system of war could with safety be directly commanded by one man.[76]

Other writers attributed this revolution as much to the organization of the Prussian army – "the greatest advance in military science that has been made in modern times," which both created a force "man for man . . . superior to any army yet put on foot at any age or by any country" and at the same time made "every citizen . . . a soldier" – as to its leadership, which was said to have "revolutionized the art of war, until it has become, not an art, but a science."[77] It was even suggested that the war of 1870–71 marked "such a complete revolution in the art of war in its various details that henceforward it will be unnecessary to study the strategy of the past."[78]

74 *Los Angeles Times*, 4 Sept. 1889, p. 4.

75 *The Times*, 23 Apr. 1867, p. 8.

76 T. D. Wanliss, *The War in Europe of 1870–1* (Ballarat, 1871), pp. 89–90.

77 *London Quarterly Review* 130 (1871), pp. 20–21 (and cf. 27 [1867], p. 153); *The St. James's Magazine and United Empire Review*, N.S., 1 (1875), p. 15; *Atlantic Monthly* 65 (1890), p. 272; cf. *Every Saturday*, N.S., 1 (1872), p. 579.

78 A view disputed in *The Temple Bar* 35 (July 1872), p. 65; similarly *Illustrated Naval and Military Magazine* 7 (1889), p. 1772; cf. *Southern Monthly* (1861), p. 632.

Once they came to be issued widely and tested in small-scale fighting, bolt-action rifles with smokeless powder cartridges were also described as potentially or actually revolutionary.[79] Although he carefully limited his claim to "so much of the art of war as relates to battle tactics," the Commanding General of the United States Army officially reported in 1888 that long-range, flat-trajectory breech-loading rifles had "brought about an entire revolution," with "fire tactics" being substituted for "shock tactics."[80] It had by then long been recognized that increases in firepower made frontal attacks unlikely to succeed even against much smaller forces: experience in the Crimean War and the American Civil War had shown as much, and Moltke had commented on it in 1865.[81] But smokeless powder rifles further magnified the advantages of the defense in two ways. First, they allowed for more rapid, more accurate, and longer-range fire than any previous rifles. Second, as Col. Henry Elsdale argued in an important article of 1892, the elimination of the smoke emitted by gunpowder rifles would leave the defender in effect much better concealed than he had been in the past, which in turn (especially in combination with the greatly increased size of armies) would make it much more difficult for a general on the offensive to use the advantages of the initiative to compensate for the defender's advantages in protection and lethality of fire. This, Elsdale predicted, meant "an absolute revolution in all our present established systems of tactics and strategy," that would immediately come clear in the next great war. "When a few enterprising and ambitious generals in command of divisions have followed the old lines and the tactics which, on the German side, proved so successful in the war of 1870," and seen their commands "cut all to pieces and reduced to a complete wreck in half an hour, then . . . they will retire much sadder and wiser men."[82]

As one might expect, this proliferation of revolutions resulted in growing skepticism. "It is wonderful how often that phrase ['revolutionize the art of war'] occurs in inventors' claims for their designs," editorialized the *St. Louis Globe-Democrat* in 1883. "The art of war jogs on pretty steadily, notwithstanding, and refuses to be revolutionized."[83] Various writers expressed doubts concerning the revolutionary nature of Moltke's generalship, of machine-guns, and of smokeless powder, to say nothing of the more far-fetched putative revolutions.[84] Col. Elsdale's prophetic article on the revolution in favor of the defense was promptly rebutted by the military historian Spenser Wilkinson, who concluded that "there has been no revolution in tactics or strategy"; rather, recent changes meant only

---

79 E.g. *San Jose Mercury News,* 22 Oct. 1889, p. 3; cf. *The Bystander* 2 (1881), p. 148.

80 *Annual Report of the Secretary of War* (Washington DC, 1888), p. 109.

81 Quoted in S. Wilkinson, "Evolution not Revolution in Modern Warfare," *Contemporary Review* 62 (1892), 417.

82 H. Elsdale, "The Coming Revolution in Tactics and Strategy," *Contemporary Review* 62 (1892), pp. 245, 257.

83 1 Aug. 1881, p. 4.

84 See note 78 (Moltke); *The Times,* 23 Aug. 1884, p. 10; *Atchison Daily Champion,* 19 Oct. 1889, p. 2.

that "certain modifications long since realised have become more pronounced."[85] In 1902, readers of *Munsey's Magazine* were informed that the military changes of the preceding two generations, great though they had been, were "evolution . . . not revolution": "Periodically, some wiseacre arises to announce that frontal attacks are forever out of date, owing to the terrible deadliness of modern weapons, or that offensive warfare in general will shortly become impossible; but such sweeping conclusions are utterly premature. The fundamental rules of war are likely to remain unchanged for ages to come."[86] In 1906, historian W. C. H. Wood scoffed at "people who are all agog with every newspaper nonsense-tale about the something or other that is always going to 'completely revolutionize' the art of war," and thought that any sensible person should realize that strategic principles were timeless, while "tactics have always been undergoing the same slow process of evolution."[87] Humorist Gelett Burgess, in 1898, mocking the same group, described a fictional scientist who had invented a paint that turned invisible whatever it was applied to. The inventor presented his invention to a soldier: "'I submit this remarkable means of revolutionizing the art of war! Your uniforms, coated with this paint, would be invisible to the enemy, and your manoeuvres would be accomplished unseen.' The Soldier blushed, and with an apologetic gesture toward the lady replied, 'The spectacle of battalions of naked men marching upon the foe does not accord with the accepted traditions of civilized warfare.'"[88] More seriously, but to the same point, Admiral Coughlan in 1908 declared, "The only man who can revolutionize war is He who will abolish it, and we are waiting for His second coming."[89]

The skeptics, however, were far outnumbered by those who accepted the validity of the concept of military revolutions. While one might properly be doubtful of the revolutionary potential of particular new technologies, the military developments of the nineteenth century were so clearly rapid and radical that it was difficult to sustain the position that the basic pattern of military change was more evolution than revolution. The revolution-skeptics could offer no good response to the *Kansas City Star*'s observation in 1898: "The branch of warfare which it is most confidently stated has been completely revolutionized is naval action. A glance a picture of an old-fashioned war vessel, Napoleon's *Victory* or our own *Constitution*, and a modern steam war ship, the *New York*, for instance, will reveal the changes. The old walls of oak and the modern iron, or rather steelclad, differ as much in every detail as in their motive powers, wind and steam, muscle and electricity . . . A ship of the white squadron looks no more like one of the monitors

---

85 Wilkinson, "Evolution," pp. 416, 414.
86 R. H. Titherington, "The Development of the Soldier," *Munsey's Magazine*, Jan. 1902, p. 504 (citing Wilkinson).
87 *The Fight for Canada* (New York, 1906), p. 270.
88 *The Lark* 23 (March 1897), n.p.
89 *United States Congressional Series Set*, 60th Cong., 1st Sess. (Washington DC, 1908), p. 3:363.

or gunboats of the civil war than the latter looked like Noah's Ark."[90] And if war at sea could be revolutionized, why not war on land?

Moreover, by the end of the century new developments were promising another round of military transformation comparable to the introduction of steam. The most significant were those that took war from two dimensions to three. As early as 1896, an American lieutenant named J. K. Cree set forth many of the ideas that would be developed by airpower enthusiasts in the next century:

> [T]he invention of a successful air ship will cause an entire revolution in the art of war more stupendous than that caused by any invention since that of gunpowder, and even surpassing that, since it only increased the distance between the lines of combatants, while the principles of attack and defense, strategy and supply, remained unchanged, or were only slowly modified. A flying-machine, however, will nullify strategy, make vital changes in the principles of attack and defense, diminish the importance of navies and seacoast fortifications, and by bringing the theater of operations to the doors of palaces and Legislatures, render speedy settlement of national grievances imperative.

In the same year, perhaps even more remarkably, newspapers reported on the development of "aerial torpedoes" – early cruise missiles, lifted by hydrogen, guided by compass and barometer, and armed with a ton of dynamite each, which the inventor believed would "altogether revolutionize the art of war." "An ordinary man-of-war could carry several hundred of these torpedoes," he claimed, "and from a distance of forty or fifty miles they could let loose sufficient to destroy the whole city of London in a few hours."[91] As with the 1896 article on airships, this one recognized that the technology was still some way from being ready to effect a military revolution. But its potential was clear; all that was needed was technical refinement, and all the "unceasing progress" that had occurred so rapidly, especially since 1850, made those refinements, and more military revolutions, expected as a matter of course.

The ultimate military revolution, of course, would be the one that would end war. "It is safe to predict," argued the *Chautauquan* in 1890, "that in the near future [war] will perish through the sheer impossibility of withstanding the havoc which explosives deal. The temerity of any nation that ventures to risk attack with guns carrying quarter-ton missiles thirteen miles and shot loaded with dynamite, will be repaid with utter extinction. Peace finally will reign because men can no longer fight. The cost of war will banish war."[92] It did not, however, take 150 years

---

90 *Kansas City Star*, 1 May 1898, p. 4.
91 *The Daily Inter Ocean*, 27 Aug. 1896, p. 8, for one model of several developed almost simultaneously. Another version was wire-guided: *Atchison Daily Globe*, 5 Aug. 1897. Cf. also *Los Angeles Times*, 7 June 1891, p. 4; *Daily Picayune* 20 Aug. 1897, p. 8.
92 *The Chautauquan* 10 (1890), p. 45.

of thought about the power of technology to transform warfare for that concept to develop. "I hold," wrote Jakob Mauvillon "that he who invents a means of certainly and inevitably destroying with a single blow an entire army, or an entire province, will render the greatest service to humanity. War would cease at once." The date of publication of his essay was 1788.[93]

93 *Essai sur l'influence de la poudre à canon dans l'art de la guerre moderne* (Leipzig, 1788), p. 130.

CAROLINGIAN CAVALRY
IN BATTLE

The Evidence Reconsidered

There is a story about early medieval warfare that was first told in the eighteenth century. It was further elaborated between the mid-nineteenth and mid-twentieth centuries, and it still is often repeated in Western civilization textbooks and sometimes in military history surveys and reference books, though specialists in medieval warfare have lately tended to dismiss it as a "myth." The story goes like this:

When the Roman Empire in the West collapsed in the fifth century, the migrating peoples who crossed the Rhine frontier were warrior nations in which all free men fought, the aristocrats on horseback but the common folk on foot. The Franks, who settled in northern France and the Low Countries, were especially infantry-focused. Over time, however, cavalry came to play a larger role in their armies. This process took a great leap forward under the first Carolingian ruler, Charles Martel. For a variety of reasons, perhaps including the need to counter fast-moving Muslim raiders from Spain and (in Lynn White's version) his recognition that the recently introduced stirrup greatly increased the effectiveness of mounted warriors engaging in shock combat,[1] Charles made every effort to increase the force of armored horsemen at his disposal, going so far as to confiscate large amounts of Church lands, which he granted out to select warriors – "vassals" – in the form of estates held by precarial tenure ("benefices," or "fiefs"). The Franks were strongly averse to Roman-style taxation, and the monetary economy had much dwindled, but these estates allowed for the direct support of the soldiers, which was found

---

1 Lynn White used the term "shock combat" in a particularly narrow sense (and one he seems to have invented), limiting its meaning to a style of fighting in which a lance is held motionless under a rider's arm, in a "couched" position, with the power (or more precisely, the energy) of the impact on the target coming from the velocity of the horse and the combined mass of horse and rider, not the muscle power of the rider's arm. The earlier and still more common sense of the term is essentially equivalent to "hand-to-hand combat" as opposed to "missile combat." E.g. see G. Truemelet-Faber, 'Fundamental Principles and Analytical Tactics of Night Attacks', *Journal of the United States Infantry*, 5, no. 1 (1908), p. 72; Stephen Morillo and Michael Pavkovic, *What is Military History?* (Cambridge, 2006), p. 83. Unfortunately, the authors who have written about medieval "shock" cavalry since White have not always been clear as to which definition they mean. I use the term in its traditional meaning.

DOI: 10.4324/9781003399971-4

to be more efficient than collecting and distributing scarce coin. Each fief generally provided revenues sufficient at least for the man holding it to sustain himself without personally engaging in farming or commerce, and to supply him with the horses, gear, and other supplies that would allow him to campaign at his own expense. This launched a "complete revolution in the art of war," and amounted to the first steps towards the feudal system that structured much of medieval politics, warfare, and economics, and the chivalric ethos that infused much of medieval art, literature, song, ethics, and aristocratic culture generally.[2]

If it were a true story, this tale could hardly be bettered as an example of the importance of warfare and military affairs in explaining the broad sweep of general history, and hence as evidence for the profound importance of the subfield of military history. Recently, however, Susan Reynolds has made a forceful argument against the prevalence of anything much like "feudalism" among the Carolingians,[3] and, for a much longer period now, medieval military historians, with Bernard S. Bachrach of the University of Minnesota leading the charge, have questioned the claim that the Carolingians focused their military system on the heavy cavalry.

Indeed, Professor Bachrach does more than merely "question" the idea that cavalry was the principal arm of the Carolingian military: he utterly rejects that proposition, both holistically and with respect to each component argument that might go to supporting it. "An entirely new state of the question must be formulated regarding the military in medieval Europe," he argues. "Knights, heavy cavalry. . . [and] small numbers of effectives, . . . must be swept away as the dominant themes. Continuity from the later Roman empire through the Middle Ages is the proper focus."[4] As Bachrach's former student Peter Burkholder writes in *History Compass*, "the main impetus to question the cavalry dominance paradigm" is the "centrality of siege warfare in the Middle Ages," which leads to the "necessary conclusion" that cavalry was not the dominant arm, since, in Bachrach's words, "There was no place for the warhorse in the sapper's mine, the artilleryman's battery, or the crossbowman's belfrey."[5]

---

2 The earliest telling of this story I have been able to trace is in George Elis's preface to Pierre Jean Baptiste Le Grand d'Aussy, *Fabliaux or Tales . . .* (London, 1796), p. xi, but Elis does not give the impression that he considers his views original. The quoted phrase is from Elis. For Lynn White's version (and a discussion of the earlier historiography), see his *Medieval Technology and Social Change* (Oxford, 1962), chapter 1.

3 Susan Reynolds, *Fiefs and Vassals: The Medieval Evidence Reinterpreted* (Oxford, 1994).

4 Bernard S. Bachrach, 'Medieval Siege Warfare: A Reconnaissance', *Journal of Military History*, 58 (1994), p. 133.

5 Peter Burkholder, 'Popular [Mis]conceptions of Medieval Warfare', pp. 512–3; Bachrach, 'Medieval Siege Warfare: A Reconnaissance', p. 125. Cf. Steven Fanning, 'Cavalry', in *Medieval France: An Encyclopedia*, ed. William W. Kibler and Grover A. Zinn (New York, 1995), p. 184: "The decisive military forces thus were the engineers and footmen who weakened and stormed the defensive works, with horsemen playing virtually no role in this sort of fighting." John France, along the same lines, writes of Carolingian warfare that "siege warfare, very much the affair of infantry, is dominant." John France, 'The Military History of the Carolingian Period', *Revue belge d'histoire*

This line of argument rests on the belief that the ebb and flow of Carolingian warfare was determined by the conduct of sieges, and also on the questionable assumption that the success of individual sieges was normally determined by the tools of assault rather than by a competition over which side could keep itself fed longer[6] – since in a long siege, the role of cavalry in protecting or attacking foragers and supply convoys could easily be decisive. It also, at least implicitly, rests on a low assessment of the importance of cavalry in open battles. There were plenty of medieval wars resolved without an open battle taking place, it is true, but it is very difficult for an invader to conduct an effective siege-based campaign unless he is able to defeat the defenders in battle, or else deter them from initiating one. Therefore, if cavalry were generally the decisive arm in battle, superiority in cavalry would normally be a practical requirement for a successful offensive aimed at conquest, even if the campaign consisted of a series of sieges and *no battles were actually fought*.[7]

But Bachrach and his followers minimize the importance of cavalry in open battle, as well as in siege operations.[8] "Even in battles in the open field," writes Steven Fanning, "cavalry was usually ineffective against well-trained and disciplined foot soldiers in prepared defenses"; hence aristocrats "usually dismounted and fought on foot throughout the Merovingian, Carolingian, and post-Carolingian periods."[9] Bryce Lyon goes even further, stating, "Not in a single significant

---

militaire, 26 (1985), reprinted in John France and Kelly DeVries, *Warfare in the Dark Ages* (Aldershot, 2008), p. 9. In that article France lines up with Bachrach to dispute the traditional view of the importance of Frankish cavalry in the eighth century. Since my purpose is to argue for a quite different conclusion, I will often cite and dispute his conclusions here. That may seem a strange practice in a *Festschrift* dedicated to him, but in fact it is a tribute to my respect for his open-mindedness and his willingness to engage in courteous debate.

6 Cf. Bernard S. Bachrach, 'Charlemagne's Cavalry: Myth and Reality', *Military Affairs*, 47 (1983), p. 184: "It is inconceivable that such massive fortifications as those at Pavia, Barcelona, or Tortosa could have been taken without a siege train of significant size and sophistication."

7 In Charlemagne's wars in Saxony, for example, there were few if any battles between Franks and Saxons after 784. However, it is likely that this was because the Saxons, having long since learned the lesson that they lost open combats far more often than they won them, avoided battle at all costs. (As early as 758, Pepin "inflicted bloody defeats" on the Saxons; in 774, three Frankish detachments fought the Saxons, each victoriously; in 775 the Saxons had some slight success with a surprise attack on a Frankish camp, but then were defeated in three combats; in 778 the Saxons were defeated near Leisa; etc. *Carolingian Chronicles*, pp. 42, 50–51, 52–3, 58.) But avoiding open battle made it very difficult for them to prevent the loss of their fortifications, one by one.

8 Indeed, Bachrach goes so far to state that in medieval battles, "mounted attacks . . . rarely succeeded." 'On Roman Ramparts 300–1300', in *The Cambridge Illustrated History of Warfare*, ed. Geoffrey Parker (Cambridge, UK, 1995), p. 84. This follows the statement, on the same page, that "it is difficult to find medieval battles where men fighting on horseback formed the tactically dominant element."

9 Fanning, 'Cavalry.'

battle or campaign did cavalry play a tactically decisive role."[10] Lyon seems to be echoing earlier remarks by Bachrach, who writes, "From the many campaigns of Charles [Martel], Pepin, and Carloman described by contemporaries and near contemporaries there is not a shred of evidence to suggest that heavily armed horsemen engaging in mounted shock combat were the decisive element of their armies,"[11] and that in the next generation, in Charlemagne's reign, "not a single significant battle or campaign has been cited in which the cavalry can be shown to have played the tactically decisive role."[12] Now, in these two quotations, Bachrach is actually saying more about the *absence of evidence* for the tactical role of mounted troops in this period than he is about cavalry's importance or effectiveness. In most cases contemporary descriptions of Carolingian battles are extremely brief and do not allow us to draw any firm conclusions about the roles or relative importance of cavalry and infantry. Bachrach could have made the point that any theory of the social implications of the dominance of Carolingian cavalry rests on shaky foundations, because of that lack of evidence, and left it there. That would have been a valuable service to the fields of medieval and military history. But in fact, he goes on to state not just that cavalry "cannot be shown to have played the tactically decisive role" in any Carolingian battle, but, more positively, that "the decisive arm of the military forces of Charles Martel, and his sons, *was not* cavalry."[13] How misleading Bachrach's logical slip can be is well

10 Bryce Lyon, 'The Role of Cavalry in Medieval Warfare: Horses, Horses All Around and Not a One to Use', *Mededelingen van de Koninklijke Academie voor Wetenschappen, Letteren en Schone Kunsten van België*, 49, Nr. 2 (1987), pp. 88, 81. The context is somewhat ambiguous here as to whether he means for these blanket statements to be taken as true of the Middle Ages as a whole; but he does seem to believe that cavalry suffered from "ineffectiveness" from the time of Gregory of Tours (d. 594) right through Gattamelata (d. 1443) and Bosworth Field (p. 88), or at least that cavalry was not the dominant arm "from the Carolingian period into the late thirteenth century," (p. 78) and *a fortiori* by the later thirteenth century (p. 83) or the early fourteenth, by which point pikemen and longbowmen were "much more effective" than cavalry, as well as being less expensive (p. 86).

11 Bernard Bachrach, 'Charles Martel, Mounted Shock Combat, the Stirrup, and Feudalism', *Studies in Medieval and Renaissance History*, 7 (1970), p. 57. Similarly, ibid., p. 55: "of the actions about which information survives, none gives the slightest hint that heavily armed horsemen engaged in mounted shock combat were the decisive arm of the armies of Charles Martel's sons." France, 'Military History', p. 92: "there is nothing to suggest that the Franks depended upon an elite force of shock cavalry."

12 Bachrach, 'Charlemagne's Cavalry', p. 181. Similarly Carroll Gillmor, 'Cavalry, European', in Joseph Strayer (ed.), *Dictionary of the Middle Ages*, 13 vols (New York, 1989), vol. 3, pp. 200–8, at p. 202: "not a single battle or campaign provides evidence that the cavalry played a tactically decisive role."

13 Bachrach, 'Charles Martel', p. 75 (emphasis added). John France, somewhat similarly, within a few pages goes from saying "the vagueness of the sources makes the construction of any theory of a military revolution extremely difficult" to saying "all the evidence is that there wa[s] no military revolution under Charles Martel." France, 'Military History', pp. 86, 97.

Charles Bowlus, for example, quotes Bachrach on this subject, says his conclusions "are convincing," and seems to accept that "mounted shock warriors were not the most important tactical element in Carolingian warfare." Charles R. Bowlus, 'Warfare and Society in the Carolingian

illustrated by Victor Davis Hanson's completely unsupportable claim that "there is not a single major Carolingian engagement in which infantrymen were not the dominant force on the battlefield."[14]

In fact, a good case could be made that shock cavalry most likely *was* the decisive arm of the Carolingian military. But I will not make that case here, since it would require a much longer essay than this one, among other reasons because of the problems inherent in even defining what it means to be a "decisive arm." I will therefore not review the evidence from the capitularies and other contemporary texts that shows how much importance Charlemagne placed on recruiting and finding fodder for the horsemen of his army. John France has in any case already done that very well, in the course of making the point that those documents only offer circumstantial support for, not hard evidence of, a key tactical role for cavalry, since horses could have been considered of crucial importance for reasons of logistics and operational mobility even if their riders were accustomed to dismounting and fighting as infantry, rather than as cavalry.[15] Instead, in the limited space available this essay will focus on re-examining the few contemporary descriptions of combat that are sufficiently detailed to allow us to assess the role of cavalry per se in Carolingian battles. The earliest of these post-Poitiers battle descriptions date to the 780s.

Frankish troops fought battles against Saxon forces in 782, twice in 783, and in 784. Although the tactical detail that has come down to us about the last of these four engagements, as is typical for Carolingian battles, amounts to less than a single sentence, it is still enough to contradict the idea that there is "not a single significant battle or campaign . . . in which the cavalry can be shown to have played the tactically decisive role." Since the combat of 784 was described by the *Annales qui dicuntur Einhardi* as an *equestri proelio*, "a battle of cavalrymen," it is hard to imagine how anyone other than cavalry can have played the decisive tactical role in it.[16]

Regarding the second battle of 783, fought near the River Hase, we are told only that Charlemagne led his full army to victory and killed many Saxons; no comment can be made about the respective importance of horse and foot. For

---

Ostmark', *Austrian History Yearbook*, 14 (1978), pp. 4–5. He goes on to add "there is nothing in the records which suggests that mounted forces were the decisive tactical element in Carolingian armies" and that, with reference to the Agilulfinger and early Carolingian Ostmark, "Cavalry units were no doubt tactically significant in pursuing enemy forces fleeing from their fortifications, but we have no reason to believe that they [armies?] were made up of mounted shock warriors." Ibid., pp. 11–12.

14 Victor Davis Hanson, *Carnage and Culture: Landmark Battles in the Rise of Western Power* (New York, 2002), p. 158; see p. 475 for derivation from Bachrach's work.

15 France, 'Military History.'

16 *Annales qui dicuntur Einhardi*, in *Annales regni Francorum inde ab a. 741. usque ad a. 829: qui dicuntur annales laurissenses maiores et Einhardi*, MGH SS rer. Germ., vol. 6, pp. 69; in the *Annales regni Francorum* (in the same volume), p. 68, Charles's force is described as a "scara," and it is noted that he initiated the battle ("*inierunt bellum*").

the battle before that, however, we have a bit more information: we know that King Charles advanced to Detmold with a "only a few Franks"; that the Saxons "prepared for battle in a plain"; and that the Franks "charged into them" or "rushed upon them" "in the usual way" [*solito more super eos inruentes*], upon which the Saxons fled. The fugitives were pursued vigorously and slaughtered mercilessly so that "only a few escaped by flight."[17] The Frankish forces are not stated to be horsemen, but there is nonetheless good reason to presume that this was another example of a battle won by a cavalry charge, considering four things. First, that a small force sent deep into enemy territory would likely be composed entirely of mounted men, and indeed of elite, armored cavalrymen (since such a mission would seem unduly dangerous for any type of troops that did not have an asymmetric advantage over the Saxons). That assertion is supported by the case of 782, when a detachment or *scara* with a very similar mission was composed entirely of cavalrymen, as we will see. Second, that men with horses who are accustomed to charge [*inruere*] as their "usual way" of fighting[18] are more likely than not to have done so on horseback, since rapid charges by infantry at the start of a battle normally create disorder and lead to defeat.[19] Third, that only a force attacking on horseback is likely to pursue so effectively as to allow few enemies to escape alive, especially when the pursuers are few in number and the fugitives are on their own home ground. And fourth, that the annalist specifies the attack was made in the "usual way," in between two battles, those of 782 and 784, in which we know the Franks involved fought entirely on horseback.[20]

As already noted, we know the later of the bracketing battles, the combat of 784, was won by Frankish cavalry because the engagement was described as a "cavalry battle." The case of 782, however, is more complicated. Early in that year Charlemagne sent a force of East Franks towards the eastern frontier of his realm, apparently a small force since it was described as a *scara* and its target was characterized as "a few rebel Slavs."[21] While this contingent was en route, its leaders learned that the Saxons had rebelled. The East Franks promptly abandoned their previous mission and moved instead against the Saxons, meeting on their

---

17 *Annales regni Francorum*, p. 64: "*cum pauci Francis ad Theotmali pervenit. Ibi Saxones praeparaverunt pugna in campo, qui viriliter domnus [sic] Carolus rex et Franci solito more super eos inruentes et Saxones terga vertentes . . . cecidit ibi maxima multitude Saxonum, ita ut pauci fugam evasissent.*"

18 The same verb is used in 782. Ibid., p. 60: "*inruerent super Saxones.*"

19 Clifford Rogers, *Soldiers' Lives through History: The Middle Ages* (New York, 2007), p. 178; idem, 'The Offensive/Defensive in Medieval Strategy', in *From Crecy to Mohács: Warfare in the Late Middle Ages (1346–1526). Acta of the XXIInd Colloquium of the International Commission of Military History (Vienna, 1996)* (Vienna, 1997), and reprinted in idem, *Essays on Medieval Military History* (London, 2010), pp. 158–161.

20 Hence I think we do at least have "clues" to how the battle on the Hase was fought; cf. France, 'Military History', p. 330.

21 *Annales regni Francorum*, p. 60.

way with a substantial force that Charlemagne's kinsman Theodoric had raised in the closest Frankish region. The two elements planned a pincer movement against the Saxon camp on the opposite side of the Süntel mountains, but in the event the East Frankish *scara*, reportedly reluctant to have the glory of the victory fall to Theodoric, attacked without waiting for his troops to move into position. "Therefore," according to the *Annales qui dicuntur Einhardi*, "they decided to engage the Saxons without him. They took up their arms, and each and every one of them charged just as fast as his horse would carry him, individually straining for the greatest possible speed, towards the place outside of the Saxon camp where the Saxons were standing in a battle array – as if they were pursuing fugitives who had turned their backs and seizing booty, rather than facing enemies standing in formation. The fighting was as bad as the approach. As soon as the battle began they were surrounded by the Saxons and slain almost to a man." The Frankish dead included Charlemagne's own chamberlain, his constable, four other counts, and twenty other noblemen, along with many of their men; of the top leaders, only Charlemagne's count of the palace escaped.[22]

Those who doubt the importance of cavalry in the Carolingian period see this episode as supporting their view. Bernard Bachrach, for example, writes with reference to this combat that in the "only real battle narrative which exists . . . for the times of Charles Martel, Peppin III and Charlemagne," "far from the Frankish cavalry playing the decisive role, it was . . . decisively defeated by Saxon foot soldiers."[23] In Bachrach's view, moreover, the description of the fight in the *Annales* has important implications for our general understanding of the role of horsemen in eighth-century warfare: "These mounted forces were . . . expected to chase down fleeing enemies and to gather booty. Indeed, such activities would seem to have been their main assignments. However, it is equally clear that from a tactical point of view, horsemen, unsupported by other troops, were not regarded as effective in breaking a prepared Saxon shield wall through the force of an uninhibited charge."[24]

I read the text and its implications for the overall military system and situation quite differently. First, a *scara* headed by three of Charlemagne's most important ministers was composed *entirely* not just of men with horses, but more specifically of cavalrymen, who were capable of fighting on horseback, since "each and every one of them charged just as fast as his horse would carry him" into battle. Second, although the Saxons were clearly much more numerous than the *scara* (since the footmen were able to "surround" the horsemen in the fighting), some of Charlemagne's most experienced military leaders clearly

<hr>

22 *Annales qui dicuntur Einhardi*, pp. 61, 63. John France may be technically correct to say that "the *Annals* do not say that they fought on horseback, merely that they approached the enemy position in this way," but the meaning of the text is nonetheless quite clear. France, 'Military History', p. 90.

23 Berard S. Bachrach, 'Verbruggen's "Cavalry" and the Lyon-Thesis', *Journal of Medieval Military History* 4 (2006), p. 145. See also Lyon, 'The Role of Cavalry', p. 81.

24 Bachrach, 'Charlemagne's Cavalry', p. 183.

expected that they would be able to defeat them decisively with a cavalry charge – the annalist specifically states they decided to attack without waiting for Theodoric because they did not want to share the "honor of *victory*."[25] Third, the annalist also seems to have shared the commanders' expectation that Frankish cavalry ought to be able to defeat Saxon infantry with a direct attack. He clearly feels that the defeat needs to be explained – that is why the annals have more detail on this battle than on others – but does not explain it simply by observing that the Frankish cavalry made a head-on charge against arrayed Saxon infantry. Rather, he emphasizes the fact that they charged *badly*, without maintaining formation. His complaint that they attacked with each man charging as fast as he could, as if riding down fugitives, rather than in the proper way for an attack "against an enemy standing in formation" means that in his mind there *was* a proper way for Frankish cavalry to charge "an enemy standing in formation," presumably with the intent of "breaking a prepared Saxon shield wall through the force of [a] charge," though indeed not "through the force of an *uninhibited* charge."[26] If Charlemagne's constable, majordomo, chamberlain, and court historian shared the presumption that a *well-conducted* cavalry charge, unsupported by other troops, could expect to achieve "victory" against a numerically superior army of infantry, ready in battle formation, then we should demand positive evidence, not just an absence of evidence, before accepting the conclusion that they were wrong.[27] Moreover, as we have already seen, the example of the battle of 784 shows that Carolingian cavalry could also be "decisive" when fighting enemy cavalry.

The point that Carolingian cavalry could be decisive in battle is supported by another of the few Carolingian engagements for which we have at least a little tactical detail, fought at Andernach in 876 between Emperor Charles the Bald and Louis the Younger, king of East Frankia. Charles saw an opportunity when he learned that "nearly all of [Louis's] army was dispersed in various locations in order to collect forage for the horses" – a detail which indicates the large role of mounted men in Louis's army.[28] Seizing his chance, Charles attempted a surprise attack on the small force still remaining with the king, apparently during a truce. According to Bachrach's own narrative of the battle, Charles opened the fighting with a cavalry charge. However, his attack failed to break through the enemy lines, and his men were counter-attacked on both flanks by Louis's cavalry, and routed. Admittedly, Bachrach's description of the tactics of this combat goes rather far

---

25 *Annales qui dicuntur Einhardi*, p. 63 ("*fama victoriae*").

26 Cf. John France's suggestion that "the passage could be read to suggest that the use of the horse in the face of the enemy was unusual." 'Military History', p. 90.

27 In 833, according to Nithard, a "small number" of Lothair's troops defeated a "large army" of the emperor's men because the latter force was disorganized, whereas Lothair's men "moved as one man." *Nithard's Histories*, in *Carolingian Chronicles*, tr. Bernhard Walter Scholz with Barbara Rogers (Ann Arbor, 1972), p. 135.

28 *Annales Fuldenses sive Annales regni Francorum orientalis, MGH SS rer. Germ.*, vol. 7, p. 88.

beyond what is securely in the sources, but the *Annals of Fulda* do imply that "all Charles's army" fought on horseback, and the *Annals of St-Bertin* do state that his squadrons (*scaris, cuneis*) opened the battle with a charge (*irruentibus*).[29] Hence we have another instance where the sources indicate that one Carolingian commander *expected* to be able to win a battle with a cavalry charge, and where – if Bachrach's reconstruction of the tactics is correct – the other side *did* win with cavalry counter-charges.[30]

There is one more Carolingian battle that can be discussed in some detail, and that casts valuable light on the question at hand: the battle of the Dyle in 891. This combat is famous in the history of tactics mainly because the author of the *Annales Fuldenses* notes in passing that at this time the Franks were unused to attacking *pedetemptim*. This passage was read by Heinrich Brunner and Lynn White as indicating that the Franks by 891 were unaccustomed to fighting on foot.[31] Bachrach has rightly pointed out that *pedetemptim* does not actually mean "on foot," but rather "step by step," and implied that this correction undermines the case for a cavalry-dominated Carolingian army. This is a bit of a red herring, however. First, if the Franks were not used to fighting "step by step" but *were* used to fighting on foot, that would imply their infantry normally either attacked with a rush, or fought standing still, i.e. defensively. The latter interpretation is improbable, since we know the Franks often did use offensive tactics (as Charles did at

29 Bachrach, 'Verbruggen's "Cavalry"', pp. 152–4; *Annales Fuldenses*, p. 89; *Annals of Fulda*, tr. Timothy Reuter (Manchester, 2000), p. 80: "Without doubt God fought against Charles in these battles, for, as the prisoners who were led away reported, when Louis and those who were with him appeared, such fear filled all Charles's army that they thought themselves defeated before battle had begun, and, what is still more remarkable, though they scored and gashed the flanks of the horses on which they sat with their spurs, these remained standing as if tied to posts." *Annales de Saint-Bertin et de Saint-Vaast*, ed. C. Dehaisnes (Paris, 1871), p. 251.

30 Bachrach, 'Verbruggen's "Cavalry"', pp. 154, 154n, emphasizes the "key role" of the Saxon infantry who met Charles' initial charge: "First, they attracted the full force of the enemy's uninhibited mounted attack. Then, as an indication of their sound training and tactical skill, the foot soldiers gradually withdrew, i.e. refused the center, while under attack . . . it must be emphasized that in the last phase of the battle, the Saxon foot held fast, and were unbroken." This view, however, is difficult to reconcile with the statement of the *Annales Fuldenses* (p. 88) that "The Saxons, who were positioned as the first line against the enemy, first began the fighting, but terrified by the multitude of their adversaries, they turned their backs *parumper* ['for a little space' or 'for a little while' or 'in a short time/quickly']. But the Eastern Franks [re-]engaging strongly from both sides, and Charles' banner-bearers being killed, they compelled rest [of Charles' men] to flee." ("*Saxones autem, qui in prima fronte contra hostes positi erant, primum iniere certamen, sed multitudine adversariorum territi parumper terga verterunt. Franci autem orientales ex utraque parte fortiter [re]pugnantes ac signiferis Karoli occisis ceteros fugere compulerunt.*") Even assuming the author was using *parumper* to mean "for a little while" rather than the more poetic usage of "*in* a little while" (which seems to make more sense in context), turning their backs in terror and then rallying after the enemy is hit on the flanks is not the same as making a dogged fighting retreat *à la* Cannae.

31 White, *Medieval Technology*, p. 3; Heinrich Brunner, 'Die Reiterdienst und die Anfänge des Lehnwesens', *Zeitschrift der Savigny-Stiftung für Rechtsgeschichte, germanistische Abteilung*, 8 (1887), p. 2.

Andernach fifteen years before the Dyle), and indeed in that the eighth century a vigorous charge was their "accustomed way" (*solito more*) of fighting. That the normal Frankish tactical practice was a charge on foot is not impossible, but it is improbable, since it was a medieval tactical maxim that when infantry forces clashed, the key to success was good order, and that therefore an attacking force (and even more so a rapidly charging force) was at a very large disadvantage.[32] Hence, it seems that, in the passage in question, the *Annales* are at least implying that the Franks *were* unused to fighting on foot.

That impression is strengthened by the context. The first thing to note is that, in order to catch the highly mobile Vikings who had been raiding his kingdom, Arnulf moved against them with only his fast-moving troops, all of whom were mounted.[33] In other words, he was seeking a battle with the invaders, and for that purpose considered his horsemen sufficient and his infantry unnecessary. After he caught up with the Danes, he "considered joining battle immediately," but "hesitated to risk so large an army, *because* with a marsh blocking the way in one area, and the bank of the river on the other side, *the terrain did not allow for an attack by the cavalry.*"[34] In other words, 'plan A' would have been a cavalry charge, and it was only because that was not feasible that he hesitated, took counsel, and decided – anxiously, the chronicler notes – to have his men dismount, prepare themselves to fight on foot, and attack step by step. This does show that one reason the Frankish cavalry was such a valuable element of Carolingian armies was because it was so flexible, but it also reinforces the conclusions already reached in this essay based on the events of 782, 783, 784, and 876: that the Franks of this period generally preferred to fight battles by attacking on horseback, and that they expected such attacks to succeed. In other words, the Franks considered their cavalry to be the decisive tactical arm of their military forces.[35]

Since I have only very briefly addressed Bachrach's points that sieges rather than battles were the most strategically important struggles in medieval warfare, and that in siege warfare horsemen "had at best a minor role to play," I will not claim to have shown that cavalry *was* the decisive arm of the Carolingian military.[36] However, I hope that I have demonstrated that the little tactical evidence

32 See note 19.

33 *Annales Fuldensis*, 120, and also Regino of Prüm, *Chronicon*, pp. 138–9.

34 *Annales Fuldensis*, p. 120: "*Transito igitur celeriter eodem fluvio nec mora meditatum est proelium applicari. Cunctanti namque regi, ne tamen valida manus periclitaretur, quia interiacente palude ex parte una, ex altera circumfluente ripa non donatur facultas equitibus aggredi, oculis, cogitatione, consilio huc illuc pervagabatur, quid consilii opus sit, quia Francis pedetemptim certare inusitatum est, anxie meditans.*"

35 Guy Halsall reaches similar conclusions in his analysis of this battle. Guy Halsall, *Warfare and Society in the Barbarian West, 450–900* (London, 2003), pp. 186–8. It is worth noting in passing that John France has pointed out that earlier in the same year as the Dyle, "a *Frankish* infantry force was defeated by a *Viking* cavalry charge." John France, 'Military History', p. 84.

36 Bachrach, 'Medieval Siege Warfare', p. 126. I do plan to address this point in a subsequent article that will focus on cavalry in the High Middle Ages.

available tends to support, rather than undermine, that proposition, at least at the tactical level. The paucity of the source material available to us makes it impossible to reach really solid conclusions about the relative importance of cavalry and infantry in Carolingian armies, but the idea that an advantage in heavy cavalry was key to Frankish military success and expansion in this period should not be cavalierly dismissed, as it often has been in recent writing.

3

THE DEVELOPMENT OF THE LONGBOW IN LATE-MEDIEVAL ENGLAND AND "TECHNOLOGICAL DETERMINISM" (EXPANDED VERSION)[1]

The Longbow and the Infantry Revolution

In 1993, as part of an article on the "Military Revolutions of the Hundred Years War," I included in my analysis of the fourteenth-century Infantry Revolution some discussion of the English strand of that phenomenon, to which England's famous longbow archers made a substantial contribution.[2] I did not expect that section to be particularly controversial; though it was not entirely derivative, it was also not exactly revisionist. The idea that the development of the English or Anglo-Welsh longbow in the late thirteenth and early fourteenth centuries had "revolutionised" warfare was a commonplace already in nineteenth-century historiography, and my own position was more moderate than that: "In the case of the English, the development of the six-foot yew longbow, substantially more powerful than the approximately four-foot Welsh elm bows of the early thirteenth century, *played an important role*."[3] I connected this point to the broader Western European infantry revolution by noting that victories won by the English in the 1330s and 1340s, to which longbowmen indisputably made very great contributions, "combined with the partial success of the Flemish pikemen, encouraged others – including but not limited to the Swiss – to develop effective infantry armies."[4]

---

1 My thanks to the *Journal of Medieval History*'s anonymous reader for helpful comments on the original version of this chapter. The text presented here is substantially expanded from the previously published version.

2 "The Military Revolutions of the Hundred Years' War," *The Journal of Military History*, 57 (1993), 241–278. Reprinted with revisions in Clifford J. Rogers (ed.), *The Military Revolution Debate* (Boulder, 1995); subsequent citations are to this version.

3 Rogers, "Revolutions," p. 59. Emphasis added.

4 Ibid., 60.

   DOI: 10.4324/9781003399971-5

Although I thought these statements were unobjectionable, that did not prove to be the case. In fact, with reference to the longbow, three main criticisms have been leveled at my thesis. Two were first raised in a 1997 article by Kelly DeVries, who claimed that I had greatly exaggerated the effectiveness of the longbow. DeVries further argued that in doing so I (like David Eltis, Geoffrey Parker, and others) had described "premodern military technology in . . . deterministic terms."[5] I have responded elsewhere to Professor DeVries's suggestion that the longbow was not a lethally effective weapon in the fourteenth and early fifteenth centuries, and will not repeat my evidence or arguments here.[6] His other claim has more recently been taken further by John Stone, who devoted a full-length article in the *Journal of Military History* to "unearth[ing] the problematic character of [my] conceptual foundations" and showing how I "endow technology – and more latterly technique – with deterministic qualities," leading to an analysis that is "unsatisfactory in many regards."[7] I will return to that topic briefly at the end of this article, but first I must address another, more serious, argument, one recently developed by Matthew Strickland.

The New Orthodoxy: The "Myth of the Shortbow"

In the distinguished volume on the *Great Warbow* he co-authored with Robert Hardy, Strickland – building on earlier work by Jim Bradbury and Sir James Holt[8] – argued that the shift from the shortbow to the longbow cannot have had any consequences whatsoever, for the simple reason that it never happened. His conclusion is that "the shortbow as a specific category of weapon forming the impotent forerunner of the longbow simply did not exist."[9] He accepted Jim Bradbury's statement that "longbows in the eleventh century might have been

---

5 Kelly DeVries, "Catapults are Not Atom Bombs: Towards a Redefinition of 'Effectiveness' in Premodern Military Technology," *War in History* 4 (1997), 454–470, at 454–5; on p. 462 he lists me along with Jim Bradbury as historians who "have tried to resurrect the old claim of longbow effectiveness" – though in fact I was not trying to "resurrect" a claim that I did not think had ever been seriously wounded, much less killed off – and implies that I subscribe to the idea of the "invincibility" of the longbow, which I certainly do not.

6 Clifford J. Rogers, "The Efficacy of the Medieval Longbow: A Reply to Kelly DeVries," *War in History* 5, no. 2 (1998), 233–42.

7 John Stone, "Technology, Society, and the Infantry Revolution of the Fourteenth Century," *The Journal of Military History* 68 (2004), 361–380, at 368, 380. Surprisingly, Stone also in effect accuses Kelly DeVries himself of technological determinism, using the typically indirect formulation that "DeVries' readership might be forgiven for inferring that technology is, after all, the driving force behind the nature of medieval warfare." Ibid., 366. Similarly Stone says Geoffrey Parker's work "invites the inference that [the Military Revolution's] developmental trajectory was governed by a logic inherent in the technology itself" then on p. 367 refers to Parker's "technological determinism" as if it were something present in Parker's work rather than something that could be inferred from it.

8 Matthew Strickland and Robert Hardy, *The Great Warbow: A History of the Military Archer* (Stroud, 2005) [hereafter cited as *Warbow*]; Jim Bradbury, *The Medieval Archer* (Woodbridge, 1985), pp. 12–15, 71–75, et passim, but especially p. 75; James C. Holt, *Robin Hood* (London, 1982), pp. 78–9.

9 *Warbow*, p. 37.

about 5ft in length, gradually increasing to become about 6ft by the fifteenth century"[10] but nonetheless believed (like Bradbury) that "the self-bow employed by the archers throughout the medieval period was essentially the same weapon."[11] Kelly DeVries and John France have concurred with the view that longbows were the normal form of self-bow throughout the Middle Ages, rather than being a development of the late thirteenth or early fourteenth century.[12] Strickland made his case sufficiently convincingly that no less an authority than Michael Prestwich, in his review of *The Great Warbow*, wrote, "The myth that [the longbow] was a new weapon developed in the later middle ages, replacing an earlier 'shortbow,' is effectively and surely finally demolished."[13] Still, having carefully considered Strickland's evidence and arguments, I contend that his revisionist stance is incorrect, and that the longbow, which first saw widespread use only in the fourteenth century, was indeed a different and significantly more effective weapon than the shorter bows common before then.

More Than Semantics

The dispute about whether the longbow was a "new" weapon that replaced the older shortbow in the late Middle Ages is in part (though, as we will see, far from entirely) about semantics rather than substance. There are two basic approaches to defining what constitutes a "longbow." One is basically meant for modern archery competition and is intended to include the style of bow characteristic of the Victorian revival of archery, which can be quite different from the medieval weapon.[14] The other, as found for example in the *Oxford English Dictionary*,

---

10 Ibid., p. 36.

11 *Warbow*, p. 34 (a statement in the form of a rhetorical question); cf. p. 36, quoting Bradbury: "probably the ordinary bow in England did have a longer stave in the later middle ages, but this should not be emphasized to the point of treating the weapon of the fifteenth century as something quite divorced from that used at Hastings"; 48, "the longbow itself was a constant."

12 Kelly DeVries, "Longbow Archery and the Earliest Robin Hood Legends," in *Robin Hood in Popular Culture. Violence, Transgression and Justice*, ed. Thomas Hahn (Cambridge, 2000), pp. 39, 51; John France, *Western Warfare in the Age of the Crusades* (Ithaca, 1999), p. 26. See also Robert D. Smith, "Weapons, Missile," in *The Oxford Encyclopedia of Medieval Warfare and Military Technology*, ed. Clifford J. Rogers et al. (New York, 2010), vol. 3, pp. 441–2.

   David Whetham, in a variation on the theme, does not dismiss the shortbow's existence, but does still discount the idea that the longbow is a late-medieval development. He claims instead that between Hastings and Edward I's reign "the bow, both long and short, remained popular not only in Britain but elsewhere." "The English Longbow: A Revolution in Technology?" in *The Hundred Years War (Part II), Different Vistas*, ed. L. J. Andrew Villalon and Donald J. Kagay (Leiden, 2008), p. 220; note also 232.

13 *English Historical Review* 122 (2007), 469–71. Prestwich's willingness to accept Strickland and Hardy's case so completely is surprising given his own astute observation a bit further along in the same review that "it is curious that casualty levels appear to have increased as defensive armour became more sophisticated in the later middle ages, particularly if, as is argued here, there was little development in bow technology."

14 For the British Longbow Society definition see Whetham, "Longbow," p. 215 n. 4.

basically defines the longbow as the weapon characteristically used by English archers from the fourteenth through the sixteenth centuries.[15] I will use the word in the latter sense. But I do not intend thereby to *define* it as something new; I fully accept that when bows with the same characteristics were employed before the fourteenth century, they were still longbows.

In the fourteenth and fifteenth centuries, the bows used by English archers were usually referred to just as "bows," but sometimes as "longbowes" and sometimes (especially when contrasted with bows used elsewhere) as "English bows."[16] The latter term was for example distinguished from the "Turkish bow"[17] and from the Irish bow (which was only about half the length of the English bow).[18] For Sir John Smythe, in the sixteenth century, "English Bowes" and "Long-bowes" were synonymous, distinct because they "exceed and excell al other Bowes used by all forren Nations not only in substance & strength, but also in the length & bignes of the arrowes," as testified by "such as have travailed in many parts of *Europe, Affricke,* or *Asia.*"[19]

It is generally accepted that the warbows used by the English in the fourteenth and fifteenth centuries were, when strung, at least equal in height to the archer, and often (especially in the fifteenth century) rather taller. They were "straight" bows (not recurved, reflexed, or double-convex) made of a single stave of wood, usually

15 *Oxford English Dictionary*, s.v.: "The name given to the bow drawn by hand and discharging a long feathered arrow (and so distinguished from CROSS-BOW), the national arm of England from the 14th c. till the introduction of firearms." The *Merriam-Webster* and *Collins* definitions are similar.

16 *Calendar of Inquisitions Miscellaneous (Chancery)*, 8 vols. (London, 1916–2003), vol. 4 (1377–88), p. 200, for a 1386 record of 100 bows "called 'longbowes'" having been provided, earlier, for the defence of Pembroke castle. My thanks to Brent Hanner for this reference. Note also *The Parliament Rolls of Medieval England 1275–1504*, ed. Chris Given Wilson et al., 16 vols. (Woodbridge, 2005), vol. 12, p. 54 (Feb. 1449, m. 5, col. b ["longe bowe"]; vol. 14, p. 464 (Jan. 1483, m. 2 col. B); vol. 15, p. 383 (Nov. 1487, m. 12); *Statute Rolls of the Parliament of Ireland*, ed. Henry F. Berry et al., 5 vols. (Dublin, 1907–2002), vol. 2, pp. 646–48; *The Paston Letters, A.D. 1422–1509*, ed. J. Gairdner, new edn., 6 vols. (London, 1904), vol. 2, p. 101 (1449?).

17 The meaning of this term is unclear. Despite the natural presumption, it does not seem to have meant a Turkish composite bow, since the one in the Esnyngton case, discussed p. 66 below, is described as made of Spanish yew. They were also not necessarily shortbows: Chaucer refers to one "Turke bow" as "long" (*The Romaunt of the Rose* [Montana, 2004], p. 28), and Gaston Phébus seems to recommend the same length for Turkish and English bows (see n. 104). The word may perhaps have been used for bows with double-convex forms.

18 See the following.

19 John Smythe, *Certain Discourses . . . Concerning the Formes and Effects of Divers Sorts of Weapons* (London, 1590), unpaginated introduction, 39 (quotation), 49v. English bows were described by a sixteenth-century Italian writer as *"ingentes"* (enormous). Paulus Jovius, quoted George Agar Hansard, *The Book of Archery* (London, 1841), p. 373. The essential correctness of Smythe's observation can be confirmed by examination of the extensive ethnographic collection of bows in the British Museum. Based on a sample of 100 weapons, the majority of the wooden selfbows in the collection are under 5' and many are under 4'. The few that are longbow-length are generally quite a bit less stout than *Mary Rose* bows, and even somewhat slighter than Migration Era bows recovered at Nydam, for example, meaning they would only be at most half as strong as a median *Mary Rose* bow. Long Paraguayan, Philippine, and Polynesian bows tested by Saxton Pope drew only 48–60 pounds. Saxton T. Pope, *A Study of Bows and Arrows* (Berkeley, [1934]), 28–29.

yew cut so as to include both elastic sapwood (on the outside or "back" of the bow) and strong heartwood (on the inside or "belly"), and not backed with sinew. They were thick bows with a so-called "D-shape" or "deeply stacked" cross-section and nocks near the end of the wood (often with horn caps to protect the soft wood from the string). Because of their length between the nocks, they could accommodate long arrows, drawn to the ear, without snapping or "setting" (losing their springiness due to cellular damage to the wood caused by over-strain) – something which, it is crucial to note, shorter selfbows of the same cross-section could not do.[20] Based on both iconographic and documentary evidence, it is clear that already by the early fourteenth century longbows were similar to those excavated from the wreck of the *Mary Rose* (1545),[21] which have thickness:width ratios of right about 1:1.1, with the thickness or "depth" at the center of the bow ranging from roughly 3.25 to 3.5 cm, and draw weights in the area of 90–180 lb. at 30", most commonly (according to one analysis) in the 150–160 lb. range.[22] The exceptional thickness of these bows is crucial, because the draw strength of a straight bow rises

20 A full draw to the ear, in the classic English archer stance, is around 30–33" for a man of normal height (e.g. 32" for me, a man of 5'11" height). Although a bow can potentially be used with an arrow fully half its length (so that a 5' bow could theoretically shoot a 30" arrow), this is dangerous and also bad for the bow – especially with a powerful bow. Fourteenth-century texts call for bows 2.2 to 2.5 times the length of the arrowshaft. (See note 104). One modern manual oriented towards archers using 50–80-pound longbows equal in height to the archer recommends arrows of just 24" for bows of under 5' and arrows of 28" only for bows at least 5'10". *Archer's Manual, or the Art of Shooting with the Long Bow* (Philadelphia, 1830), pp. 19–20; slightly differing from its source, Thomas Waring, *A Treatise on Archery* (London, 1824), p. 26. Saxton Pope notes that it would be practically impossible for a 4' bow to be used with a 26" arrow unless backed with sinew and that a twentieth-century English target bow will "invariably" fracture if drawn to 30"; that a bow of around 5'6"–5'8" will break if drawn to 31"; and that it would take a 6'6" longbow (which happens to be the average length of the sixteenth-century longbows recovered from the wreck of the *Mary Rose*) to be able to withstand the strain of a 36" draw. Saxton T. Pope, "A Study of Bows and Arrows," *University of California Publications in American Archaeology and Ethnology* 13, no. 6 (1923), pp. 354–5; *Hunting with the Bow and Arrow* (New York, 1923), pp. 42, 45; idem., "Study," p. 366. In other words, a shoulder-high D-section bow cannot be drawn fully to the ear without snapping, whereas a head-high one can. This fundamental point seems to be missed by Bradbury, *Archer*, p. 73, and Strickland, *Warbow*, p. 35, though it was appreciated by earlier writers such as Oman and Morris. See also Jim Hamm, ed., *The Traditional Bowyer's Bible* [hereafter cited *TBB*], 4 vols. (Guilford, CT, 1992–2008), vol. 1, pp. 61–2.
21 For the documentary and iconographic evidence, see the following.
22 *Warbow*, pp. 17–18: this was validated by comparing newly-made replica bows to a mathematical model's predictions and finding that the latter returned "the exact draw-weights of each bow at various draw-lengths." Moreover, this match was to a replica made with Oregon yew; subsequent tree-ring study of the *Mary Rose* bows suggests that may have led to *understatement* of the actual bows' original strength. Robert Hardy et al., "Assessment of the Working Capabilities of the Longbows," in *Weapons of Warre: The Armaments of the* Mary Rose, ed. Alexandra Hildred, Archaeology of the Mary Rose 3, vol. 2 (Portsmouth, 2011), 2:627. Another method gives a range of 90–180 lb., with the mode and median in the 110-lb. range. *Weapons of Warre*, 2:616, fig. 8.27. Confirmation of the great power of late-medieval longbows is provided by a fifteenth-century treatise that says good archers normally do target shooting at 300 *pas* (275 English yards), while the best archers shoot at 400 *pas* (366 yards). *L'Art d'archerie*, ed. Henri Gallice (Paris, 1901), p. 29.

proportionally with its width, but proportionally with the *cube* of its thickness, so that, for example, 30% greater thickness equates to 120% greater power.[23]

How many of these characteristics does a bow need to share before it can be called a "longbow"? I would say nearly all of them, since the word "longbow" was brought into common usage specifically to describe the English bows that contributed so significantly to the English victories of the fourteenth and fifteenth centuries. Still, bows with somewhat less circumference and draw-weight should be accepted as longbows, since the *Mary Rose* bowmen were elite archers, and would have drawn mightier bows than many of their contemporaries. Strickland says, "It is the combination of length and depth which distinguishes the longbow from other types of self-bow such as the 'flat bow.'"[24] I essentially agree with that statement, but then the questions become *how long* and *how deep* (both absolutely and relative to width) does the bow have to be to qualify as a longbow?

Since it is generally accepted that the bows shown in the Luttrell Psalter are "longbows," and the artist clearly thought of the normal height for these bows (when strung) as equal to the height of the archer, I will accept that standard as sufficient to warrant the name longbow, even though documentary and iconographic evidence suggests that in the late fifteenth century the norm was for longbows to be some 4"–6" or more taller between the nocks than were their users.[25] The *Mary Rose* bows, with just a few outliers, range from 6'3" to 6'8".[26] Strickland, however, accepts as "longbows" bows of "about 5ft in length."[27] This I cannot agree with, because the reason to have a separate name for a "longbow" is to distinguish it from other sorts of bows by its length, and, as already noted, we know from iconographic and documentary evidence that already before the time of Halidon Hill (1332) and Crécy (1346) English archers were using bows with staves significantly longer than their own height.[28] So, in this article I will reserve the word

---

23 *TBB*, vol. 1, p. 66; Bob Kooi, personal communication; *Weapons of Warre*, 2:624–5. Hence, most of the Neolithic and pre-Christian Germanic longbows that have been found had only around 50–60% of the strength of a median *Mary Rose* longbow. By calculation, using the Kooi method with data from *Warbow*, p. 39.

24 *Warbow*, p. 40.

25 *Statute Rolls of . . . Ireland*, vol. 3, p. 292. For examples of fifteenth- and sixteenth-century depictions of longbows significantly above the height of the user, see Bodleian Mss. Douce 332 f. 18v; Douce 353 f. 154; New York, Pierpont Morgan Library Mss. G 23 f. 93v; G 55 f. 140v; M 285 f. 249v; M 1053 f. 191; *Warbow*, pp. 363, 392; Bradbury, *Archer*, pp. 37, 47 161. For fourteenth-century examples, see London, British Library [hereafter BL], Ms. Royal 10 E IV fols. 93v, 168, and Egerton 189 f. 14v; Koninklijke Bibliotheek, the Hague [hereafter KBH], Ms. MMW 10 B 23 f. 317. See also note 104.

26 Clive Bartlett et al., "The Lonbow Assemblage," in *Weapons of Warre*, fig. 8.9, p. 595.

27 *Warbow*, p. 36 (quoting Bradbury). On p. 40 Strickland accepts as a "longbow" the Chessel Down bow, which the excavation report described as "about five feet in length." C. J. Arnold, *The Anglo-Saxon Cemeteries of the Isle of Wight* (London, 1982), p. 66.

28 The Luttrell Psalter artist (BL Add. Mss. 42130, fos. 147v, 45v) clearly depicts the archers whose strung bows are the same height as they are (meaning the bows unstrung would be around 2" longer), as does (to give one example among many) the artist of BL, Ms. Royal 6 E VI f. 303v

"longbow" for deeply stacked selfbows at least equal in height (when strung) to their users. I will use the phrase "fully developed longbow" more specifically to designate *Mary Rose*-equivalent yew bows, comfortably long enough to support a draw of 30" or more, and with thicknesses of at least 3 cm and draw weights of 100 lb. or more.

Traditionally, since the popularization of the term "longbow" to refer to the English bow of the fourteenth century, the term "shortbow" has been used in context to designate bows of lesser stature, including those in the 3'–4' range (when strung) and those in the shoulder-high class (that is, for a 5'8" tall archer, in the region of 4'9"–4'11" when strung, or around 4'11"–5'1" in total length).[29] Since it will be valuable to distinguish between those two classes of bows, I will refer to the former (up to breastbone height, or around 45" on a 5'8" man, when strung) as "shortbows," and designate bows longer than that, but no more than shoulder high, as "medium bows." The term "ordinary bows" will include both short and medium bows. "Shortbows" will mean wooden selfbows, including those with sinew backing, unless qualified by the designation "composite." The reader should note that this is a change from the terminology I (following the historiographic tradition) used in the past, when "shortbow" was used more broadly also to include bows in the shoulder-high class. In this article I will use the terms "near-longbow" or "transitional bow" for bows that when strung stand above the archer's shoulder line, but below his head-top line.[30]

Given these definitions, a longbow is a clearly a significantly different weapon than a shortbow. Moreover, although it is less intuitively obvious, a head-high bow is also sufficiently different from a shoulder-high bow of similar width, thickness, and general design to be classed as a distinct weapon. The reason was given by Oman and Morris in the nineteenth century: a longbow can be used with a long arrow drawn to the ear, which makes it possible to launch an arrow with substantially increased power, while a shoulder-high bow cannot. It is a bowyer's rule of thumb that a thick, unbacked wooden selfbow should be 2.2 to 2.4 times the length of the arrow it is to be used with, because drawing an arrow more than 42–45% of the bow's length is likely to damage or break the bow.[31] Thus, a 4'8" bow with a longbow-style "D-shape" cross-section cannot safely be drawn even to 28," whereas a 5'8"-tall archer's fully developed longbow is (optimistically) just robust enough to be drawn a full 32" to the ear, about the longest

---

(c. 1360–75). The Skeffington murder case (discussed later) describes a bow 6' in height, which would be at least 4" taller than the average Englishman of the time. The *Mary Rose* crew averaged about 5'7" tall.

29 This was the usage of, for example, Oman and Morris, though Bradbury argued the word "shortbow" should be used only for composite bows.

30 I realise that some may dislike these categories, or the names given to them, but I know of no other clear-cut definitions of the terms, and one must start somewhere.

31 Bradbury seems to miss this key point when he writes that "there is nothing to prevent one drawing the shorter bow to the ear." *Medieval Archer,* 73.

draw biomechanically possible anyway for someone of that height.[32] Furthermore, because of its longer draw and more favorable string-angle, a long bow stores substantially more energy than an ordinary bow with the same draw-weight. This means that a 6' bow at full draw can store around half again as much energy as an otherwise-similar 4'8" bow.[33] The additional kinetic energy and momentum a longbow can impart to an arrow make a large difference in range, and a huge difference in the terminal effect on an armored target.[34] It is less significant, but also worth noting, that a longbow releases an arrow more smoothly and with less "hand shock" than a shorter bow, making it more consistently accurate.[35] All this justifies the designation of the longbow as a distinct and different weapon.

My more restrictive definition of what constitutes a "longbow" will account for some portion of the disagreement between myself (among others) and Strickland (among others) as to the "newness" of the longbow in the fourteenth century. The real difference between our positions, however, is primarily substantive rather than semantic. It is my contention that the bows normally used in warfare in the British Isles (and the rest Western Europe, with the possible exception of Scandinavian areas) between the eleventh century and the fourteenth were of the shortbow and medium bow classes, as defined earlier. Strickland, on the contrary, seems to believe that the military bows of that era and geographical range were a mix of true longbows and what would by my definition be "near longbows," and to believe that bows of less than 5' (which I think were the norm) were never widely used in warfare.[36]

Evidence for the Transition from Ordinary Bow to Longbow

Aware that he is going against a long- and widely accepted conclusion by denying that the longbow was a new development in the reign of Edward I (1272–1307), Strickland wisely begins by explaining how the opinion he considers to be wrong came to be so commonly held.

The main reason for belief in a shift from ordinary bow to longbow originally came from iconographic evidence. Art from before (and during) Edward I's

---

32 Note 20, and Jim Hamm, *Bows and Arrows of the Native Americans* (Guilford, CT, 1992), p. 28; *Warbow*, p. 418 n. 42.

33 Rogers, "Revolutions," 59–60; *Warbow*, p. 37. The force/draw curves comparing a 4' bow to a 6' bow, each tillered to draw 50 lb. at 28" in *TBB*, vol. 1, p. 52, indicate that the longer bow stores roughly 22% more energy at the 28" draw, and (by extrapolation) around 76% more energy at a 32" draw (compared to the shorter bow at 28" draw, about the maximum it could handle without breaking, even if sinew-backed). If the 48" bow were unbacked and drawn only to a safer 24" (still long for that bow) it would have a draw weight of around 33 lb. and would store only about 40% as much energy as the longer bow at the longer draw.

34 See the following.

35 *TBB*, vol. 1, pp. 70, 74–5, 78.

36 *Warbow*, p. 36, 48.

reign usually shows bows that when strung and resting on the ground would be at most shoulder-high on the archer, and which are drawn, with two fingers on the string, back to the center of the chest (most commonly), or to the chin, or to the cheek in front of the eye. Depictions of English longbowmen in the late middle ages, by contrast, show bows as tall as, or taller than, the archers, drawn to the ear (a longer, more powerful draw) with three fingers on the string (indicative of the greater power, because the pressure of a powerful longbow's string is too great to be at all comfortable unless spread over three fingers.)[37] (See Figure 3.1.)

The other reason for the belief that the longbow developed during the reign of Edward I (or later) is the clear importance of military archery in late medieval England – testified to in accounts of battles and in administrative documents – compared to its relative insignificance earlier, especially when used against armored men rather than horses or unarmored Scots. We know from detailed accounts that the longbow of the fourteenth and fifteenth centuries was a highly effective weapon and played a major role in nearly all the English battles of the period;[38] the fact that before the fourteenth century archery is usually not given much prominence in battle narratives (except when used against horses or unarmored men) suggests that the bows in use must have been less powerful than the bows of the later era, when archers' contributions are often emphasized, even though defensive armor had in the meantime become much stronger and more complete.[39] A variation on this point is that in England through the 1260s, crossbows and crossbowmen were more highly valued than selfbows and archers, whereas by the early fourteenth century the English monarchs were using huge numbers of archers and few crossbowmen in their armies, and by the 1340s (particularly at Crécy in 1346) it had become clear that English longbows were significantly more effective weapons than crossbows.[40]

Strickland nicely summarizes the second of these broad points, in order to disagree with it: "The clear dominance of the crossbow in the armies of Henry II, Richard and John," thought Sir Charles Oman and J. E. Morris, "could only be accounted for by the inferiority of the shortbow: It was impossible to believe that a commander as gifted as Richard the Lionheart ... would have favoured the crossbow if he had known the military value of the true longbow of the fourteenth century."[41]

37 We know this mainly from modern experience and iconographic evidence. In the fifteenth-century *L'Art d'archerie,* pp. 26–27, the three-finger draw with two fingers below the arrow-nock is assumed.

38 Rogers, "Efficacy."

39 Morris, *Welsh Wars,* 26, writes: "The powerful longbow and the heavy couched tilting lance of Poitiers would have annihilated Harold's men [at Hastings] in an hour." This is a strong statement, but I am not sure it is wrong. See also 28: "Mr. George has shown on the clearest negative evidence that [the selfbow] played no part in the crisis of Lewes and Evesham."

40 Ibid., pp. 28, 100.

41 *Warbow,* p. 35.

Figure 3.1 Ordinary, Short, and Long Bows. The Bayeux Tapestry clearly shows medium bows drawn to the center-chest with two fingers on the string. The next two images (British Library Board, BL MS Royal 10 E IV, fos. 120 and 92v, generously released into the public domain by British Library) are the work of one English artist of the mid-fourteenth century. The first shows a shortbow with a two-finger draw to the chin. The other shows a longbow drawn with three fingers to the ear. Note how the *front* of the fletchings are in roughly the same place on the longbow-man as the arrow-nock is on the shortbow archer. The detail of the Bayeux Tapestry (eleventh century) is reproduced by permission of the City of Bayeux.

The Revisionists' Objections: The Argument of Efficacy

Strickland responds to this second line of argument with a reassessment of the importance of archery in the period before Edward I's reign, with emphasis on the Anglo-Norman ambit, arguing that bowmen with selfbows were important in armies right through the Middle Ages. He claims that "there is nothing to suggest that the bows of the royal archers at Bourgthéroulde [1124] or of the Yorkshire levies at the Standard [1138] were any less powerful than . . . those bent by their fourteenth-century successors in the Hundred Years War."[42] If (Strickland argues) the archers did not often effect the same sort of slaughter in this period that they would in the fourteenth century it is partly because they were often ordered to shoot at horses rather than riders, for reasons of chivalry, and also because "massed archery was not deployed on many twelfth-century battlefields."[43] Then, from the mid-twelfth to the late-thirteenth century, the role of archers with selfbows was limited by "the increasing dominance of the crossbow" and by the decline of the tactical use of archery in combination with dismounted knights, a pairing that had flourished in the Anglo-Norman period and would again in the fourteenth century.[44] Nonetheless, "archers continued to form a key element in Angevin armies under Henry II and Richard I. If in the second half of the twelfth century it is hard to trace the tactical role of the longbow in major battles in England and Normandy, its importance is strikingly revealed in other theatres of war. Archers were to form an essential element in the forces of Richard I during the Third Crusade," and "Anglo-Welsh warfare and the operations of Anglo-Norman marcher lords in Ireland from 1169 saw longbowmen playing a crucial part in warfare on the peripheries of the Plantagenet lands, coming into their own in operations in difficult terrain."[45]

The Efficacy of the Bow in Western Warfare Before Edward I

In fact, however, both the accounts of Henry of Huntingdon and of Ailred of Riveaulx suggest that the bows used at the Battle of the Standard *were* much less powerful than fully developed longbows. Henry of Huntington specifies that the English arrows penetrated the Scots "who were, evidently [or 'doubtless'], unarmoured" [*eos nimirum inermes penetrabant*].[46] Since the armor of the day

---

42 *Warbow*, p. 76; note also p. 78.
43 *Warbow*, pp. 78–9.
44 Ibid., pp. 79–80.
45 Ibid., p. 83.
46 Henry of Huntington, *Henrici archidiaconi Huntendunensis Historia Anglorum*, ed. Thomas Arnold (London, 1879), pp. 264–5. Other chroniclers understood the same point. William of Newburgh says the battle was won when the "light-armored men" (*levis . . . armaturae homines*) were put to flight by the English arrows. In Richard Howlett, *Chronicles of the Reigns of Stephen, Henry II and Richard I*, 4 vols. (London, 1884–1889), vol. 1, p. 34. Henry's expectations match with

was simple mail, such a qualification would have been quite unnecessary had the bow in use been the fully developed longbow, which could drive an arrow through mail armor with ease even from long range. Indeed, the chronicler's use of *nimirum* implies that he would expect arrows to be ineffective against mail-armored soldiers; the same goes for John of Hexham's contrast between the "naked" (i.e. unarmored) Scots who were "pierced and transfixed" with arrows, and the "mail-armored [*loricatos*] and therefore invulnerable" Anglo-Normans.[47] The expectation that mail would render its wearers largely secure against arrows, in turn, implies the bow with which these authors were familiar was not any sort of longbow, much less the equivalent of an Agincourt bow or one of the hefty weapons recovered from the wreck of the Tudor warship the *Mary Rose*.[48]

Henry also describes an attack by the Scottish king's armored household knights on the intermixed archers and dismounted knights of the English force. The attack was unsuccessful, and the cavalry retired "with wounded horses," but there is no mention of dead horses or of wounded or killed riders.[49] Ailred, moreover, describes Galwegian soldiers "bristling all round with arrows," "like a hedgehog with its quills," yet "none the less brandishing [their] sword[s]."[50] Given that the Galwegians were generally lacking metal armor and at the very most would have been armored with mail, this too is strongly indicative of bows much weaker than the weapons of Crécy or Agincourt, which even at long range could inflict severe wounds against unarmored or mail-armored men.[51]

---

the suggestion in version C of the *Anglo-Saxon Chronicle* that an English archer did no good in attempting to dislodge a mail-shirted Norwegian who was guarding Stamford bridge in 1066, and thereby preventing King Harold from pursuing the defeated invaders. *Anglo-Saxon Chronicle*, ed. and tr. Benjamin Thorpe. 2 vols. (London, 1861), vol. 1, p. 339; vol. 2, p. 68. The Old English text is unfortunately ambiguous because the verb "sciten" can mean either shot or thrown, and the noun "flan" can refer either to a javelin or an arrow, though it more often means the latter. (See Joseph Bosworth, *An Anglo-Saxon Dictionary*, ed. and enlarged T. Northcote Toller, Supplement [Oxford, n.d.], s.v. flán.) Either way, however, the episode is a testimony to the lack of effective archery: either there was a bowman who failed to do anything against a mailed target, or else there was apparently no bowman available even to try.

47 John of Hexham, in *Symeonis monachi opera omnia*, ed. Thomas Arnold, vol. 2 (London, 1885), p. 294.

48 Even a relatively weak longbow in the 65–72 lb. draw range is, if used at full draw, highly effective against mail. See Clifford J. Rogers, "The Battle of Agincourt," in Villalon and Kagay, eds., *The Hundred Years War (Part II)*, n. 22, pp. 44–45, and Appendix I of that article, and notes 60 and 62–64 of this chapter.

49 Henry of Huntington, *Historia Anglorum*, p. 265.

50 Ailred of Rievaulx, in *Chronicles of the Reigns of Stephen, Henri II and Richard I*, vol. 3, p. 196; trans. from Alan O. Anderson, *Scottish Annals from English Chroniclers, A.D. 500 to 1286* (London, 1908), p. 203.

51 For abundant evidence of the ability of the fourteenth-century longbow to inflict large numbers of deaths on horses and armored men-at-arms alike, see Rogers, "Efficacy," especially 240–1. For the longbow's effectiveness against mail, see note 48 of this chapter. By contrast, in the Morgan Picture Bible of ca. 1250, arrows are usually shown penetrating their (mail-armored) targets only to the depth of the head or a bit more: e.g. *Warbow*, pp. 108, 80. In fifteenth-century manuscripts,

Although archery was valued and useful in eleventh- and twelfth-century battles, that does not prove either the presence of the longbow or the absence of the shortbow. Even the shortbow could be lethal against unarmored men or horses[52] (just as it was hunting game), and it is against just such targets that accounts from this period do describe arrows as being effective.

Yet bowmen in the armies of Henry II, Richard, John, or Henry III did not have anything like the *relative* importance that they had under Edward III or Henry V.[53] As already noted, Strickland suggests that the bow did not have as large a battlefield role in the earlier period because archers were not used in mass, and because of the dominance of the crossbow. But that raises two questions: *why* were they not used in mass, and *why* was the crossbow dominant if the longbow was available?

Against the mail-armored enemies of the Anglo-Norman kings, archers would have evidently been *even more effective*, by a substantial margin, than they later were against the mail-and-plate armored men-at-arms of Crécy or Poitiers, much less the full-plate armored soldiers of 1415 – if, that is, they were indeed using "powerful longbows" that were "essentially the same as those of their fourteenth- and fifteenth-century counterparts."[54] Since the English archers of the fourteenth and fifteenth centuries were *extremely* effective,[55] in the twelfth and thirteenth

---

arrows from fully developed longbows are frequently shown having penetrated (judging by the amount of shaft still visible) right through human targets, including sometimes plate-armored targets. E.g. *Warbow*, pp. 210, 236, 266; Bibliothèque Nationale, Paris [hereafter BnF] Ms. fr. 2663, f. 61; Bodleian Ms. Douce 353, f. 154. The point is not that these images should be taken as photographic evidence of terminal effects, but rather that they reflect what artists and their audiences *expected*, or at least found within the boundaries of willing suspension of disbelief. The hedgehog metaphor is also used in descriptions of Crusaders peppered with arrows from Eastern composite shortbows, e.g. Ambroise, *L'estoire de la guerre sainte*, ed. Gaston Paris (Paris, 1897), ll. 11,625–11,630, which describe Richard sallying out into the midst of a crowd of Turks and returning without harm [*perte*], though he and his horse were so covered with arrows as to look like a hedgehog. The same metaphor is also used in a longbow context with reference to the Germans at Stoke Field in 1487, who were "broken and defeated, shot up and covered with arrows like hedgehogs." The latter, however, despite being half-armored in plate, clearly did not escape without harm. Jean Molinet, *Chroniques*, ed. J.-A. Buchon, vol. 3 (Paris, 1828), pp. 155–6.

52 See note 84.
53 *Warbow*, p. 36. Strickland acknowledges that Bradbury himself "believed there was a development in the weapon itself from 'the ordinary wooden bow' to the longbow"; but Strickland then adds "though this transformation was gradual rather than the result of some technical revolution." And on p. 38 Strickland notes we do not "find any chroniclers or contemporary observers remarking on a major technical improvement in bow design or a military 'big bang' in the weaponry of the English armies during the decades spanning the end of the thirteenth century." My linkage between the development of the longbow and the Infantry Revolution has to do with a revolution in the weapon's *effectiveness*, and does not disallow the possibility of that revolutionary change being the result of an evolutionary development; indeed, I make it clear in other work (including Chapters 4 and 5 of this volume) that evolutionary change eventually reaching a tipping point and producing or contributing to revolutionary effects is a common pattern in military history, e.g. with the Artillery Revolution of the fifteenth century resting on a century of evolutionary progress.
54 *Warbow*, pp. 79, 78.
55 Rogers, "Efficacy."

centuries they would, if using longbows, have been so effective, and so *obviously* effective, that it is hard to imagine any consideration that would have kept military leaders from employing them in mass, or that would have led rulers to treat bowmen less favorably than crossbowmen. It is possible that, as Strickland suggests, leaders "generally eschewed" using massed longbows against their enemies because of "chivalric mores [of] the twelfth century, which emphasised the desirability of capturing noble opponents where possible rather than slaying them," and that the reason many Galwegians were killed by arrows at the Battle of the Standard, while few or none of the Scottish king's household knights were, is that the English leaders were happy to shoot down men regarded as "scarcely human . . . barbarians," but unwilling to allow the targeting of Earl Henry's knights, "considered [by the Anglo-Normans] as their own."[56] That is possible – but not, in my judgment, at all probable. I think it far more likely that (as Oman and Morris argued long ago) the reason archers of this period are never mentioned as inflicting large numbers of casualties on armored men is that they were not able to do so, because their bows were nowhere near as powerful as fully developed longbows, and that that in turn is why they were not usually deployed in mass. Western selfbows of this period were probably less powerful than Byzantine composite shortbows, yet in 1107 Alexius Comnenus "considered shooting at [Western knights] useless and quite senseless," because he "knew that the Franks were difficult to wound, or rather, practically invulnerable [to arrows], thanks to their breastplates and coats of mail."[57]

Strickland recognizes that "the spread of horse armour in the form of mail barding" "go[es] some way to explaining the predominance of cavalry" and (by implication) the declining role of battlefield archery in the second half of the twelfth century.[58] This implies that mail barding was quite effective against arrows. It is true that mail, in combination with quilted soft armor, could offer a very good level of protection from arrows fired from the sorts of bows used in Western Europe before the fourteenth century. The early twelfth-century chronicle of Galbert of Bruges describes the exploits of an exceptionally powerful archer named Benkin, whose arrows could be distinguished from those of other bowmen because they were able – in addition to the more ordinary result of seriously wounding unarmored men – to drive off mailed knights, "forcing armoured men whom he hit to flee, stunned and bruised though not wounded."[59] If, as is not unlikely, Benkin was using a particularly strong version of a bow similar in form to the ones depicted

---

56 *Warbow*, pp. 78–9.
57 Charles Oman, *A History of the Art of War in the Middle Ages*, 2nd edn., 2 vols. (London, 1924), vol. 2, p. 58; Morris, *Welsh Wars*, pp. 26, 34, 100–2. Anna Comnena, *Alexiad*, trans. E. A. S. Dawes (London, 1928), p. 341. For a different translation and other examples to the same point, see *Warbow*, pp. 100–101, and later in this chapter.
58 *Warbow*, p. 81.
59 Galbert de Bruges, *Histoire du meurtre de Charles le Bon*, ed. Henri Pirenne (Paris, 1891), p. 59: "persequebatur armatos sine vulnere contunsos, stupefactos in fugam vertebat." Galbert does, however, mention one *armiger* killed by an arrow to the heart in an earlier assault (along with many killed by stones). Ibid., p. 54.

in Stephen Harding's Bible (from Cîteaux, dated 1109), this is just the result we would expect. The bows in question appear comparable in length, thickness, and shape to a replica Apache bow that was found in testing to impart a mere 25.8 joules to its arrow; at least half again as much energy that would be required to be effective at any range against mail and padding.[60] The *Chronicle of Colmar*, as late as 1298, describes men wearing mail armor over thick gambesons as impossible to wound with arrows from bows.[61] The chronicler was not employing hyperbole. Using a bow that was probably a good approximation for a late thirteenth-century English bow,[62] even at close range David Jones had 0 of 10 test arrows make *any* penetration against mail and a heavy gambeson of one design; another design allowed 3 of 10 arrows to penetrate, but none more than 1 cm.[63] That was with a narrow bodkin head. With a broader head, even against mail backed by a lighter, unstuffed gambeson, only 1 of 10 arrows penetrated at all, and that one only to a depth of less than 1 cm.[64]

By contrast, however, it is clear from both narrative sources and modern tests that such armor provided little protection against the heavy arrows of fully developed longbows, and could even be defeated by relatively weak longbows. This is because longbows impart much more energy and momentum to their arrows than do most sorts of selfbow shortbows. For example, a 6'4" yew longbow with an 80-lb. draw – far less than the median *Mary Rose* bow, and only about half the most powerful ones – drawn to 32" can impart to an arrow well more than *double* the energy and momentum of the arrows from the 64-lb. at 27" bow in the Jones tests.[65] That does not prove that the archers known to Galbert of Bruges or the Col-

---

60 There is a nice reproduction of the Harding Bible scene showing archers (BM Dijon, Ms.14, f. 13v) on p. 77 of *Warbow*; compare to the photograph of the replica Apache bow on p. 664 of C. A. Bergman et al., "Experimental Archery: Projectile Velocities and Comparison of Bow Performances," *Antiquity* 62 (1988). The Apache bow was very light (with a 37.8 lb. draw at 24"), but it did impart more kinetic energy than did a Sioux bow of similar size and heavier draw (54.7 lb), which, according to Bergman et al., imparted only 13.5 joules (J) to its arrow. Ibid., p. 663. It would take something like 37–45J for an arrow of moderate weight to penetrate mail and padding and inflict a serious wound. Rogers, "Agincourt," appendix 1; David Jones, "Experimental Tests of Arrows Against Mail and Padding," *Journal of Medieval Military History* 18 (2020), 143–172.

61 *Les annales et la chronique des Dominicains de Colmar*, ed. and tr. Charles Gérard and L. Liblin (Colmar, 1854), p. 344: "perque nulla sagitta arcus poterat hominem vulnerare." The context is a battle against Hungarians and Cumans.

62 It was a longbow with a 64 lb. draw-weight (appropriate for a thirteenth-century medium bow) to a draw of 27" (significantly longer and therefore more powerful than a Waterford or Skeffington type bow), imparting about 31J to an arrow on the heavy side for the thirteenth century (52.6g arrow with a 9.5mm socket).

63 A heavier bow, with a 72 lb. draw weight at 27" and imparting around 37J, did significantly better, but still had only 2 of 10 arrows penetrate more than 2 cm, and none reach 4 cm.

64 For this and the previous two notes, see Jones, "Experimental Tests of Arrows Against Mail and Padding."

65 Bergman et al., "Experimental Archery," 663 (up to 83J); cf. Rogers, "Agincourt," appendix I, 110–11.

mar chronicler were necessarily using ordinary bows rather than near-longbows or longbows, but it does strongly suggest it.

Similarly, the relatively heavy continued use of archery in warfare in Wales, Ireland, and Scotland noted by Strickland can be explained by the relative scarcity of metal armor among the peoples of the "Celtic fringe," and does not imply the use of the longbow. On the contrary, it tends to support the idea that the bows of the period were strong enough to be very useful against unarmored targets, but not strong enough to be highly effective against mail-armored ones – which again tends to argue against the longbow being in widespread use.

Selfbows, Crossbows, and Composite Bows Before Edward I

Strickland's two final points under the rubric of battlefield effectiveness – (1) that "archers were to form an essential element of Richard I's army during the Third Crusade," where they "strikingly revealed" the continuing importance of the longbow into this period, and (2) that the lack of massed selfbow archery in Western battles from the mid-twelfth to the late thirteenth century was due in part to "the increasing dominance of the crossbow" – are interrelated.[66] They both are linked to a point made by the Morris-Oman school that is noted, but to my mind not satisfactorily rebutted, by Strickland: that it is "impossible to believe that a commander as gifted as Richard the Lionheart . . . would have favoured the crossbow if he had known the military value of the true longbow of the fourteenth century."[67]

As Strickland well demonstrates, the archers Richard I took with him on the Third Crusade were certainly valued and effective; Richard of Devizes's description of their role at Messina, where the defenders abandoned the walls because they "could not look out without getting an arrow in the eye" shows this clearly. More than was commonly appreciated before Strickland and Hardy published their book, during his campaigns in the Holy Land Richard used infantry archers along with crossbowmen to keep Muslim horse-archers at a distance.[68] But this does not mean that Morris and Oman were wrong to conclude that Richard "much admired" the crossbow, and that crossbowmen, rather than archers, were the most important missile troops in England through the mid-thirteenth century.[69] The crusaders of the Third Crusade generally employed crossbowmen in conjunction with archers, and the evidence strongly suggests that the crossbows significantly outperformed the selfbows in overall effectiveness. At

---

66 *Warbow*, pp. 83, 79.

67 *Warbow*, p. 35; also Oman, *Art of War*, vol. 2, p. 58.

68 *Warbow*, pp. 104–7; *Chronicon Ricardi Divisiensis de rebus gestis Ricardi Primi regis Anglie*, ed. Joseph Stevenson (London, 1838), pp. 23–4. For a twelfth-century English example of effective archery during a siege, see *Gesta Stephani*, ed. K. R. Potter, 2nd edn. (London, 1976), pp. 182–3.

69 Oman, *Art of War*, vol. 2., pp. 58–9. Indeed, Strickland fully accepts that "in the later twelfth and thirteenth centuries it is clear that it was the crossbow that reigned supreme as the missile weapon of choice not only in much of Europe but also in the lands of the Angevin kings of England." *Warbow*, p. 113; see also p. 119.

Jaffa, when Richard prepared his defense against the Turkish horse-archers, he made an outer line of sergeants with shields and lances, then placed a crossbow-man between each two of them, and another man behind the latter to reload the crossbow. Although selfbow archers were present, Ambroise (who notes the deployment just described) does not bother to mention them, saying instead that the Turks did not dare linger in front of the crossbowmen, who "struck them, into [*es*] the body and into the horses." Ralph of Coggeshall does record the presence of bowmen, but only in passing, whereas the crossbowmen are singled out as worthy of the highest praise for the "incomparable *virtute* [manliness, value, excellence, or valor]" they showed in the combat.[70] The *Itinerarium* likewise emphasizes the "slaughter" wrought among the Turkish horses by the king's crossbowmen.[71] At Acre, when Richard constructed a covered firing-position from which to harass the defenders on the walls, he had it manned with "his most skilful crossbowmen."[72] The king himself, moreover, preferred to use the crossbow rather than the selfbow.[73]

He was not alone in this. His rival Philip Augustus also used a crossbow, and like Richard and Richard's successor John he paid crossbowmen much better than other sergeants.[74] In the thirteenth century, according to Ibn Said, even the Muslims of Andalusian Spain generally equipped their infantry with crossbows rather than Eastern composite bows. The basic reason for all this seems to be the greater power of the crossbow, more particularly its ability to penetrate armor and inflict serious wounds at significant ranges. Arrows from Byzantine and High Medieval Turkish composite shortbows, like Western ordinary bows but unlike crossbow bolts,[75] seem to have been only marginally effective against men wearing mail over padded gambesons. Western, Muslim, and Byzantine sources agree on this point. Western chroniclers of the First and Third Crusades speak of their knights as "fearing an arrow about as much as a straw," and being made by their armor as invulnerable as "grey rock."[76] Beha ad-Din says Muslim arrows "made

---

70 Ambroise, *Estoire,* ll. 11,455–11464, 11,502–11,505; Ralph of Coggeshall, *Chronicon anglicanum,* ed. J. Stevenson (London, 1875), p. 50.

71 Richard of Holy Trinity, *Itinerarium peregrinorum et gesta regis Ricardi,* ed. William Stubbs (London, 1864), p. 410.

72 Ibid., p. 225.

73 Not just at Acre, where he was ill and his use of the weapon could be attributed to weakness, but also at Jaffa. Ibid., pp. 225, 408. By contrast, Edward III (1327–77) seems to have been a powerful longbow archer: in 1333, the wardrobe paid 10s. to an archer in recompense for a bow broken by the king: The National Archives, Kew, E 387/9, under 4 July.

74 *Warbow,* pp. 109, 118; Frederick C. Suppe, *Military Institutions on the Welsh Marches: Shropshire, A.D. 1066–1300* (Woodbridge, 1994), p. 22.

75 *Alexiad,* pp. 255–6; *Warbow,* p. 121.

76 Clifford J. Rogers, *Soldiers' Lives through History: The Middle Ages* (Westport, 2007), p. 172. Ottokar von Steiermark, similarly, has a German knight, in order to encourage his comrades to fight the Hungarians, say that he would willingly let the four best archers of the latter take a shot at him, as long as he had his armor and helmet on. *Ottokars österreichische Reimchronik,* ed. Joseph Seemüller, vol. 1 (Hannover, 1890), p. 334.

no impression on" Frankish footsoldiers who were protected by strong mail and thick felt vests.[77] Anna Comnena, as already noted, claims Frankish mail "repels arrows and keeps the wearer's skin unhurt" so well that archers find the Franks "difficult to wound, or rather, practically invulnerable."[78] All this should not be taken entirely literally, but it does seem that while High Medieval composite-bow arrows could penetrate Western mail-and-padding armur, at least at short range, they used up most of their force in doing so, and generally inflicted little injury.[79] There is no reason to think that these weapons were inferior in range or power to Western selfbows of the same period – if anything, the evidence suggests the reverse.[80] This tends to reinforce the conclusion reached already: that Western selfbows before the fourteenth century were generally rather ineffective against

77 *Warbow*, p. 101.

78 *Alexiad*, p. 341.

79 Jean de Joinville, *Histoire de Saint Louis*, ed. Natalis de Wailly, 2nd edn. (Paris, 1874), pp. 132, 212–14; *Warbow*, pp. 100–1. For exceptions, see *Warbow*, p. 100, and cf. n. 81 below. Bergman et al. tested two types of replica Eastern composite bows and found that the replica of an ancient Egyptian bow could impart a maximum of 46J to its arrows, while the replica Crimean bow could impart a more formidable 63J. The former is just enough to do significant damage through mail and a heavy gambeson at short range; the latter would be more effective, especially at long range. Bergman, "Experimental Archery," 663 (ke); Rogers, "Agincourt," Appendix I (requirement). Ottoman Turks in the fourteenth through sixteenth centuries seem to have used substantially more powerful bows, with draw-weights comparable to English longbows. However, since these were used with shorter draws and lighter arrows, they still did not have equal terminal effects. A middling 110-lb.-at-28" Ottoman bow using a heavy Turkish war arrow might impart to an arrow around 95J of kinetic energy [ke] and 2.76 kg*m/sec momentum, whereas a middling 150-lb.-at-32" *Mary Rose* longbow might impart 136J and 5.1 kg*m/sec, i.e. 43% more ke and a formidable 84% more momentum. Even an arrow from a very heavy Ottoman bow (180 lb. draw-weight), though it might have slightly more kinetic energy (145–147J) than one from a 150-lb. longbow, would still have 1/3 less momentum. Adam Karpowicz, "Ottoman Bows – An Assessment of Draw Weight, Performance and Tactical Use," *Antiquity* 81 (2007), pp. 681–2; *Warbow*, pp. 409–411.

80 Joseph O'Callaghan, *Reconquest and Crusade in Medieval Spain* (Philadelphia, 2003), p. 141; James F. Powers, *A Society Organized for War: The Iberian Municipal Militias in the Central Middle Ages, 1000–1284* (Berkeley, 1988), p. 133 n. 92; *Alexiad*, p. 398; *Warbow*, p. 98 (*Gesta Francorum*), p. 109 (Richard I). Composite shortbows, like short selfbows, were recognised as weaker than crossbows (e.g. Ibn Said, quoted *Warbow*, p. 109; *Le Livre des fais du bon messire Jehan le Maingre, dit Bouciquaut*, ed. Denis Lalande [Paris, 1985], 235–6: "leur trait ne fu mie pareil ne de tel force"). In the fifteenth century, by contrast, English longbows were noted as shooting as far as Italian arbalests – which at that date might well be steel-bowed weapons – and since English arrows were at least as heavy as crossbow bolts, this indicates approximately equal power. Dominic Mancini, *The Usurpation of Richard the Third*, tr. C. A. J. Armstrong (London, 1989), pp. 8–9. Also in the fifteenth century, Bertrandon de la Broquière's *Voyage d'Outremer*, ed. Charles Schefer (Paris, 1892), pp. 226–8, 271 (and quoted *Warbow*, p. 101) mainly with English longbowmen in mind, treated Christian bows as equivalent to crossbows with crannequins in their superiority to Muslim composite bows, in both range and hitting power. The experienced soldier Jean Waurin, indeed, described the shot of English archers as more valuable than that of Liégeois crossbowmen at Othée in 1408: *Recueil des croniques d'Engleterre*, ed. William Hardy, 5 vols. (London, 1864–91), vol. 2 (1399–1422), p. 124.

mail-armored opponents.[81] They were therefore probably not longbows, and certainly not fully developed longbows.

The Revisionists and the Archaeological Record

Although the evidence of the limited battlefield effectiveness of the selfbow in the Anglo-Norman period, compared to the crossbow of the time or the longbow of a later era, does suggest the use of short- or medium bows rather than longbows, it is not definitive. Bradbury and Strickland accept that bows of the eleventh century were shorter (at somewhat above five feet, they suggest) than those of the fourteenth century. As the Jones tests discussed earlier show, the relatively low level of efficacy suggested by the narrative sources for archery in the period could have been achieved by relatively slight bows with medium draws rather than (as I believe to have been the case) reflecting the use of various sorts of stout but shorter bows with short draws.

Strickland argues that iconographic evidence (to which we will return later) supports his claim that longbows were in use right through the medieval period, whereas the shortbow was a "myth." But the main impetus behind his belief in that conclusion seems to be the archaeological record. As he rightly notes, there is plenty of archaeological evidence to show that longbows or reasonable approximations thereof were used by Neolithic and early Iron Age humans, by Migration-era Germans, and by Vikings, from the third millennium BC through the tenth century AD. Archaeological evidence for the shortbow in pre-Christian Europe, by contrast, is relatively sparse, though certainly not non-existent.[82]

---

81 There is on the other hand the isolated statement of Giraldus Cambrensis that describes a twelfth-century Welsh archer's arrow penetrating two layers of mail, a man's thigh, and a saddle to kill a horse. This is so obviously exaggerated, however, that it can be disregarded entirely, just as can be the same author's story of beavers castrating themselves to avoid hunters who desired their testicles for medicinal uses. Giraldus' other examples of the power of the Welsh bow – that it could drive an arrow through mail to pin a soldier's leg to his horse, or through an oak door as thick as a palm (which, rather than "a palm [4"] thick," is probably what Gerald meant by "palmalis . . . spissitudinis"; if not, the story can be dismissed as a tall tale) so that the arrowhead protruded on the other side – do indicate a strong bow, with penetrating power comparable to a contemporary crossbow (as he explicitly indicates). However, it is clear that his text is meant to be surprising, indicating that the normal expectation would be that an arrow from a bow could *not* do such things. It is also clear that Giraldus is retelling soldiers' tales of feats understood to be exceptional even for Welsh archers. Giraldus Cambrensis, *Opera*, ed. James F. Dimock and G. F. Warner, 8 vols. (London, 1861–91), vol. 6, p. 54.

82 *Warbow*, pp. 38–41; for more detail, J. G. D. Clark, "Neolithic Bows from Somerset, England, and the Prehistory of Archery in North-western Europe," *Proceedings of the Prehistoric Society* 29 (1963); G. Rausig, *The Bow. Some Notes on Its Origins and Development* (Lund, 1967). Prehistoric shortbows: one approximately 1.25 m long was excavated in Slovenia, for example, and Neolithic bows from northern Europe outside the Alps are mostly "rather small" flatbows, e.g. one from Saxony at c. 134 cm. Boštjan Odar, "The Archer from Carnium," *Arheološki vestnik* 57 (2006), 262; Clark, "Neolithic Bows," p. 64; note also p. 84.

However, even if the absence of evidence for the military use of shortbows among pre-Christian Germanic peoples were accepted as evidence that they were not used, that would not be anywhere near sufficient reason to presume that the peoples of Celtic or Anglo-Saxon Britain, or of High Medieval France, Italy, or Iberia, likewise used the longbow. In other areas of the globe, longbows are far from universal. Among the peoples of Latin America, North America, and Africa who traditionally used selfbows, some used longbows, but shortbows or medium bows were more common. Native American tribes of the Great Plains in the nineteenth century, for example, commonly used bows of only around 3'6".[83] More appositely, it is clear from a variety of sources, textual as well as iconographic, that Irish warriors of the fourteenth, fifteenth, and sixteenth centuries normally used bows in the shortbow class, described by a Catalan diplomat who visited Ireland in 1394 as half the length of English bows. Probably with some exaggeration, an Englishman in the 1590s described the Northern Irish as using short "Scottishe bowes, not past 3 quarters of a yard long," with even shorter arrows.[84]

We know from a variety of sources that the Welsh and Anglo-Norman archers who came to Ireland from England in the late twelfth century were highly effective, and Gerald of Wales indicates that the Irish, who previously did not much use archery in warfare, trained with the bow in response to the threat they posed.[85]

---

83 *Archer's Manual*, p. vi; such short bows could be used even to hunt bison (Pope, *Study*, 19). See also note 19 of this chapter.

84 Ramon de Perellós says "alguns se aiudan darcs, que son ayssi petitz com miech arc d Anglaterra; he si fan aysi gran colp com los angleses" ("some [Irish soldiers] make use of bows, which are as short as half an English bow; but they hit as hard as the English [bows]"). See "Viatges del Cavaller Owein y de Ramón de Perellós al Purgatori de Sant Patrici," in *Llegendes de l'altra vida*, ed. R. Miquel y Planas (Barcelona, 1914), p. 144. Strickland suggests this must refer to composite bows, but I think he is too credulous of Perellós's claim that the Irish bows were as powerful as English bows (something which, unlike the simple length of the bows, Perellós could not easily judge.) Especially given the Waterford evidence of short selfbows we now have (see the following), we can presume it was selfbows Perellós encountered. 1590s: Edmund Spenser, *Veue on the Present State of Ireland*, in his *Complete Works*, ed. Alexander B. Grosart et al., 9 vols. (London, 1882–4), vol. 9, p. 93. Spenser adds that though the arrows are short and the strings slack, so that the missiles "are shott forth weakely," nonetheless "they enter into an armed man or horse most cruelly" because of their extremely sharp and slender steel arrowheads. (There may be an error here in the use of "armed" [i.e. armored] since the overall sense makes the inclusion of that word improbable; perhaps the author wrote or meant "unarmed." An earlier edition of Spenser's work omits the word. *The Works of Edmund Spenser, with Observations on His Life and Writings* [London, 1850], p. 496.) A Spanish ambassador in 1529 likewise described Irishmen using "short bows." James Anthony Froude, *History of England from the Fall of Wolsey to the Death of Elizabeth*, vol. 2 (New York, 1895), p. 269. Early Modern artistic depictions confirm Irish and Scottish use of shortbows. E.g. see *Warbow*, p. 95; Geoffrey Parker, *The Military Revolution* (Cambridge, UK, 1988), p. 50. So does medieval iconographic evidence, e.g. the 1316 seal of Carlisle.

85 Giraldus Cambrensis, *The Conquest of Ireland*, tr. Thomas Forester, rev. and ed. Thomas Wright (Cambridge, Ont., 2001), pp. 76–7; note also 48, 81. However, there are indications that pre-Norman-invasion Irish did make at least some use of the bow in war, assuming that the *Caithreim Cellachain Caisil* is indeed a late-eleventh-century text, and that references in it to arrows are not either mistranslations from the Irish or later interpolations (the earliest manuscript being from the fifteenth

Gerald's claim that the Irish made little use of bows before the Norman invasion is supported by archaeology, which has found that although "the crannogs have preserved a remarkably full picture even of the more perishable aspects of the Celtic material equipment of the period" they have yielded only a single bowstave, which from its provenance was almost certainly a Viking weapon.[86] Even more telling is the rarity of arrowheads: the excavation of Lagore crannog, a royal residence from the seventh to the tenth centuries, for example, "produced a wealth of weapons, but among numerous swords, scramasaxes and spears, and a few throwing axes, not a single arrowhead and, despite the excellent preservation of wood, not a trace of bow-stave or arrowshaft came to light."[87]

It would thus be unsurprising if the bows used by the Irish in the late twelfth and thirteenth centuries were similar to those used in the same period by the Anglo-Normans and the Welsh archers they employed. Fortunately, recent archaeological work in Ireland has now begun to fill in the "serious hiatus in securely datable bow finds between the tenth-century Hedeby bow and the *Mary Rose* bows" that Strickland lamented.[88] Excavations undertaken at Waterford now allow us to

---

century.) See Alexander Bugge, ed. and trans., *Caithreim Cellachain Caisil: the Victorious Career of Cellachan of Cashel* (Christiania [Oslo], 1905), pp. xvi, 64, 99, 101, 104, 105. Strickland notes some clear evidence of twelfth-century Irish use of bows: *Warbow*, p. 94. In *The War of the Gaedhil with the Gail*, however, the only mention of archery I could find was a comment on the effectiveness of the Scandinavians' "sharp, swift, bloody, crimsoned, bounding, barbed, keen, bitter, wounding, terrible, piercing, fatal, murderous poisoned arrows," which seem (even allowing for the style of Irish poetry) to have made a great impression. By contrast, bows and arrows are notably absent from the roster of military equipment of the Gaels, unless perhaps (though not likely) the "terrible sharp darts with variegated silken strings; thick set with bright, dazzling, shining nails, to be violently cast at the heroes of valour and bravery" are arrows rather than javelins. *The War of the Gaedhil with the Gaill*, ed. and tr. James Henthorn Todd (London, 1867), pp. 159–163.

86 This is the well-known Ballinderry bow, probably of the tenth century and "indisputably a Viking weapon," which, though not as thick as a typical *Mary Rose* bow, should be considered a longbow. For details, see Andrew Halpin, "The Ballinderry Bow: An Under-Appreciated Viking Weapon?" in *The Vikings in Ireland and Beyond: Before and After the Battle of Clontarf*, ed. Howard B. Clarke and Ruth Johnson (Dublin: Four Courts Press, 2015), 151–60.

87 Clark, "Neolithic Bows," 88; see also Andrew Halpin, "Military Archery in Medieval Ireland: Archaeology and History," in *Military Studies in Medieval Europe – Papers of the "Medieval Europe Brugge 1997" Conference*, vol. 2, ed. Guy De Boe and Frans Verhaeghe (Zellik, 1997), p. 51.

88 *Warbow*, p. 40. Like the Balinderry bow, the Hedeby or Haithabu bow is a Viking longbow. Given that Strickland himself recognises the lack of datable longbows between the tenth and the sixteenth centuries, it seems clear that he overstated his case when he wrote (p. 48) that the archaeological evidence "demonstrates the existence of longbows . . . from the Neolithic period *onwards* and unequivocably disproves the supposed technical leap forward of the late thirteenth century." [Emphasis added.] One more bow has been identified that was likely a Viking longbow, and may possibly be younger than the Hedeby and Balinderry bows, though more likely it is older. It was found near Wassenaar in the Netherlands, and with a radiocarbon dating of 789–879, with a calibrated estimate of 800–950 (Lanting) or 770–950. Jürgen Junkmanns, *Pfeil und Bogen. Von der Altsteinzeit bis zum Mittelalter* (Ludwigshafen, 2013), p. 364; J. N. Lanting, B. W. Kooi, W. A. Casparie and R. van Hinte, "Bows from the Netherlands," *Journal of the Society of Archer-Antiquaries* 42 (1999), 9. The surviving part is 150 cm; Junkmanns estimates its original length at 170–190cm, i.e. from transitional to *Mary Rose* length.

confirm that shortbows of a bit under 3'6" to a bit over 4' long (unstrung) were indeed in common use there in those centuries. As Andrew Halpin of the Irish National Museum reports:

> Bows are rarely found at excavated sites but excavations in the town of Waterford have produced one complete bow and fragments of six others. All are simple yew bows dating between the mid-12th and mid-13th centuries. . . . The only complete example is 125 cm [49.2"] long and the other surviving bowstaves were probably of much the same length. The arrowheads found with the bows are . . . strongly military in nature . . . so it is impossible to argue that these bows were for hunting and that longer bows were used in war. These are military bows and, what is more, most of them seem to be Anglo-Norman . . . the Waterford evidence surely demonstrates that Anglo-Norman archers of the late 12th and early 13th centuries were not using bows which would be recognised as longbows.[89]

Along with these bowstaves, one complete arrowshaft has been found at Waterford, which so far as I know is the only surviving Western European shaft that can securely be dated to the High or Late Middle Ages.[90] In diameter it was clearly typical for its time and place, as the sockets of the relatively numerous surviving arrowheads show. The fragmentary arrowshafts found at Waterford (including one nearly complete one) suggest that the one complete shaft was also typical in its length, which was around 60 cm (a bit under 24"), a perfect match for the length of the complete bowstave.[91] In addition to the Waterford bows, a nearly complete

89 Halpin, "Military Archery," pp. 56–7. See also Andrew Halpin and Aidan O'Sullivan, "Bowstaves," in Maurice F. Hurley et al., *Late Viking Age and Medieval Waterford. Excavations 1986–1992* (Waterford, n.d.), pp. 546–52. Halpin does go on to say that the Waterford bows "are essentially identical to the later medieval longbow in all respects except length," and, in consonance with the Holt-Bradbury-Strickland line of thought, adds that "any attempt to distinguish the longbow from other wooden self bows purely on grounds of length must always be entirely arbitrary and ultimately unsustainable." However, as explained earlier, length *does* matter. Moreover, the Waterford bows, judging by Halpin's drawings, also differ from proper longbows in the higher width:depth ratios of the upper and lower sections of their limbs. And even if the thickness of the bow at the center were *proportionally* equivalent to a fully developed longbow (which actually it is not, with a 1:1.25 thickness:width ratio vs. a 1:1.1 typical for a *Mary Rose* bow), it would be *absolutely* much less, so it would be far less powerful. The complete bow is 2.5 cm wide by 2 cm deep at the center (Andrew J. Halpin, *Archery and Warfare in Medieval Ireland: A Historical and Archaeological Study*, Ph.D. diss., University of Dublin, 1998, vol. 2, p. 9) vs. 3.5 by 3.2 cm for one of the slighter *Mary Rose* bows. A replica Waterford bow drew a surprisingly high 60 pounds, but because of the short draw imparted only a low level of kinetic energy (probably around 20J) to its arrows: Jeremy Spencer, "A Short War-Bow Examined," *Journal of the Society of Archer-Antiquaries* 60 (2017), 100–102; my calculation of energy based on an estimated arrow weight of 40g.
90 The famous Westminster Chapter House arrow could well be Tudor rather than medieval.
91 The shaft is 9mm in diameter, corresponding to the most commonly found arrowhead socket diameter; this is only about 75–80% of the diameter of typical late-medieval military arrowheads. Halpin and O'Sullivan, "Bowstaves," pp. 549–50. This makes a substantial difference in mass

weapon of similar form, which was probably only around 39–41" in length when intact, was found at Desmond castle in Limerick in 1864.[92]

Outside the British Isles, archaeology has yielded an additional three complete or nearly complete selfbows probably dating from the eleventh through the fourteenth centuries, and none is long enough to meet my definition of a longbow. At 47" unstrung, a fourteenth-century bow from southwestern Poland is at the top end of the shortbow class.[93] A 49" elm bow from Pineuilh in southwestern France is at the low end of the medium-bow range.[94] The third, a twelfth-century yew bow from Burg Elmendorf in northwestern Germany, would by my classifications technically be a transitional bow, though with only a 58" nock-nock distance, it would be *functionally* equivalent to a medium bow.[95] Even that weapon, though easily the most powerful of the three, would only have had at most *one third* of the power of a median *Mary Rose* longbow.[96]

---

because the volume (and hence mass) of a cylinder increases with the *square* of the radius, and proportionally with length. If made with the same type of wood, a 32" arrow shaft with an 11–12mm diameter would have 100–137% more mass than the Waterford shaft. The second most common socket diameter among the Waterford finds is 8mm (ibid.), which corresponds almost exactly to the one-inch circumference of the arrow mentioned in the 1298 murder indictment discussed later in this chapter. An 8mm shaft of 24" would have exactly *one third* the mass of a 12mm by 32" shaft (or 42% of an 11mm by 30" shaft). Arrowheads probably from the English siege of Caerlaverock in 1300, and a large number of military arrows found at Urquhart probably also from the time of Edward I, were also mostly 8mm to 9mm external socket diameter. Jones, "Tests of Arrows," 156.

92 This yew bow was 88cm long, with an estimated 10–15cm missing from one end. Etienne Rynne, "Was Desmond Castle, Adare, Erected on a Ring Fort?" *North Munster Antiquarian Journal* 8 (1961), 199–202. My thanks to Brian Hodkinson for providing me with a copy of these pages. I have not included for consideration an even shorter (67.8cm long) complete bow of the thirteenth century excavated in Dublin because it is so short that it must have been for a child. Halpin, *Archery and Warfare*, vol. 2, p. 7. A 50cm bow fragment probably from the early fourteenth century that was excavated at St. Andrews in Scotland suggests the use of similar weapons there. The original bow was definitely slightly built, but the fragment is not large enough to judge the original length of the weapon. Colm Moloney, Louise M. Baker et al., "Evidence for the Form and Nature of a Medieval Burgage Plot in St Andrews: An Archaeological Excavation on the Site of the Byre Theatre, Abbey Street, St Andrews," *Tayside and Fife Archaeological Journal* 7 (2001), 70–71.

93 Marian Głosek and Leszek Kajzer, "Łuk średniowieczny znaleziony w Brzegu," *Silesia antiqua* 19 (1977), p. 242. This bow (under 47" tip to tip; 43.6" between the nocks unstrung; standing perhaps 45" strung) would have been quite a weak bow. Two bows similar in form (one complete and one nearly so) dating from the tenth to the twelfth centuries were excavated at Opole, but both the artifacts and the records concerning them were lost in the Second World War. As far as can be judged from a published drawing, however, they were more likely medium rather than shortbows. Włodzimierz Hołubowicz, *Opole w wiekach X-XII* (Katowice, 1958), 231; Łukasz Urman, pers. comm.

94 Pierre Lansac, "Les arcs de Pineuilh (Gironde)," *Lettre des amis du musée de l'archerie et du Valois* 7 (Feb. 2008), 2–8. This dates from 979 to 1060.

95 It is nearly 64" in total length, so would have been at most 62" strung. Junkmanns, *Pfeil und Bogen*, 377–8. By fifteenth-century standards (as noted later) longbows were supposed to be a fist longer, between the nocks, than the height of the archer.

96 At the middle the Burg Elmendorf bow is only 2.7cm wide by 2.3cm thick, and because of its short nock–nock distance would not have taken an arrow of more than 24–26" draw. Junkmanns, *Pfeil und Bogen*, 377–8.

Even taken together, five bows, a number of bow fragments, and one arrowshaft make a scanty data set from which to make any broad generalizations or firm conclusions about military archery in High Medieval Europe. Still, it is noteworthy that all the limited archaeological evidence now available supports, rather than undermines, the traditional view that the bows of Europe during the High Middle Ages were normally ordinary bows, employed with relatively short arrows (and therefore short draws), and were nowhere near as formidable as fully developed longbows.

Documentary Evidence for Sizes and Types of Bows

The first document to provide specific information regarding the length of a Western bow is the record of a murder indictment dating to 1298. The text is well known, but its interpretation has been mistaken.[97] In the Latin text of the document, the bow used to kill Simon "de Skeftington" on 22 March 1297 is described as made of yew, and the *"longitudinis arcus unius ulne et dimidie, grossitudinis in circuitu sex pollicum,"*[98] with an arrow *"longitudinis trium quarteriorum unius ulne et grossitudinis in circuitu unius pollicis."*[99] The translator rendered this as a bow one ell and a half long, with an arrow three quarters of an ell in length and one inch thick. A minor problem with this translation is that the arrow is actually described as an inch in circumference, i.e. under one third of an inch thick – about the same as the Waterford shaft – rather than an inch thick. A major problem is that "ulna," in this context, means a 36" measure, and so would have been better translated as "yard," rather than "ell."[100] An *ulna* or ell could in other contexts be of a variety

---

97 S. H. Skillington, "The Skeffingtons of Skeffington," *Transactions of the Leicestershire Archaeological Society* 16 (1929–31), p. 113.

98 Note that a six-inch circumference is extremely thick; a typical *Mary Rose* longbow is only about 4.7" in circumference. *Pollicum* here may be used in its original sense of "thumbs" (which might then be a circumference of as little as 4.5"–4.9"); the first use of the word in the document, describing the length of the arrowhead, specifies "pollicum hominis."

99 TNA De Banco Roll 123, Easter, 26 Edward I (1298), m. 107, photograph in Bradbury, *Archer*, p. 81; transcript in F. Cottrill, "A Medieval Description of a Bow and Arrow," *Antiquaries Journal* 23 (1943), 54–5.

100 Ronald E. Zupko, *British Weights and Measures: A History from Antiquity through the Seventeenth Century* (Madison, 1977), pp. 20–21, and idem, *A Dictionary of Weights and Measures for the British Isles*, s.v. "ell," noting particularly the citations to 1272 ("Et xij pollices faciunt pedem; et tres pedes faciunt ulnam") and c. 1400 ("nota quod tres pedes regii faciunt ulnam Regis.") That the *ulna* of 36" rather than the *ulna* of 45" is meant in this context is further confirmed by the fairly frequent references to arrows an *ulna* or an "elle" long, e.g. in a purchase order by Edward III: *Calendar of Patent Rolls Preserved in the Public Record Office*, 16 vols. (London, 1891–1916), vol. 4: *CPR 1338–40*, p. 124; *Robin Hood Garlands and Ballads*, ed. John Mathew Gutch, vol. 1 (London, 1850), p. 165. 45" is absurdly long for an arrow, while 36," though very long for an arrow, matches later references to English arrows of a "clothyard" (37"), a "full yarde" (Edward Hall, *The Unyon of the Two Noble and Illustre Famelies of Lancastre and Yorke* [London, 1550], Henry VII, f. 43; intended as exceptional) or "bicubitales" (around 36"; in Paulus Jovius' *Descriptio Britanniae* [1548], quoted in Hansard, *Book of Archery*, 373) long. Hansard also notes two English arrows in the 38" range still extant in his century. Ibid., p. 378.

of lengths, and in this case it has generally been taken by scholars as equivalent to the late-fifteenth-century English ell of 45 inches.[101] This led historians to believe that this document provided the first clear documentary evidence of the English longbow as early as 1297, with a weapon 5'7½" long shooting an arrow of 33¾"; whereas in fact it does not show that, but rather demonstrates conclusively that weapons of significantly lesser stature than longbows were then in use. Indeed, the 4'6" bow of the Skeffington record is almost the same size that I used in my 1993 article as a baseline "shortbow" against which to measure the advantage of a longbow, and was provided with an arrow of 27" – which would be drawn only around 24", allowing for the stated length of the arrowhead.[102]

The first document that *does* unambiguously describe a longbow is another murder indictment, purporting to describe the killing of Robert de Esnyngton on 3 July 1314. This document again mentions two assailants using bows of a yard and a half in length (which again has often been misinterpreted in terms of 45" ells). One of these bows, specified as being of the type "called Turkeys" [Turkish] and made of Spanish yew, shot an ash "Welsh arrow" of three quarters of a yard (27"), including an iron head six thumbs long. The other medium bow was of Irish yew and shot a "Scottish" arrow. Another of the attackers, however – John, son of Roger de Swinnerton – is described as employing a bow two yards in length and four thumbs in circumference. Its "clotharewe" was a yard long, with an iron head four thumbs long and two thumbs broad. A fourth attacker, John Charles, is also described as using a very stout six-foot bow, and again using an arrow a yard in length.[103]

The next clear textual evidence for bows of longbow length we have is in a French hunting treatise written between 1354 and 1377, which says a bow should be 22 *poignés* (fists) in length between the nocks, with an arrow 10 *poignés* in length from the nock to the start of the barbs. Although the fist's length would vary from individual to individual (just as the length of the bow should), this might typically be around 72" for the bow between the nocks and 33" for the arrowshaft,

---

101 E.g. Prestwich, *Armies and Warfare*, 131; DeVries, *Medieval Military Technology*, 34. Although Strickland does in an endnote (n. 59, p. 419) acknowledge that various different "ells" were in use, he nonetheless states positively in the main body of his text that "the ell was . . . gauged in England at 45 in . . . making this bow some 5ft 7½ in," and that this record "provides a firm *terminus post quem* for the existence of heavy and powerful longbows." *Warbow*, pp. 41, 43.

102 Note that a 1" circumference (=8mm diameter) and a length of around 24–25" makes the arrow described in this document very close to the Waterford shaft. See note 91.

103 The National Archives, Kew, Coram Rege rolls, KB27 no. 220, membranes 105–107; calendared with some errors and many omissions in "Extract from the Coram Rege Rolls and Pleas of the Crown, Staffordshire, of the reign of Edward II, A.D. 1307 to A.D. 1327," in George Wrottesley, ed., *Collections for a History of Staffordshire*, vol. 10, part 1 (London, 1889), pp. 16–19. In this case "ulne regis" are specified, as is the "pollicum hominis." See Clifford J. Rogers, "New Details on the Bows and Other Weapons of the Esnyngton Murder Indictment (1315)," *Journal of Medieval Military History* 22 (2024).

i.e. around 36" for the arrow including the head, the same as Swinnerton's and Charles's longbows and arrows in the Esnyngton case.[104]

Again, this is too small a data set from which to draw any strong conclusion or to make any broad generalization. It does however show unequivocally that ordinary bows were in use in central England in 1297 and 1314, and that by the latter date proper longbows[105] were in use alongside them. In conjunction with the iconographic evidence discussed next, these records suggest that the longbow may still have been practically unknown at the end of the thirteenth century, then have begun to become common within just fifteen years.[106]

104 Henri de Ferrières, *Le Livre du roy Modus et de la royne Racio*, ed. Elzéar Blaze (Paris, 1839), fol. 53. The *poigné* (fist or handful) is not a standard measurement, and does not appear as a unit of length in any dictionary I have consulted. The author presumably means the bowman should measure the shaft with his own fists, hand over hand, so that the length of the bow will vary with the size of the archer, just as English prescriptions called for. However, for our purposes, it is likely that the fist would in general be either equivalent to the *paume* (palm) or slightly larger, and the standard French *paume* was three French inches, or just under 3.3 English inches. Ronald E. Zupko, *French Weights and Measures before the Revolution: A Dictionary of Provincial and Local Units* (Bloomington, 1978), s.vv. *paume, pouce*. I have small hands for my height, but for me measuring fist over fist 22 times equaled my own height (under 6') almost exactly. For my wife, who is 5'5" tall, the 22-fist measure was 5'6". It is unlikely that the *poigné* was meant as an equivalent of the English "hand" (four inches), since an arrowshaft of 33" is already quite long, and an arrowshaft of 40" would be very improbable. Gaston Phébus, though clearly drawing on Ferrières, reduced his recommended lengths to 20 fists for the bow and 8 for the arrowshaft not including the head (which he says should be 5 fingers, or around 3.25 French inches/3.56 English inches), i.e. perhaps 5'6" and 27"/30.5". Note that in both cases this recommendation is for a hunting bow, which (as Phébus specifically states, and modern experts agree) should be a relatively light bow, so that it can be held in the drawn position for relatively long periods. Gaston Phébus, *The Hunting Book of Gaston Phébus. Manuscrit français 616. Paris, Bibliothèque nationale*. Facsimile; intro. Marcel Thomas and François Avril, commentary by Wilhelm Schlag (London, 1993), fols. 111v-112; the fifteenth-century *L'Art d'archerie* (pp. 12, 27) repeats the *Roy Modus* recommendation, altering it slightly: "ton arc, se to voeulz quil te dure, doit avoir deux petites poignies plus de long que la longeur deux fois de la flesche"; "ung bon achier selong lusage doibt tyrer dix plames du trait."

105 The *Mary Rose* bows range from just over 6' to just under 7', in total length (i e. unstrung). *Warbow*, p. 18. The six-foot measure is also a good match for bows required in a 1465 Irish statute to be, measured between the nocks, "at the very least" the height of the archer plus a *fistmele*. *Statute Rolls . . . of Ireland*, vol. 3, p. 292. Since an "inch" is in Latin (and, by origin, in concept) a "thumb," the four thumbs thickness (*grossitudo*) probably means a circumference of around 4" at the grip, which fairly closely approximates one of the lighter *Mary Rose* bows (e.g. longbow 80A0445 at 110mm= 4.33"; pers. comm., Andrew Elkerton, Mary Rose Trust, 3 March 2010.) The yard-long arrow would be somewhat longer than those found on the *Mary Rose*, which have shafts an average of 30.5," corresponding to an arrow of around 33.5" when fitted to an arrowhead of the long bodkin or heavy armor-penetrating "quarrel" types. This is not surprising considering that Smythe, writing a bit later in the sixteenth century, noted that the archers of his day used bows (and therefore, one might infer, arrows) shorter than those used in the medieval heyday of English archery. Smythe, *Discourses*, fol. 19v.

106 Another piece of evidence, though only suggestive evidence, that the real advent of the longbow was later than previously thought is that the earliest inventory of arms at Dover Castle, from 1320, includes 47 crossbows but no bows. In 1323, however, 60 bows were added, and from then on were stored in larger numbers than crossbows. Dan Spencer, "The Armaments of Dover Castle, 1320–1437," *Journal of Medieval Military History* 22 (2024).

The Iconographic Evidence

So, from a combination of archaeological and documentary evidence we know for certain that within the British Isles ordinary bows were in military use in the period around 1200; that medium bows were in use in 1297; and that both medium bows and longbows were in use in 1314. We also know that by the mid-fourteenth century, it could be presumed that English military archers would use "longbowes" and yard-long arrows.[107] Two questions remain open, however: were longbows in use in England alongside ordinary bows all along, and if not, when did they first come into use? The one class of evidence that we have not yet laid under contribution for answering these questions is iconographic evidence.

From the time of the first scholarly discussions of the development of the longbow, artistic depictions of bows, for example those in the main panel of the Bayeux Tapestry, were cited as one principal reason for concluding that the Anglo-Saxons, the Normans, and the thirteenth-century Welsh employed "shortbows" (including what are in this article termed medium bows).[108] Not only do the artistic depictions of the eleventh and twelfth centuries seem to show only bows of the short and medium classes, they also almost always show arrows being drawn to the breastbone, or occasionally to the chin,[109] which, as Oman and Morris rightly understood, indicates a relatively short draw, which could be accommodated by an ordinary bow.

Strickland, however, has argued that such evidence is so unreliable as to be almost useless before the late thirteenth century, because of the small skill and limited concern with naturalistic use of proportion of medieval artists before that time, when there was a sort of artistic revolution that has misled those who, like me, succumb to "too literal and too unwary" a reading of the iconography. "In reality," Strickland concludes (and DeVries agrees), "what changed from the thirteenth century was not the bow itself but rather the manner in which many artists depicted bows, with an increasing stress on proportion and naturalism."[110] "As

---

107 See notes 16 and 100.

108 The other, which as noted previously is recognized but not (in my opinion) effectively rebutted by Strickland, is, in Oman's words: "it is impossible that [Richard I] should have so much admired [the crossbow], and taken such pains to secure mercenaries skilled in its use, if he had been acquainted with the splendid longbow of the fourteenth century. It is evident that the bow must always have a great advantage in rapidity of discharge over the arbalest: the latter must therefore have been considered by Richard to surpass [the crossbow] in range and penetrating power. But nothing is more certain than that the English longbow at its best was able to rival the crossbow on both these points. The conclusion is inevitable that the weapon superseded by the arbalest was merely the old shortbow, which had been in constant use since Saxon times." *History of the Art of War*, vol. 2, p. 58.

109 E.g. the Bayeux Tapestry main panel archers; the *Eadwine Psalter* (Trinity College Cambridge Ms. R.17.1), fos. 19, 54v, 100, 230v, 242, 244); Morgan Library Ms. 736 f. 14; St. Johns College Cambridge Ms. K 30, f. 6; University of Glasgow Hunterian Library, Ms. Hunter U 3 2 (229), f. 6; BL Ms. Arundel 91, f. 28v; Kencot St. George (Oxfordshire), tympanum.

110 *Warbow*, 38; DeVries, "Longbow Archery," 50.

more artists became skilful in the use of proportion and perspective and broke free more readily from existing exemplars, so the longbow and the draw to the chin or ear becomes recognisable in more illustrations."[111] Moreover, Strickland argues, the historians who relied on iconographic evidence to argue for a transition from shortbow to longbow used that artwork not only naively, but also selectively, since, he says, "depictions of bows which are proportionally longbows occur before the period of the supposed technical development under Edward I."[112] Kelly DeVries has reached the same conclusion: that "the English were well aware of the longbow prior to the thirteenth century, and that a misreading of iconographical evidence has distorted the length of the early bow, making it short, when it was not."[113]

The interpretation of medieval art, especially before the fourteenth century, does indeed require caution. It is, in particular, dangerous to draw any conclusions from individual sculptures, embroideries, carvings, or illuminations. Those seeking to derive useful information from medieval artworks must be alert for the possibility that particular images are modeled on earlier *exempla*, in which case they may represent something closer to the arms, armor, and clothing of the time of the latter rather than the former. One must also take into account the possibility in particular cases that an image has been distorted in order to fit a particular space, to make some sort of symbolic point, or because of aesthetic conventions that did not stress realism. Kings, for example, are often depicted as larger than other men to emphasize their importance, and humans are often shown radically out of scale with buildings or trees in the same images. When it comes to depictions of bows, more specifically, there are particular difficulties in judging the length of the weapon in an image. The posture of the archer, the angle of view, and the extent to which the bow is bent all must be taken into account by the artist and by the viewer, and the latter cannot presume the former has done so according to the correct rules of realistic artistic depiction.

Still, iconographic evidence can be used effectively. To do so the historian must be conscious of the problems inherent in its interpretation, and must if possible examine a large number of depictions, preferably across a variety of media. This makes it feasible to discard, or place little weight on, images that are suspicious or unclear for any of the reasons just given. Fortunately, depictions of bows in medieval art, even before the fourteenth century, are common enough that it is quite possible to do so without obscuring the overall pattern.

Historians who have argued for the prevalence of ordinary bows rather than longbows in Britain from Anglo-Saxon times past the middle of the thirteenth century have often referred in support of this view to the images of the Bayeux Tapestry, and

---

111 *Warbow*, 48. Note that draw to the ear and draw to the chin should not be equated; a draw to the chin is actually similar in length to a draw to the breastbone, around 5" shorter than a full draw to the ear.

112 *Warbow*, 46.

113 DeVries, "Robin Hood," 50.

Figure 3.2 English Archer with Ordinary Bow, 11th Century. British Library MS Cotton Claudius B. IV, fo. 41v. The bow in this manuscript illumination is clearly at the low end of the medium bow class, not a longbow. The arrows appear to be about 24" long or less.

to the depictions of Welsh bowmen in the *Littere wallie* margins, as Strickland notes.[114] But those are merely examples of the iconographic evidence, not the sum of it.

What is needed, but still lacking in the literature, is a thorough and well-documented survey and assessment of the artistic evidence for the development of the bow in the Western European Middle Ages. In order to write the present article I began that project, and have now collected around two thousand images of representations of bows in medieval art.[115] To report properly on the methodology and

114 *Warbow*, p. 44.
115 The majority of these have been taken from online copies of medieval manuscripts found in the BL's Catalogue of Illuminated Manuscripts, the Bodleian Library, the Morgan Library, the BnF, and various other French libraries contributing to the Enluminures website of the Institut

Figure 3.3 A Typical Artistic Depiction of a Medieval Shortbow. From Palencia, Spain, c. 1175–1200. Metropolitan Museum of Art, New York. Author's photograph.

de recherche d'histoire des textes (<www.enluminures.culture.fr>). Others have been taken from other online collections (including the Koninklijke Bibliotheek in The Hague; the Imareal database of the Institut für Realienkunde des Mittelalters und der frühen Neuzeit of the Österreichische Akademie der Wissenschaften; the Bildindex database of the Deutsches Dokumentationszentrum für Kunstgeschichte; the Bridgeman Art Library collection; the Liber Floridus website; the Hill Monastic Library; the Digital Scriptorium project hosted by Columbia University; the Parker Library of Corpus Christi College, Cambridge; the KBR in Brussels; the Rylands library, University of Manchester; the Fitzwilliam Museum, Cambridge; the Bayerische Stadtsbibliothek; and the Biblioteca Ricciardiana), from photographs of illuminations found in books, from photographs of museum objects, cathedral sculptures, wall-paintings, etc., and from British Library manuscript images not included the BL database, particularly those in Additional and Cottonian manuscripts.

results of that study will require more space than is available here. I hope to do so in a separate article in the future.

Meanwhile, a few general results of this inquiry can be reported here. A wide survey of medieval images of bows shows that the bow as depicted in art matches quite closely to the pattern suggested by the limited archaeological and documentary evidence. Before the fourteenth century there are many clear depictions of various sorts of shortbows;[116] a good number of renderings of medium bows;[117] and a few depictions of bows that might be (but probably are not) representations of near-longbows or longbows.[118] Unambiguous depictions of near-longbows or longbows between the tenth century and the fourteenth are exceedingly rare: I have been able to find several hundred clear-cut portrayals of short-to-medium

116 Among the best and most typical depictions of shortbows used by men on foot are, by period, as follows. Pre-eleventh century: BnF, Mss. Latin 8085, f. 58; Latin 8318, f. 53; Württembergische Landesbibliothek, Cod. Bibl. Fol. 23 (Stuttgart Psalter), fos. 21, 90v; note also Corpus Christi College Cambridge Ms. 023 f. 22v. Eleventh century: BL, Ms. Claudius B IV f. 36v; BL, Ms. Cleopatra C VIII f. 22v; BM Angers Ms. 4, fos. 205, 270v; BnF Ms. Latin 6 (1) f. 6; most of the bows in the Monreale capitals; the Anglo-Saxon archer in the main panel of the Bayeux Tapestry; St. Gallen Cod. Sang. 863 f. 77. Twelfth century: BL, Arundel Ms. 91, f. 28v; BnF, Ms. Lat. 5134 f. 20; the Copenhagen Oliphant (National Museum, Copenhagen); stone capital from Palencia displayed at the Metropolitan Museum of Art, New York (Fig. 3 of this chapter); BM Epernay Ms. 1, f. 15v; BL Add. Ms. 2799, f. 194v; BM Moulins Ms. 1, f. 136; Bib. Mazarine Ms. 3, f. 4v, and Ms. 8, f. 13; BnF Ms. Fr. 403, f. 7v; BL Ms. Harley 2799, f. 194v; the *Livro das Aves* of Lorvão, f. 2v (printed in *Exposição: Pera guerriar. Armamento medieval no espaço Português* [Palmela: Câmara Municipal de Palmela, 2000], 78; my thanks to João Gouveia Monteiro for a copy of this excellent exhibition catalog); Lamech capital, Cathedral Museum of St. Lazare, Autun (Bridgeman image 146849); Biblioteca Nazionale, Turin, Inv. J. II. 1, ff. 189v-190, as reproduced in Helen Nicholson and David Nicolle, *God's Warriors* (New York, 2005), 32. Thirteenth century: BL, Ms. Harley 3244 f. 41; Bodleian Ms. Bodley 602 f. 64v; Bodleian Ms. Hatton 77 f. 1; BnF, NAF 1625, fos. 92 and 101v; BnF, Mss. Latin 3630 f. 77v; BM Besançon Ms. 54, f. 15; BM Carpentras Ms. 106, f. 2; BM Amiens Ms. 21, f. 183; Limoges gilded copper St. Sebastian, Musée du Moyen Age (Cluny), Paris; Österreichische Nationalbibliothek codex 507, f. 3; Huntington Library Ms. 3027 f. 81; Berlin, Staatsbibliothek, Ms. Germ. Fol. 282, fos. 2, 32v, 42, 46.

117 One of the best, which approaches the near-longbow class, was (if the nineteenth-century drawings of the destroyed manuscript can be trusted, and if it is presumed to be in scale with the human figure shown next to it) in the Strassbourg *Hortus deliciarum*, discussed (and Fig. 3.4) later in this chapter. Other good examples (including some that might instead be classed as shortbows) are the following: Pre-1000: the Ruthwell Cross; Sueno's Stone (per an 1861 drawing); Württembergische Landesbibliothek, Cod. Bibl. Fol. 23, fos. 46v, 74v. Eleventh century: the Norman archers in the main panel of the Bayeux Tapestry; BL, Ms. Cotton Claudius B V f. 41v. Twelfth century: BL, Add. Ms. 11695, f. 223r; BM Dijon, Ms. 14, f. 13v (*Warbow*, 77); Albert Boecker, *Das Stuttgarter Passionale* (Augsburg: Dr Filser Verlag, 1923), p. 47; the larger bows in the capitals of Monreale cathedral cloister; Lamech panel, Modena Duomo (Cortauld image B67/1395); Lescar Cathedral pavement (Bridgeman image 172540). Thirteenth century: BL, Add. Ms. 35166, fos. 36; Morgan Library Ms. M. 736, f. 14; BL, Add. Ms. 21926, f. 12; Add. Ms. 48985, fo 95v; Tours cathedral, Lamech windows; Bodleian Library, Mss. Bodley 270b f. 139v; Bodley 533, f. 5; Douce 188, ff. 95, 155v, 167; Westminster Abbey chapter house (*Warbow*, 137); Österreichische Nationalbibliothek [hereafter ONB] codex 832, f. 18.

118 Several of these that have been identified by other scholars as High Medieval longbows are discussed at length later in this chapter.

Figure 3.4 A Long Bow, But Not a Longbow. The long bow in the twelfth-century *Stuttgarter passionale*. Author's photograph from the public domain 1923 facsimile edition.

bows from that era, but only one clear-cut depiction of a longer bow. Moreover, the one exception (from the twelfth-century *Stuttgarter passionale*) does not have the form of a longbow; because of its heavily deflexed arms it would have been no more powerful than a more typical contemporary medium bow. [Figure 3.4] It is in any case shown with a short arrow and a two-fingered draw to the center-chest.[119]

There are also some bows that look to be longbows in the lower margin of the Bayeux Tapestry, but these examples are not clear-cut: the marginal figures are small and their proportions are quite distorted. The artists of the Tapestry included six depictions of bows in the main panel, where proportions are more carefully and consistently rendered, and these are all clearly ordinary bows. Moreover, the bows on the Tapestry, including in the lower margin, are shown as drawn with two fingers, and to the center-chest.

Despite the traditional view that the rise of the longbow was closely related to Edward I's Welsh Wars (1277–1292), even in the last quarter of the thirteenth century there is no alteration in the artistic depiction of bows. Both English and Continental manuscript illuminations continue to depict a variety of shortbows and medium bows, but offer no unambiguous portrayals of longbows, or even

119 *Stuttgarter passionale*, ed. A. Boeckler (Augsburg, 1923), pl. 99 (Ms. f. 34v), and Fig. 3.4 of this chapter. Note that the same artist's other four archers, on pl. 47 (Ms. f. 16v), have bows of at most medium stature.

near-longbows.[120] That changes rather suddenly and dramatically in the early fourteenth century. Although ordinary bows continue to appear frequently in English manuscript illuminations, we can now find alongside them clear images of both transitional bows and true longbows.[121]

Some of the manuscripts containing these illuminations cannot be dated more precisely than "first quarter of the fourteenth century," but what may be the earliest [Figure 3.5, middle image], dated c. 1305, already shows a very stout bow, probably a true longbow, employed with a very long, three-fingered draw. The Tickhill Psalter, which can confidently be dated between 1303 and 1314, includes the earliest unambiguous depiction of a longbow that I have found.[122] Into the

120 For thirteenth-century English depictions of ordinary bows catalogued as dating within or possibly within Edward I's reign, see most usefully Fitzwilliam Museum, Ms. 46–1980 (1291); BL Add. Mss. 48985, f. 45v; 21926, f. 12; Harley 3244 fos. 41, 42v; Egerton 1151, f. 95v; Harley 3487, f. 140v; Royal 1 D I, fos. 1, 259; Add. 24686, f. 11; BnF Ms. Latin 3630, fos. 77v, 78; Rylands Library French Ms. 142 f. 28v (though possibly fourteenth-century); *Warbow*, 144; also BL Add. 42555, f. 11v; Royal 14 B V, m. 2; Bodleian Mss. Douce 366, f. 12; Rawlinson C117, f. 152 (c. 1300); BL Egerton 1151, fos. 38, 118; St. John's College, Cambridge, Ms. K26 f. 6; *Warbow*, 85. I have also identified over three dozen English artistic depictions of bows c. 1250–1275; with one problematic exception, all, in my judgment, show only ordinary bows. The problematic exception is that one of three archers on Lambeth Ms. 209, f. 51 (ca. 1267), is shown (in what seems to be a case of badly executed art) stringing a misshapen longbow-length weapon. The other archers on the same folio (and the demonic one on f. 53) have stout ordinary bows, one shown with a two-finger draw to the eye and another with a two-finger draw to the chest. Continental artistic representations of bows from the period of Edward I's reign, which are quite numerous, likewise show only ordinary bows, again with one possible exception. E.g. BM Besançon Ms. 457, f. 61; BnF Ms. Fr. 95 fos. 177v, 13v, 113v, 321; Fr. 412, fos. 128, 52v; Fr. 776, f. 9v; Lat. 15158 f. 56v; BM Marseille Ms. 111, fos. 18, 73v, 138; BM Reims Ms. 679, f. 93; BL Ms. YT 8, f. 45v; St. Gallen Ms. Vad. Slg. 302, f. 169v; Huntington Library Ms. HM 3027, fos. 81, 22v; Lausanne, BCUL, Ms. U 964, fos. 10r, 124r; Morgan Library, Ms. 183, f. 10v; Musée de Cluny, Ivory Coffret (c. 1300–1310). The possible exception is BL Ms. Egerton 747, f. 6, from Salerno. This manuscript, cataloged as dating to c. 1280 – c. 1310, shows a bow that could be a near-longbow, though the artist's poor use of proportion leaves the length of the weapon ultimately quite unclear.

121 Examples of longbows or near-longbows: BL, Ms. Harley 324, fos. 4v, 5 (c. 1305); BnF, Ms. Fr. 24364 f. 63 (c. 1308–12); BL, Ms. Royal 2 B VII (1310–20), fos. 151 (*Warbow*, 143) and probably 13; Royal 19 B XV, f. 9v (c. 1300–24); *Douai Psalter*, f. 1 (in *The Gorleston Psalter*, ed. Sydney C. Cockerell [London, 1907], pl. XVII); Bodleian Lib., Ms. Canonici Bibl. 62, f. 17 (c. 1320–30). The bow of Bodleian Ms. Douce 88, f. 115, dated c. 1300, is difficult to judge relative to the rather ill-proportioned centaur wielding it, but does appear also to be a longbow. The arrow is shown only partly drawn, but would be at least to the ear at full draw. The Queen Mary Psalter (1310–20), BL Ms. Royal 2 B VII, shows short bows (e.g. fos. 161, 238, 277 right), medium bows (fo. 277 left) and transitional bows (fos. 13, 151v).

122 New York Public Library Ms. Spencer 26 (1303–14, probably c. 1310), f. 69: an unstrung bow slightly longer than the height of the archer. Characteristically for this period (and consistently with the evidence of the Esnyngton case discussed earlier) the artist also shows shorter bows. On fos. 26v–27, depicting 1 Samuel 20, the artist inconsistently shows Jonathan first holding a medium bow, then stringing and drawing (to the ear, with an odd four-fingered draw) a longbow or near-longbow, then handing the unstrung bow, equal in length to the height of the archer, to a boy servant. Another archer on f. 38 has what appears to be a near-longbow or longbow at full draw. A bowman on f. 74 holds what measures as a near-longbow. For reproductions see

Figure 3.5 From Ordinary Bow to Longbow. English illuminations of bows, c. 1260, c. 1305, and c. 1360. British Library, Add. Mss., 35166, fo. 35; Harley 324, fo. 4v; Egerton 1894, fo. 3.

middle of the fourteenth century, in English manuscripts, ordinary bows continue to be depicted frequently – often with the same artist depicting both longbows and ordinary bows.[123] Thereafter, longbows become predominant in English art, though shorter bows do also continue to be shown.[124] In Continental art, however, ordinary bows continue to be the norm, even in scenes of warfare, at least until the second half of the fifteenth century.[125]

Thus, the iconographic evidence provides confirmation for the traditional belief that for two centuries or more following the Norman Conquest the bows used in England, like those used elsewhere, were normally short or medium bows, not longbows. Indeed, the iconographic evidence, in combination with the documentary evidence, suggests that the rise of the longbow may have been even later and more rapid than was previously believed – possibly beginning, and probably gaining real momentum, only in the reign of Edward II (1307–27), rather than that of Edward I.

Putative Longbows in Art Before 1300

In contradiction to what is stated in the last section of this chapter, several historians have claimed that there *are* at least a few unambiguous artistic depictions of longbows from before the fourteenth century. If that were correct, it would undermine my argument that the longbow was practically nonexistent in Europe (with the possible exception of Scandinavia) from 1000–1300 – though a few such images would not refute the more important point that shorter bows were the *norm* in that period, as indicated by archaeological and documentary evidence, and by the evidence of effectiveness, as well as by large numbers of artistic representations. Still, I think the question is worth addressing, because from a history-of-technology perspective,

---

Drew Donald Egbert, *The Tickhill Psalter and Related Manuscripts* (New York, 1940), plates. This is a sharp contrast with the contemporary Gorleston Psalter (dated 1310–24), BL Add. Ms. 49622, which includes over thirty depictions of bows, none of which surpasses the medium bow category.

123 E.g. in the Luttrell Psalter (BL, Add. Ms. 42130; c. 1325–35), longbows on fos. 147v, 45 (*Warbow*, 198, 172), ordinary bow on f. 75. Royal 10 E IV (c. 1340), longbows e.g. fos. 92v, 93v, 97v, 128, 168; ordinary bows e.g. fos. 120, 46, 41. Bodleian Ms. Bodl. 264 pt. 1 (c. 1338–44, likely Flemish illuminator but for Edward III), fos. 156, 88v for longbows, f. 51v for an ordinary bow. BL Yates Thompson 13 (c. 1325–49), f. 192 for a longbow being strung; ordinary bows on fos. 192, 187, 81v, 68v, 68, 79v. Both bows painted for the tomb of Sir Oliver Ingham (d. 1344) were longbows. Charles Stothard, *The Monumental Effigies of Great Britain* (London, 1817), plate 67.

124 Longbows: E.g. *Warbow*, pp. 382 (right), 332, 210; BL Mss. Egerton 1894, fos. 3, 14v; Royal 6 E VI, f. 303v; Harley 2278, f. 61; Bodleian Ms. Douce 332, f. 18v; Douce 335, f. 55; Douce 353, f. 154; Rawlinson B 214, f. 199; Morgan Library Ms. M 126, fos. 42, 126. Shorter bows: e.g. *Warbow*, p. 382 (left); BL Mss. Royal 17 B XLIII, f. 149v; Burney 323, f. 324v; Fitzwilliam Museum Ms. 1–2005 (c. 1330–40), fos. 133r, 36r.

125 E.g. (clear depictions in military contexts), *Warbow*, 200; BL Ms. Yates Thompson f. 62; BnF Mss. Fr. 48, f. 47; Fr. 59, f. 291v; Fr. 77, f. 329v; Morgan Library Ms. G 24, fos. 37v, 82v; KBR Ms. 9401–04, f. 72. By the late fifteenth century, longbows are fairly common in western Continental art, e.g. *Warbow*, pp. 12, 14, 47; Morgan Library Mss. G 23 f. 93v, M 1053 f. 191; KBH, Ms. 10 A 15, f. 25v; 74 G 35, f. 89v; 74 G 37, f. 76v.

it matters whether the widespread adoption of the longbow was the consequence of a new invention (or re-invention), rather than being simply a shift of preferences in favor of one particular type of bow out of a range of choices already in use. Readers for whom that is a moot point may wish to skip to the next section.

It is Strickland's contention that "longbows and the archer's draw to the ear are depicted in manuscript illumination and other media long before the traditionally supposed development of the true longbow in the later thirteenth century."[126] Kelly DeVries concurs, noting what he considers four examples of "clear representations of longbows in use" in English art from the eleventh, twelfth, and thirteenth centuries as evidence that "the longbow was a favored weapon in Europe ... throughout the entire Middle Ages."[127] David Nicolle likewise identifies a number of artworks from before the fourteenth century that he thinks "clearly" or with "no doubt" portray longbows.[128] Richard Wadge adds one more example of what he considers a "true longbow" in an English manuscript of c. 1260.[129]

Between them, Strickland, DeVries, Nicolle, and Wadge cite seventeen examples of what they believe to be artistic representations of longbows from before the reign of Edward I.[130] Of these sixteen, two have already been discussed. The depictions of bows in the lower margin of the Bayeux Tapestry should not be given any weight, since the larger, more careful, and less distorted images of bows in the main panel are clearly short or medium bows rather than longbows or near-longbows. The *Stuttgart Passionale* (Figure 3.4) does indeed show one archer with a relatively long bow (though not a longbow), and is acknowledged as an isolated case of a pre-fourteenth-century image of a bow that was clearly depicted as more than medium-bow length.

Of the remaining fifteen putative depictions of longbows, nine are of bows that pretty clearly are no more than shoulder-high on the archer, thus (by my definitions, though not necessarily theirs) falling into the medium bow class.[131] (Figure 3.6)

126 *Warbow*, 45.
127 DeVries, "Longbow Archery," 51.
128 David Nicolle, *Arms and Armour of the Crusading Era, 1050–1350*, 2nd ed. vol. 1 (London, 1999), nos. 10d, 100, 118, 154, 436, and note also 101, 117, 471, 571.
129 Wadge, *Archery*, 183: viz., the *Rutland Psalter*, BL Ms. Add. 62925, f. 87v.
130 See the prior two notes and *Warbow*, 46, apparently though not unambiguously including Morgan Ms. M 736 f. 14 as "proportionally" a longbow, though "slight."
131 Viz. Morgan Lib., Ms. M 736 f. 14 (*Warbow*, 58); St. Aignan centaur capital (https://commons. wikimedia.org/wiki/File:Saint-Aignan_%2841%29_Coll%C3%A9giale_Saint-Aignan_-_Chapiteau_-_10.jpg); Adel Church centaur (www.greatenglishchurches.co.uk/html/adel.html); Fitzwilliam Museum Ms. 254, f. 19 (*Warbow*, 81); Trinity College Cambridge, Ms. R.16.2, fos. 13, 40v, 30v (Fig. 3.6); BL Cotton Claudius B IV, f. 41 (Fig. 3.2); BL Ms. Add. 62925, fo 87v; Barton-le-Street church centaur (https://commons.wikimedia.org/wiki/File:St_Michael_and_All_Angels,_Barton-le-Street,_Sagittarius.JPG); the *Hortus deliciarum* bow (Fig. 3.9); and one bow in the Vezelay tympanum. The Vezelay archers mostly have shortbows, but one has a bow that is shoulder-high unstrung (though Nicolle's sketch makes it look longer), so it would be under shoulder height strung, i.e. a medium bow by my definitions, though Nicolle (no. 10) says it would "clearly be considered a longbow."

Figure 3.6 Ordinary Bows from a Thirteenth-Century English Manuscript. Although one prominent historian claims this thirteenth-century manuscript "show[s] clear representations of longbows in use," in fact (despite the problems posed by inconsistent proportions and odd postures) the weapons shown can all safely be considered ordinary bows. This is clearest for the figure at the far right of the upper panel, who is stringing a bow of less than shoulder height. Trinity College Cambridge, MS R.16.2, fo. 30v, by permission of the Master and Fellows of Trinity College, Cambridge.

The one most likely (though not definite) representation of a longbow in the group is a graffito that, though "thought to be Norman," is actually impossible to date, and could well be from the fifteenth century.[132] The remaining five images also do

132 De Vries, "Longbow Archery," 51; Bradbury, *Archer*, 33, 34 (photo); <https://twitter.com/ColMuseums/status/1272510054470758400/photo/4> (accessed 9 Oct. 2023). The weapon looks to be a longbow or near-longbow, but the proportions are too inconsistent to have any confidence in that interpretation.

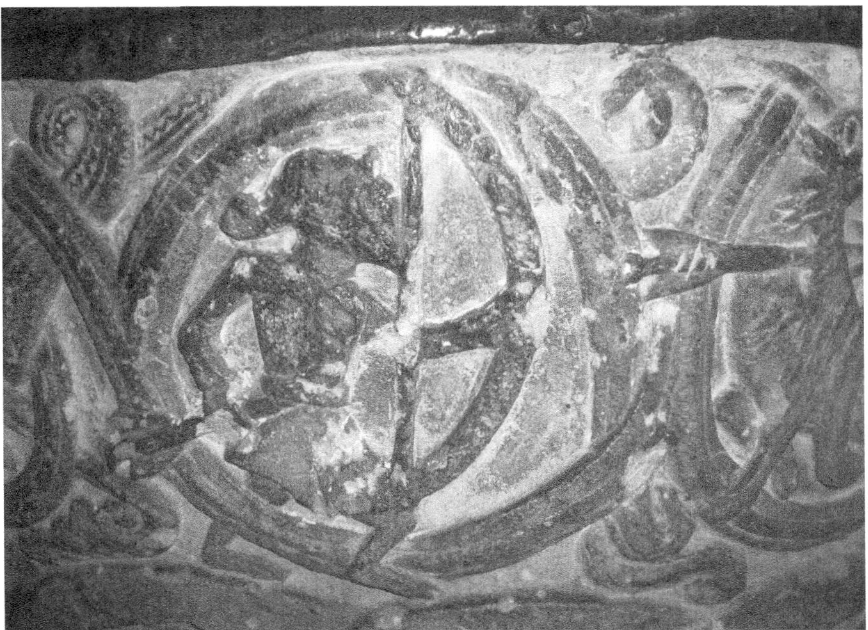

Figure 3.7 Anglo-Norman Font in St Michael and All Angels Church, Alphington, c. 1120–1140. Photograph by Gill Thorne, used by permission.

not provide good evidence of the use of longbows before the fourteenth century, but are worth more detailed examination.

Although the bows on the Alphington font, on folio 9 of the Eadwine Psalter, and in the Cambridge *Topographia Hiberniae* could perhaps have been intended by the artists to depict longbows, that identification is far from unequivocal. The last-mentioned image, described by Strickland as "showing archers holding bows as long as they are tall," is difficult to interpret. We do not see the feet of the figures, making it harder to compare the one bow depicted to the heights of the men, and furthermore it looks – by following the curve of the bow – like the artist originally sketched a bow that was relatively short, but then redid the bow as longer so that its lower end would not be obscured by hand of the figure holding it. Furthermore, the artist was not particularly skilled and his proportions are not internally consistent; if the length of the bow (even at its extended length) is judged in proportion to the sword of another figure or to the hands of the archer, it would be only a medium bow.[133] So I would throw this image out as insufficiently clear to allow any conclusion to be drawn from it.

133 Cambridge Univesity Library Ms. Ff.1.27, p. 322 (*Warbow*, 92, 46). The "extended" bow is about fourteen times the width of the hand holding it; the proportion for a longbow is, per the *Livre du roi Modus* (note 104) 22 *poignés*. Thus by this measure (as also by comparison with the main

Similar problems apply to analysis of the Alphington archer (Figure 3.7).[134] The artist who carved the Alphington font was clearly not interested in employing natural proportions. The arrow, for example, is about as thick as the man's ankle. In the figure of the archer himself, the line from his back heel to his waist is less than double the length from shoulder to head-top, rather than the more than triple it would be in natural proportions. The bow would be a near-longbow or longbow in comparison to the man's height (because of his very short legs), but nowhere near longbow length if judged relative to his head-top-to-shoulder distance, left elbow-to-fist distance, or left fist height. Thus, it cannot rightly be said that the carving "unequivocally portrays a huntsman with a powerful longbow."[135]

The putative longbow in the Eadwine Psalter (Figure 3.8) combines both sets of difficulties.[136] First, there is a problem with how the bow has been depicted; the artist has either shown it held well below the center of the bowstave, or else has failed to show all of the lower bow-arm that should be shown (i.e. that would not have been obscured by elements in its foreground). If the former is the case, it is clearly a short bow. That might not be considered likely were it not that this particular image of a bow seems to be the result of an artistic slip. In the original artwork on which the Eadwine Psalter is based, the Utrecht Psalter, the equivalent figure does not have a bow at all; rather, he is holding a switch, near the bottom of its length. But even if we assume that the artist did mean to depict a bow held in the normal position at the center of the bowstave, it is still far from clear that the intention was to depict a long bow, because the archer's own proportions (the only standard by which to judge the bow's length) are inconsistent. Compared to the archer's narrow shoulder width, it might be a long bow, but judged by the figure's shoulder-to-head-top or neck-to-left-wrist distance, it would be a medium bow. Moreover, the dozens of other bows depicted in the same manuscript are definitely *not* longbows or transitional bows, but rather shortbows or (possibly, in a few cases) medium bows.[137] So this image, too, cannot bear any weight in demonstrating that longbows were used by the English before the fourteenth century.

---

figure's sword) the bow would be in the 4' range. Only one archer is shown there; perhaps the plural was meant to refer also to p. 262 of the manuscript, but the bows of the archers there can only be identified as tall if it is presumed that they are being held with their lower ends on the ground (as is not shown and should not be presumed: cf. Morgan Library Ms. G 4 f. 37v; Hereford Cathedral Library, Ms. P. viii.II, reproduced *Warbow*, 86).

134 My thanks to Reg Williams (former Church Warden of Alphington Church) and Gill Thorne (who took the photo) for the photograph reproduced as Figure 3.7.

135 *Warbow*, 46. The relationship between the bow-length and bracing height is also far out of proportion for a longbow; a deep brace such as the one shown would greatly reduce the draw-under-tension and so the power of the bow.

136 Trinity College Cambridge Ms. R.17.1, f. 9.

137 Trinity College Cambridge Ms. R.17.1. Short: fos. 6v, 19, 21, 27v, 33v, 36v, 39v, 41v, 44v, 62v, 76, 98v, 100, 101, 108v, 141, 163, 230v, 242, 244; possibly medium: 54v, 66, 98v, 108v.

Figure 3.8 The twelfth-century artist who illuminated the Eadwine Psalter based his images on those found in the earlier Utrecht Psalter, but not so slavishly that the dozens of bows depicted take the same form as those in the original. Nearly all are shown as short selfbows; this detail shows the only archer depicted in the manuscript who could conceivably be thought to hold a long or transitional bow. Detail from Trinity College Cambridge, MS R.17.1. fo. 9. By permission of The Master and Fellows of Trinity College, Cambridge.

The twelfth-century *Hortus deliciarum* depicted Goliath standing between a quiver and a freestanding bow (Figure 3.9). Assuming the nineteenth-century drawings of the relevant illuminations accurately convey the now-lost original, the bow actually would (if its base were aligned with Goliath's nearest heel) stand just shoulder-high. Therefore this too would by my definition not be a longbow, but rather a medium bow.[138]

138 Christian Maurice Engelhardt, *Herrad von Landsperg, Aebtissin zu Hohenburg*, etc. (Stuttgart, 1818), Atlas, plate 4 (Fig. 3.9). Note that regardless of length this would not be a classic longbow because of its reduced nock-to-nock distance and its deflexed arms. Nicolle (*Arms and Armour*, no. 436) describes it as "a longbow of just under a man's height." He may be judging it relative to the figure to the bow's right (Balthazar), but since the Goliath figure is a warrior and has a quiver on his other side, presumably the bow is meant to be associated with him. The bow would also be shoulder-high relative to the figures of Caspar and Melchior. The other bows in the work (pls. 6, 9) are difficult judge but also seem to be medium bows.

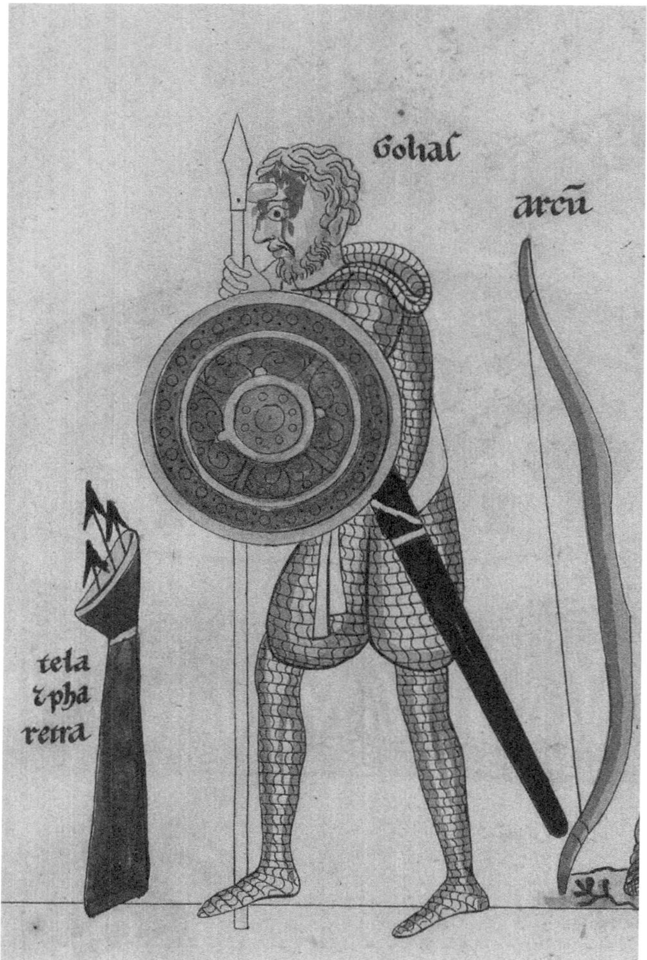

Figure 3.9 A substantial medium bow as depicted in a nineteenth-century engraving of a now-lost twelfth-century manuscript. Note that the base of the bow is not level with the "ground." Public domain.

There is one more manuscript illumination that must be considered here. Strickland writes that "any lingering doubts that . . . contemporary representations are intended to be longbows are rapidly dispelled by comparison with the contemporary psalter from St. Albans, compiled after 1246, which shows one of St. Edmund's assailants drawing a massive longbow."[139] [Figure 3.10] Although this is perhaps the most likely candidate for a pre-1300 representation of a longbow, I am nonetheless left with more than lingering doubts. In this image, the length of

139 Royal 2 B VI, f. 10. The manuscript was made before 1260.

Figure 3.10 The Martyrdom of St. Edmund. From BL MS Royal 2 B VI, fo. 10.

the bow is hard to judge due to the posture of the archer and the artist's poor use of proportion. Relative to the length of the archer's head, or elbow-wrist measurement of either of his arms, the bow would be a shortbow, not even a medium bow. Judged by the archer's left knee-ankle distance, the bow would be a medium bow. But the heel-head measure of the archer himself, compared to his bow, does indicate a longbow; so too would comparison with the archer's waist-to-left-knee distance. Thus it is impossible to determine from the relative measurements of the archer and his weapon whether this is indeed intended to represent a longbow. But what is clear is that the bow is shown with a short draw to the breastbone.[140] An archer who employed such a draw would be unlikely to use a long bow, since longer bow arms have disadvantages (making the weapon unwieldy and increasing the energy wasted in moving the arms rather than adding velocity to the arrow) with no major offsetting advantages except the crucial one of allowing a longer draw.[141]

140 Note that the archer's left arm is not even extended.
141 Nicolle writes of a bow shown in the thirteenth-century Trinity Apocalypse (Fig. 3.6) that it is "one of the earliest accurate illustrations of the longbow," but "clearly the bow has a much shorter draw than the Asiatic composite bow." This misses the point that the main purpose of a longbow's length is to allow a full draw to the ear (as also used with Asiatic composite bows), and the consequent large gain in power, so a selfbow that does not allow such a draw is not a longbow. *Arms and Armour*, no. 154.

The same point in reverse suggests that an artistic depiction that does show a selfbow drawn fully to the ear might be indicative that the artist was attempting to depict an archer with a longbow. But of the pre-1300 pictures of selfbows said to depict draws to the ear,[142] one actually shows fairly clearly a draw to the chest,[143] others show a draw to the eye or chin (equivalent to a draw to the breastbone rather than a draw to the ear),[144] and one shows an archer bent over sideways at the waist in order to shoot straight up, so that even if he were drawing to the ear (which he may or may not be)[145] it would still be no longer than a draw to the chest in a normal posture.[146]

"Only the style and technical illumination or carving distinguish these longbowmen from those of the Luttrell Psalter,"[147] Strickland suggests: but I cannot agree. Many of the images he notes do show bows similar in style and even thickness to longbows of the fourteenth century, but they are shorter weapons used with shorter draws, whereas the Luttrell Psalter clearly shows bowmen whose bows, strung, are the same height as the archers, and who use a three-fingered draw to the ear, not a two-fingered draw to the chest.

Longbows and Technological Determinism

The widespread use of the longbow in fourteenth-century England was a new development, and the weapon the English archers used at Crécy and Poitiers was substantially more powerful than the bows used earlier in England, or at the same time in other parts of Western Europe. Since the development of the longbow did happen, that means it at least could have been important, and could even have contributed to an Infantry Revolution. I have demonstrated elsewhere that, despite the skepticism of some modern historians, the longbow was indeed a lethally effective weapon, capable under the right circumstances of killing even armored men-at-arms in large numbers, and that it made a major contribution to the impressive battlefield victories won by various badly outnumbered English armies of the fourteenth and even the early fifteenth centuries.[148]

Does that amount to "technological determinism"? That concept is one that I think has been widely misused. In the case of my analysis of the longbow, for example, I did not say that its development *caused* the Infantry Revolution, or

---

142 *Warbow*, 46, 47; images pp. 48, 81. Again, Strickland seems to treat a draw to the chin as equivalent to a draw to the ear, when it is actually equivalent in length to a draw to the breastbone.

143 Fitzwilliam Museum Ms. 254, f. 19 (*Warbow*, 81).

144 Corpus Christi College Cambridge Ms. 16, fos. 56 (*Warbow*, 48), 59v.

145 This cannot be determined for certain because the position of the nock is obscured by the head. Judging by the fletchings he would seem to be drawing to the chin; judging by the string and his right wrist, he would seem to be drawing to the ear. Bodleian Ms. Ashmole 1511, f. 48v.

146 When I attempt to replicate the posture depicted, I (at a little over 5'11" height) draw only around 24" to the chin, or 27–28" to the ear.

147 *Warbow*, 46.

148 Rogers, "Efficacy"; Rogers, "Agincourt."

even the English strand of it; still less did I claim that it was *the* cause of that phenomenon. In fact, I specifically stated that "the adoption of the longbow" did *not* produce a revolution in military affairs, and "the stunning improvement in British military effectiveness from the 1330s to the 1350s was not technologically driven."[149] But even if I *had* said that the longbow caused the English to be successful at Crécy and Poitiers, it would still not actually have been technological determinism. When we speak of one thing "causing" another, we are using ambiguous language; we may mean that it is a *necessary* cause – without which the effect would not have occurred – or we may mean that it is a *sufficient* cause – something that guarantees or "determines" the effect, so long as the cause is present. Almost always, unless we specify the latter meaning, we mean the former usage, and intend to highlight one salient cause (typically one that is new or unusual) among the various necessary causes.[150] That does not mean we are denying the existence of other necessary causes and claiming that the cause we are drawing attention to was the only thing that affected a given result – that by itself it "determined" the outcome.

It is clear that the longbow was a *necessary* cause of the startling and one-sided English victories at Poitiers or Agincourt: *ceteris paribus*, if the English archers had all been equipped with shortbows or medium bows, with their lesser range and armor-penetrating ability, the English almost certainly would not have won those battles at all, and if they had, they quite certainly would not have won so crushingly. To make that statement is not an "attemp[t] to identify a single cause of the Infantry Revolution," (or even a single cause of the outcome of the battle), or to deny that "war is powerfully shaped by the societies that wage it," nor is it a failure to recognize that "military-technical innovation is [at least in part] a consequence, as well as a cause, of broader sociopolitical change."[151] John Stone claims that I "vie[w] technological innovation as a prime mover in the Infantry Revolution" and that in my work "the notion that the character and conduct of warfare are shaped by, as much as they shape, human society is undervalued."[152] All important historical phenomena have both causes and effects; my 1993 article, which has been accused of "technological determinism," was about the effects of the Infantry Revolution, rather than its causes.

The development of the warbow of the fourteenth century is certainly an example of a technological change that had important effects. It was a fundamentally

---

149 Clifford J. Rogers, "'As If a New Sun Had Arisen:' England's Fourteenth-century RMA," in *The Dynamics of Military Revolution, 1300–2050*, ed. MacGregor Knox and Williamson Murray (Cambridge, Eng., 2001), pp. 20, 22. I should note that a major reason for my reaching that conclusion in that article was that at that time I thought the longbow had become common during the reign of Edward I. I would now guess that most of Edward II's archers at Bannockburn were not longbowmen. Still, I would stand by my formulation that the longbow contributed to but did not drive the reversal of England's military fortunes under Edward III.

150 Marc Bloch, *The Historian's Craft*, tr. Peter Putham (New York, 1953), pp. 190–192.

151 Stone, "Infantry Revolution," 379; abstract, p. 361; 362.

152 Ibid., 370, 368.

different weapon than its predecessor of a century earlier because it was far more effective against armored troops than the shortbow or medium bow. Although it was not a "deterministic cause of" the English strand of the Infantry Revolution, it did make a very significant contribution to that phenomenon. The English military successes of the 1330s–50s and the 1410s–20s therefore cannot be satisfactorily explained without reference to the development of the longbow.

4

THE ARTILLERY AND ARTILLERY FORTRESS REVOLUTIONS REVISITED

In an article published in 1993, I proposed that the development of warfare in the West between 1300 and 1800 could profitably be viewed as a process of "punctuated equilibrium evolution," a mix of incremental, Darwinian evolution and more radical "mutation," and identified as two of the "punctuations" an Artillery Revolution in the fifteenth century and an Artillery Fortress Revolution in the sixteenth century.[1] Although this model has found some favor, it has also had its critics. The purpose of this chapter is to respond to some of their criticisms.

The Artillery Revolution thesis holds that improvements in cannon technology in the decades around 1430 effected a rapid and radical improvement in the effectiveness of these weapons. Better manufacturing techniques allowed for larger guns with longer barrels (increasing muzzle velocity and hence kinetic energy and accuracy, and also permitting more rapid loading), strong enough to exploit the full explosive force of wet-mixed gunpowder. In combination with an increase in the number of large guns, the synergistic interaction of these technical improvements made it, for the first time, possible to use artillery to smash down the stone walls of castles or fortified towns relatively rapidly. This great change in the tactical-level balance in siege warfare had tremendous implications for strategy also. Especially for the three previous centuries, the imbalance between offense and defense in individual sieges, and the great density of fortification in Western Europe, had created a corresponding superiority of the defense at the operational and strategic levels. Conquest was difficult, so wars most often ended with compromise, negotiated peaces rather than outright victories. Stone walls sheltered the weak from the strong, enabling them to retain their independence against stronger neighbors and overbearing suzerains alike. The Artillery Revolution changed all this, allowing the strong to efficiently employ their superior power both for external conquest and for internal consolidation. Battles, more than sieges, became the principal determinants of military success, and this gave a great advantage to wealthy, centralized states able to support standing forces like the French and

---

1 Clifford J. Rogers, "The Military Revolutions of the Hundred Years War," *Journal of Military History* 57 (1993). Note that the footnotes in this essay as originally published were kept to bare minimum due to the publisher's length requirements, and have not been greatly expanded here.

DOI: 10.4324/9781003399971-6

Burgundian *compagnies d'ordonnance*, and to pay and supply large field armies on campaign. "Great towns," wrote Andreás Bernáldez, "which once could have held out a year . . . now fell in a month."[2] Francesco Guicciardini agreed: "the siege and taking of a city became extremely rapid and achieved not in months but in days and hours," so that the art of war was "turned upside-down, as if by a sudden storm."[3] The Vegetian style of defense that had up until then frustrated many invaders lost much of is efficacy. The new artillery also made a substantial contribution towards the growing strength of the "new monarchies" relative to the sub-sovereign political actors they ruled, such as great nobles and cities. The consequences of these military-technical developments were thus of sufficient consequence and sufficient rapidity (relative to the centuries-long predominance of the defense they reversed) to merit the designation of "revolutionary."

The era of rapid sieges, and the consequent increase in the decisiveness and frequency of open battle, was however only temporary. By around 1500, Italian architects had synthesized a new style of fortress, commonly called the *trace italienne*, that could effectively resist long sieges. By the 1520s the new design was becoming widespread, and by 1530 it was used almost universally for the construction of new defenses, permanent or improvised. This brought the era of easy conquests to a crashing halt. Just as the Artillery Revolution had shortened the duration of sieges, so the Artillery Fortress Revolution lengthened them. Just as the Artillery Revolution had increased the decisiveness and frequency of great battles, so the Artillery Fortress reduced them. This helped create a new military environment in which, for example, despite his dogged efforts the mighty Philip II could not conquer little Holland; the city of Geneva could hold off the duke of Savoy in 1588 and 1602; and the duke of Mantua and Montferrat's two superb citadels could sustain his principality against the combined power of Spain and the Empire in 1628–31.[4]

In opposition to this thesis, it has been claimed that cannon, and anti-cannon fortifications, developed by slow, steady change over time: in other words, by *evolution* rather than *revolution*. Another line of argument holds that the military-technical changes in question did not effect sufficiently radical change to merit the name "revolution."

---

2 Quoted in Weston Cook, "The Cannon Conquest of Nasrid Spain and the End of the Reconquista," *Journal of Military History* 57 (1993), p. 43.

3 Francesco Guicciardini, *History of Florence*, ed. J. R. Hale, tr. C. Grayson (New York: Washington Square Press, 1964), p. 20.

4 Rogers, "The Military Revolutions of the Hundred Years War"; idem, "The Medieval Legacy," in Geoff Mortimer, ed., *Early Modern Warfare* (London: Palgrave, 2004), pp. 6–24; Thomas F. Arnold, "Fortifications and the Military Revolution: the Gonzaga Experience, 1530–1630," in Clifford J. Rogers, ed., *The Military Revolution Debate* (Boulder: Westview, 1995), pp. 201–252.

Significance of Change

In general, even those skeptical of the idea of an "Artillery Revolution" have been prepared to admit that gunpowder weapons "had, by the end of the middle ages, completely altered siege warfare."[5] Although poliorcetics is not the whole of medieval warfare, it ranks with devastation and open battle as one of the three principal modes of the conduct of war in the Middle Ages. Hence, its "complete alteration" might seem *prima facie* to be of sufficient importance to qualify as "revolutionary," provided that the other primary criterion of a revolution – rapidity of change – also is met. But experts as eminent as Michael Mallett and Philippe Contamine have endorsed J. R. Hale's view that "cannon revolutionised the conduct but not the outcome of war."[6]

This does not amount to a denial of the existence of an artillery revolution *per se*, but it does challenge the Artillery Revolution thesis as just described, which holds that the new cannon *did* radically alter the outcome as well as the conduct of wars, and *therefore* merit the description of "revolutionary." The key difference, in my view, is that wars which under the conditions of the *ancien régime* would have ended in a partial or complete failure for the aggressor might well, under the *nouveau régime*, have a very different result: the defeat and submission of the defender. This radical change in outcomes (observed and expected) allowed the "gunpowder empires" both to expand and to tighten their control over their own territory.

It is of course very difficult to evaluate the extent to which the presence or absence of cannon altered the outcome of a particular war, much less of wars in general. The question inherently raises counterfactuals – what would have happened in this siege or that war or the wars of this period if the cannon that were used had not been, or vice versa? Such questions cannot be answered conclusively, but we can address them probabilistically. For example, if a long siege involving gunpowder bombardment ended because the defenders' walls had been undermined, we can say that in all probability the siege would not have been radically different or shorter if the guns used in the attack had not been present at all. (In an earlier draft of this essay, I had in the last sentence "ended *because the garrison had run out of food* or because its walls had been undermined," but the length of time it takes to starve out a garrison depends on the amount of food it has stockpiled, which in turn depends on how long the defenders expect to be able to hold off an assault. If a fort can be taken by storm in a few weeks, why would anyone

---

5 Kelly DeVries, "Facing the New Technology: Gunpowder Defenses in Military Architecture before the Trace Italienne, 1350–1500," in B. Steele and T. Dorland, eds., *The Heirs of Archimedes* (Cambridge: MIT Press, 2005), p. 37, and idem, "The Impact of Gunpowder Weaponry on Siege Warfare in the Hundred Years War," in I. Corfis and M. Wolfe, eds., *The Medieval City under Siege* (Woodbridge: Boydell, 1995), p. 244.

6 Michael Mallett, *Mercenaries and their Masters* (Totowa: Towman and Littlefield, 1974), p. 163, quoting and endorsing J. R. Hale. In his closing remarks at the Parthenay colloquy, Philippe Contamine quoted the same passage as summarizing his own view.

bear the expense of storing supplies for three or six months, or even for a year? It would be very worthwhile to study a full series of cases of sieges ended by starvation to track the change in their duration over time, as an index or proxy for changing expectations of defensibility over time.) On the other hand, if we study the siege of a stone-walled town defended by a strong, well-supplied garrison, and find that the defenders capitulated after a week or two of bombardment had created a broad breach in the walls, it is only reasonable to conclude that the siege would most likely have lasted much longer had the guns not been present.

Similarly, if we look at an invasion that barely penetrated the frontier of the province that it aimed to conquer, due to a failed effort to capture a strategically sited border fortress, we cannot say with certainty or precision what degree of success the operation would have enjoyed if the invaders had been able to blast through the stronghold's walls in a matter of days. We can, however, conclude that it would very likely have been much more successful than it was, and might well have been very dramatically more successful.

More concretely, it is possible to compare repeated unsuccessful attempts of the French to push the English out of Guienne (e.g. in 1337, 1339, 1345–6, 1377, 1403–6, and 1442) with the *Blitzkrieg* that culminated in the capture of Bordeaux in 1451, or to compare Henry V's relatively slow conquest of Normandy with its much more rapid reconquest in 1450, or to compare the many French efforts to overrun Montfortist Brittany in the fourteenth century with the rapid occupation of the duchy in 1491, or to compare the many failed Christian assaults on the emirate of Granada in the fourteenth and fifteenth centuries with the eminently successful conquest of the territory by Ferdinand and Isabella at the end of the Middle Ages. In each case, the question of why the last effort succeeded when earlier ones had failed leads to complicated answers, which space does not allow us to explore here.[7] Still, we can already say this much: many attempted conquests of the fourteenth century failed because their momentum was broken by long sieges of positions that in a later era were (or would have been) rapidly captured by armies with strong siege trains, either because their walls were battered down or because their garrisons knew their walls *could be* rapidly demolished. Moreover, both contemporary observers and the modern historians who have studied these operations in most depth have generally agreed both that these conquests were carried out with unprecedented speed and ease, and that effective gunpowder artillery was at least *a* major cause of the change.

Professor Kelly DeVries, however, has recently argued that while "it is true that the time required to capture a fortification was greatly diminished by gunpowder weapons between 1370 and 1420 . . . in many sieges after 1420 the time required to reduce a fortification returned to that of the pre-gunpowder era."[8] Each of these two claims poses a challenge to my conception of an Artillery Revolution centered

---

7 Developed somewhat more fully in Rogers, "The Medieval Legacy."
8 DeVries, "Facing the New Technology," p. 42.

around 1430; combined, if accepted, they would completely invalidate it. In order to support his assertions he cites an excellent article by Alain Salamagne, published in the same year as my "Military Revolutions of the Hundred Years War." However, although Salamagne emphasizes a break-point in the first rather than the second and third decades of the fifteenth century, his chronology is much closer to mine than it is to DeVries's. He opens his section on the "efficacité des tirs" with "quelques contre-examples qui attesteront de la valeur conservée par la fortification au début du XVe siècle," including the siege of Dortmund in 1388 where large guns "restèrent sans effet contre les murailles." He also notes – in addition to examples I also used in my article – a number of long sieges in the 1410s and 1420s (including Vellexon 1409–1410, Coucy 1411, and Champtoceau 1420) that were finally brought to a close not when artillery breached the defenders' walls, but rather when the garrisons ran out of food, or when the defenses were demolished by undermining. Conversely, he adds to my list of fortresses captured after the walls were breached by post-Revolutionary artillery the excellent example of Avranches in 1449, where after 30 days of battery the guns brought down a stretch of wall so wide that ten carts abreast could have passed through.[9] Overall, Salamagne's examples reinforce rather than modify my general chronology: around 1400 cannon had developed to the point where they could often seriously *damage* walls, e.g. by demolishing hoardings, machicolations, or merlons, or even by shattering the towers that provided crucial flanking fire. Sometimes, particularly versus weak fortifications, they could also effect breaches, but in the 1400s and 1410s that was still the exception, not the rule (particularly when well-made stone walls were the target). In the 1420s and 1430s, guns became able to force surrenders more rapidly and more often. By the 1440s, the situation of the 1410s was reversed: some major fortifications, if provided with sufficient supplies and large, determined garrisons, could withstand regular sieges of three months or more, but this had become quite unusual, rather than the norm. Before 1420, it was often mentioned that fortresses withstood battery with little damage, but rarely explained (presumably since no explanation seemed necessary). By contrast, when the French artillery failed to demolish the fortifications of Avesnes in 1477, it *did* seem necessary to explain that the bombardment failed "because the wall was made of strong and hard stones, and the guns [*engins*] were too small."[10]

The change of expectations this represents was of great importance in the case of places that surrendered when *threatened* with battery by a strong siege train, even if the bombardment never began, or was not continued to the point of breaching the walls. A town that knew it would not be able to hold out for the length of

---

9 Alain Salamagne, "L'attaque des places-fortes au XVe siècle à travers l'exemple des guerres anglo- et franco-bourguignonnes," *Revue historique* 289/1 (1993), pp. 65–113, esp. 84, 94–6.

10 Quoted Salamagne, "L'attaque des places-fortes au XVe siècle," p. 92. In 1352 and 1400, indentures allowed the constables of Berwick and Roxburgh to negotiate for surrender only after three months of siege – an unreasonable standard by 1430. Michael Prestwich, *Armies and Warfare in the Middle Ages* (New Haven: Yale U.P., 1999), p. 241.

time it would take for a relieving army to assemble and come to its rescue had good reason to make its inevitable surrender quickly, to minimize the damage it would suffer during a siege and to secure better terms from the attackers. In such a case, the rapidity of the operation can still be said to be the result of the Artillery Revolution.[11]

In 1409, Christine de Pizan recommended that defenders prepare for sieges lasting six months,[12] and this was not unrealistic, as shown by the five- to seven-month sieges of Cherbourg, Meaux, Rouen, Montaiguillon, Torcy, and Château Gaillard between 1418 and 1429. But in the 1440s through 1470s, more typical were Charles VII's sieges of Bayonne (six weeks), Rouen (one month), or Caen (four weeks), or, in the reign of Louis XI, the capture of Lectoure in two weeks, Roye in two days, Pont de l'Arche in four days, Grandson in three days, Bouchain in sixteen hours, etc. Beauvais and Neuss, to be sure, held out successfully against Burgundian attacks, but this was due to exceptional efforts of the defenders, not the resistance of their walls. Beauvais had a quarter of its enceinte demolished within two weeks. Even a pile of rubble, however, has significant value to determined defenders, and an assault on a breach is a very perilous and difficult undertaking. Still, while a breach may be much easier to defend than an open field, it is much harder to hold than an intact wall. Hence, the ability to create a breach, while it does not guarantee the attacker's success, greatly reduces the defensive advantage in siege warfare, particularly if cannon have also demolished the towers that otherwise could have covered the breach with enfilading fire.

When a medieval ruler attempted to conquer a territory, he was playing against time. The longer the campaign lasted, the greater the risk that he would run out of money, that his supporters would desert him, that his enemies would gather the allies needed to make a vigorous counterattack, or that he would have to break off operations to respond to a dangerous threat elsewhere. A siege that lasted three or four months because the defenders had to be starved out, or because their walls had to be laboriously undermined, might deplete reserves of cash that had taken years to build, or might depend on loans that mortgaged future revenues for years

---

11 Mallett admits this: "The slowness with which siege pieces were moved into position was balanced, however, by the speed with which a heavy gun, once ready to fire, could breach the medieval fortifications of most Italian cities and castles. Thus if a siege lasted more than a few weeks, the chances were that the besieged would then quickly be forced to sue for terms." *Mercenaries and their Masters*, p. 163. But, surprisingly, Salamagne is more reserved: "Il n'est pas sûr que l'irruption de l'artillerie entraîna . . . une dimunition de la durée des sièges, celle-ci variant en fonction de multiples facteurs, contingents en présence, capacités militaires de l'attaque et de la défense, conditions climatiques, etc. qui ne sont pas réductibles à l'action de la seule artillerie." Salamagne, "L'attaque," p. 112 n. 39. The many factors at work might indeed outweigh the advantages given to the attack by the new artillery in particular cases, e.g. the siege of Calais in 1436. Still, artillery added a major weight to one end of the scale in nearly all sieges; that that weight could be counterbalanced in some instances does not change the fact that the system itself had been tilted to one side.

12 Christine de Pizan, *The Book of Deeds of Arms and of Chivalry*, ed. and tr. S. Willard and C. C. Willard (University Park: Penn. State U.P., 1999), pp. 110–11.

to come. Moreover, a campaign that made some gains but halted short of forcing the submission of the defender might leave an invader in the dangerous position of, to paraphrase Clausewitz, having jumped half-way across a ditch.[13] It was often more difficult for the invader to consolidate his gains than it was for the dispossessed defender to recover his losses through counterattacks. Conquest delayed often proved to be conquest denied.

Thus, gunpowder artillery did, starting in the 1430s, have a revolutionary impact on individual sieges, on the overall balance between offense and defense in warfare, and hence on the outcome of wars.

The same logic applies to the Artillery Fortress Revolution of the sixteenth century. From the 1520s, as before the 1430s, it might take six, seven, or eleven or more months of siege to capture a strongly fortified town, for, as Fourquevaux wrote in 1548, "the art of building [bastioned] ramparts was only brought to light a short time ago. Those places which have been walled since then . . . must be considered very difficult to take by force."[14] As a result of the new Italian style of fortifications, François de la Noue remarked in 1587, a town that formerly could have been taken in eight days now required nearly a whole season to capture.[15] Indeed, Fourquevaux concluded, anyone attempting to capture modern fortresses should expect to fail more often than succeed, "so that the conquest of a land will be from now on very difficult."[16] The defender rather than the attacker once again had great advantages at the operational and strategic levels. The era of conquests in Europe largely came to an end; field battles lost their decisiveness and became less frequent.

This reversal of the impact of the Artillery Revolution, it must be emphasized, does not invalidate the thesis that it was a revolution in the first place, any more than the fact that a Bourbon wore the crown of France in 1816 means that there was no French Revolution.

Pace of Change

The term "revolution" indicates change that is not only deep, but also rapid. The Artillery and Artillery Fortress revolutions have both been challenged on that basis also.

Kelly DeVries, for example, is *a priori* skeptical of the idea of sudden and radical change deriving from technological innovation. "The structure of technological revolution," he declares, "is not revolution at all, but evolution." Gunpowder weapons, he argues, saw "continually progressive [increases in]

---

13 Carl von Clausewitz, *On War*, ed. and tr. M. Howard and P. Paret (Princeton: Princeton U.P., 1983), 598.

14 Raymond de Rouer, seigneur de Fourquevaux, *Instructions sur la faict de la guerre* (Paris, 1548), fo. 85.

15 F. de la Noue, *Discours politiques et militaires*, ed. F.-E. Sutcliffe (Geneva: Droz, 1967), p. 372.

16 Fourquevaux, *Instructions,* fos. 85–85v.

effectiveness": "evolution slowly and over a long period of time and not a techno-logical revolution."[17] The same goes for change in fortification design: "the *trace italienne* system of artillery fortifications . . . is not something that sprang fully formed from the head of Zeus at the beginning of the sixteenth century. . . . [A]ll the characteristics which Parker identifies as original in the *trace italienne* artil-lery fortifications have medieval precedents."

Hence, in sum, "the move to the *trace italienne* signified an evolution from medieval military engineering, *not* the sixteenth-century military revolution that Parker and others have claimed."[18]

The premise with which DeVries begins is correct. But where this line of argu-ment goes from that premise is deeply problematic, for it presumes that the pres-ence of evolution disallows the possibility revolution. DeVries implies that to be a "revolution," a change must "spring fully formed from the head of Zeus," and be without "precedents." But by that standard we would have to eliminate the word "revolution" from the vocabulary of historians and political scientists. The American Revolution, the French Revolution, the Industrial Revolution, the Rus-sian Revolution: none of these appeared on the historical scene *ex nihilo*. Rather, revolutions often "turn everything upside-down" (to use Guicciardini's phrase again) with a sudden change that occurs when a basically linear advance reaches what scientists who study non-linear phenomena call a "tipping point," as when the addition of one more ounce causes the arms of a balance scale to tip, or when a bucket of cooling water turns to ice. It may be only one degree colder than it recently was, but that one degree makes all the difference. Now, if you climb to the top of a castle wall and empty the bucket onto someone's head, he will be *dead* instead of *wet*.

In both these examples, we see a revolutionary "tipping point" coming from a single process. Such non-linear breaks are more common when several related phenomena interact in complex ways. In the case of cannon, for example, we see evolutionary change in bore diameter, in barrel length, in numbers of big guns, and in the power of the charge. Each of these developments adds weight to one side of the scale that balances offense against defense; together, they cause it to tip. The same goes for *trace italienne* fortresses. Thicker walls, more use of absorptive earthwork, and better firing positions for defensive artillery all add weight to the defensive side of the scale, but it is not until the final ingredient – the *trace* per se – is added to the mix that the balance again reverses itself. All the pieces of the puzzle come together in a few places around 1500; the whole package becomes standard for new construction by around 1520. By 1530 there are enough of the new works and enough engineers who understand their principles and can impro-vise defenses for areas under attack, that the whole framework of war is greatly altered.

---

17 Kelly DeVries, "Catapults are not Atomic Bombs: Towards a Redefinition of 'Effectiveness' in Premodern Military Technology," *War in History* 4 (1997), pp. 470, 469.
18 Ibid., 466–7; idem, "Facing the New Technology," 37.

To say that these technological developments have changed the art of war greatly is not to say that nothing remains the same, or that they monocausally *determine* how long sieges will last, how wars will be fought, or who will win them. But it *is* to say that one cannot explain the differences between the wars of the fourteenth century and those of the later fifteenth century without taking the revolutionary improvements in gunpowder artillery into account. Likewise, the differences between the wars of the late fifteenth century and those of the mid-sixteenth cannot really be understood without reference to the new style of fortifications. I do not think that makes me a technological determinist, but if it does, then so be it.

GUNPOWDER ARTILLERY IN EUROPE, 1326–1500

Innovation and Impact

The Technical Development and Capabilities of Gunpowder Artillery: 1326–1415

In 1326–1327 an English manuscript illuminator provided us with the first depictions of guns in European art. (Figure 5.1) In 1327, as we know from documentary evidence, the town council of the city of Florence ordered the acquisition of a number of "engines called cannon," along with the iron balls they shot, for the defense of the city.[1]

These Florentine guns of 1327 were probably similar in scale to the one found in the ruins of Monte Varmine, which was destroyed in 1341. This forged-iron weapon, meant to be mounted on a wooden stock, was 20 cm long, had a trumpet-shaped tube expanding from 4 cm diameter at the base to 6.4 cm at the mouth, and weighed 2.7 kg. It most likely threw an iron ball of around 4.6 cm diameter weighing about 370 grams – rather larger than the acorns used as a metaphor for cannon-shot by Petrarch in the 1350s.[2] The other type of cannon in use at

---

1 Angelo Angelucci, *Delle artiglierie da fuoco Italiane. Memorie storiche con documenti inediti* (Turin: G. Cassone, 1862), 16–17n. Note that the document is given as "die 11 februarii *MCCCXXVI*," and every single secondary source I have seen referring to this document dates it to 1326, but the Florentine new year did not begin until March 25, so the text actually dates to 1327.

2 Angelo Angelucci, *Documenti inediti per la storia delle armi da fuoco Italiane* (Turin: G. Cassone, 1869), 69. My estimate for the size and weight of the ball assumes it sat just inside the mouth. Angelucci's sketch shows the ball smaller and set fairly close to the touch-hole, but that would not have allowed for a reasonable weight of powder to drive an iron ball. This is similar to the bombards used by the Papal States in 1350, which fired iron balls averaging 280 grams, so about 4.2 cm in diameter, and even closer to the one-pound (339g, 4.5cm) balls fired by four bombards purchased in 1358. By that date the Papal States had acquired dozens of guns and thousands of iron balls. Alberto Pasquali-Lasagni and Emilio Stefanelli, "Nota di storia dell'artigleria dello Stato della Chiesa ni secoli XIV e XV," *Archivo dell R. Deputazione romana di storia patria* 60 (1937), 149–189, at 150–53. See also Gustav Köhler, *Die Entwickelung des Kriegswesens und der Kriegführung in der Ritterzeit*, 3 vols. in 5 (Breslau: W. Koebner, 1886–1887), vol. 3, pt. 1, pp. 250, 226 (Petrarch). A better match for Petrarch's "acorns" would be the half-pound (170g, 3.6 cm) iron balls fired by certain Papal "bombarda magna" in 1358. Pasquali-Lasagni and Stefanelli, "Nota di storia dell'artigleria," 153. Angelucci also gives details of a now-lost gun of somewhat similar proportions that was cast with

---

DOI: 10.4324/9781003399971-7

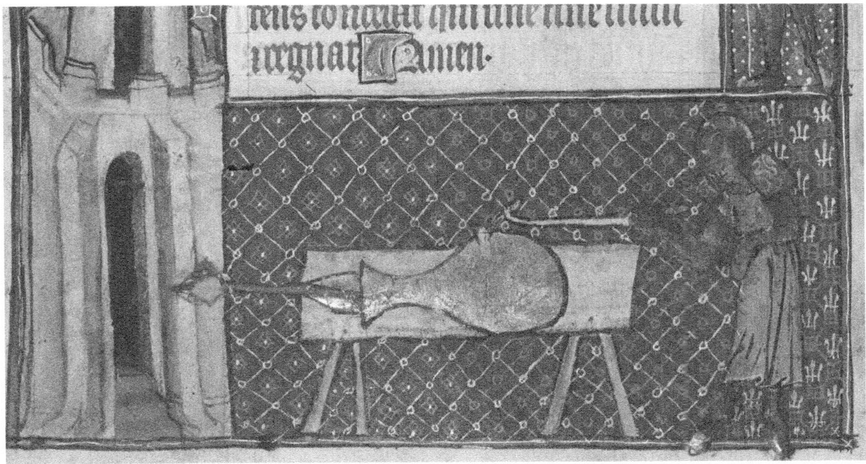

Figure 5.1 Depiction of an Early European Cannon (1327). From Walter de Milemete's *De nobilitatibus, sapientiis, et prudentiis regum*, Christ Church College Oxford MS 92, fo. 70v. Copyright the Governing Body of Christ Church, Oxford; reproduced by permission.

the time, as depicted in the 1326–1327 illuminations of Walter de Milemete's manuscripts *De secretis secretorum* and *De nobilitatibus*, was larger – a piece of artillery rather than a hand-gun – and designed to shoot either lead balls or heavy lances.

Over the next century, gunpowder artillery developed rapidly, so that the gunners of each new generation worked with weapons quite different from those their fathers had employed. Indeed, gunpowder artillery changed as much in *each* twenty-five-year period between 1325 and 1500 as it did over the following three centuries. (Figure 5.2)

But what drove this development? Some historians have suggested that the spread of and progress in gunpowder technology over its first century was propelled not by rational considerations of weapon efficacy, but rather by some sort of emotional reaction to the shock and awe generated by loud explosions and sheets of flame. Even Lynn White, who generally explained other technological changes in terms of practical utility, seems to have reached this conclusion about artillery: "Any rational technology assessment of the cannon in 1326 or for a hundred years later," he wrote, "would have concluded: 'Stick to the trebuchet.'"[3]

---

the date 1322; I have not referred to it in my main text because I am not confident that it is not a later fake. *Documenti*, 99.

3 Lynn White, Jr., "The Crusades and the Technological Thrust of the West," *Medieval Religion and Technology: Collected Essays* (Los Angeles: University of California Press, 1978), 269.

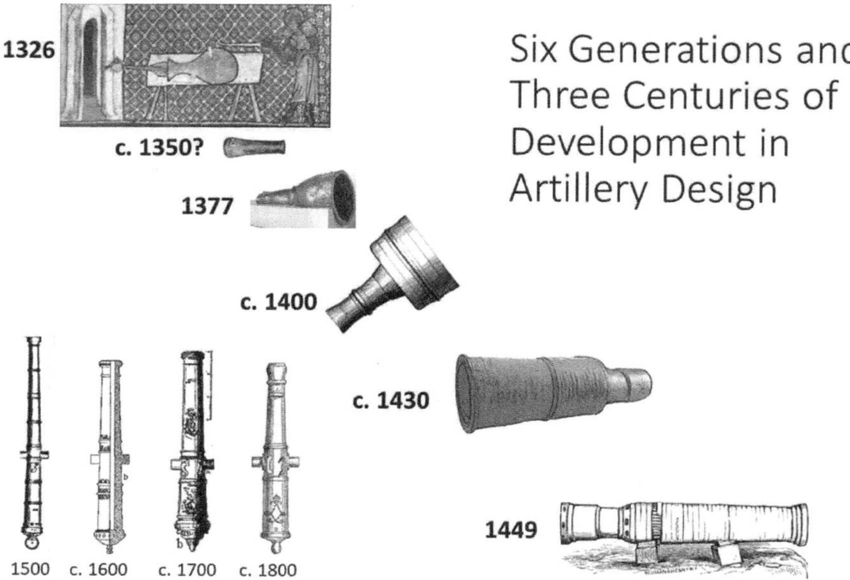

Figure 5.2 Development of European Cannon, 1326–1800. From around 1380 to around 1405, the highest recorded gun weights, shot weights, and powder charge weights rose in tandem rapidly, but in a linear fashion. Thereafter the curve became almost asymptotic until gun and shot weights peaked around end of the Hundred Years War.

White, however, was fundamentally wrong. In fact, from the very beginning guns had important advantages that made it perfectly rational to include them in the mix of weaponry used in sieges. Then from each generation to the next the largest cannon grew heavier and stronger, capable of handling larger and larger charges of gunpowder, even while that powder was reformulated to make it stronger. Well before 1426, the power of a large bombard far surpassed that of any mechanical artillery. Other changes in gun design made it possible to fire more shots per gun per day. At the same time, the number of artillery pieces in use increased steadily. These evolutionary lines of development interacted with each other in ways that exponentially increased the destructive punch a siege train could deliver against a fortress. The rapidly growing power of gunpowder artillery mattered in qualitative ways, changing the fundamental balance between offense and defense in European warfare. This fifteenth-century Artillery Revolution was an important factor in shaping the broad sweep of Western history in the fifteenth and early sixteenth centuries, facilitating the expansion and internal consolidation of the Renaissance monarchies that became the basic patterns for the modern nation-state. I have already sketched out the technological foundations of those epochal military developments, but this essay will present fuller and more quantified detail, linking technical developments to changes in their practical consequences over time.[4] This should not only help historians better understand one particular military revolution of the past, but also provide them with insight into patterns to look for and lines of analysis to pursue in examining other exemplars of the phenomena of "revolutions in military affairs" and "military revolutions."[5]

The Earliest Cannon

The first real description we have of a European gun in action might at first glance seem to support the idea that there must have been some reason other than simple usefulness for medieval governments, war-leaders, and technologists to invest in gunpowder weapons. In 1346, a seller of metal-ware from Bruges named Peter was summoned by the town council of Tournai to give a sales demonstration of one of the new weapons for them. He set up his gun to fire at a postern gate or the neighboring wall, but missed his target, even though it was much broader than the proverbial side of a barn. The projectile, a lance given extra mass by a two-pound lead ball below the point, flew out with such great velocity that the observers could not see its flight. Instead of striking the gate, it went over the both the outer wall of the town and also its inner wall (280 meters farther away) to the plaza in front of a church, where it happened to kill an extraordinarily unlucky passer-by.

As a seller's demonstration of a new technology, that was a pretty epic failure.

But one of the first things I learned from Joe Guilmartin was that medieval people were not stupid, and that – especially in the realms of war and technology – if they did something that seems stupid to you, there is a good chance that the problem is in your lack of understanding rather than theirs. Consider again Peter of Bruges's demonstration. His weapon may not have done what it was supposed to do, but in the process it demonstrated remarkable potential. Since the plaza where the victim was killed was about 440 meters from the outer gate and 160 meters from the inner wall, the total horizontal flight must have been over

4 Clifford Rogers, "The Military Revolutions of the Hundred Years' War," in Clifford J. Rogers, ed. *The Military Revolution Debate* (Boulder: Westview, 1995), 64–76; idem, "The Artillery and Artillery Fortress Revolutions Revisited," in Nicolas Prouteau, Emmanuel de Crouy-Chanel and Nicolas Faucherre, eds., *Artillerie et Fortification, 1200–1600* (Rennes: Presses Universitaires de Rennes, 2011): 75–80, reprinted as Chapter 4 of this volume.

5 For these two concepts as distinct phenomena, see Clifford J. Rogers, "'Military Revolutions' and 'Revolutions in Military Affairs': A Historian's Perspective" in Thierry Gongora and Harald von Riekhoff (eds.), *Toward a Revolution in Military Affairs? Defense and Security at the Dawn of the 21st Century* (Westport: Greenwood Press, 2000): 21–36, and Williamson Murray and MacGregor Knox, "Thinking About Revolutions in Warfare," in *The Dynamics of Military Revolution, 1300–2050*, ed. MacGregor Knox and Williamson Murray (Cambridge: Cambridge U.P., 2001). For the long tradition of looking at military developments as actually or potentially "revolutionary," see Clifford J. Rogers, "The Idea of Military Revolutions in Eighteenth and Nineteenth Century Texts," *Revista de História das Ideias* 30 (2009): 395–415, and chapter one of this Variorum volume.

600 meters[6] – and since Peter was targeting the gate, he was likely using a low trajectory (as we'd expect from the Milemete illustration), which would mean the cannon-shot could have carried much farther still if it had been elevated to a higher launch angle.[7]

On the other hand, despite its relatively high velocity – probably at least triple the speed of a crossbow-bolt[8] – the projectile would probably not have been able to smash open the gate even if it had hit, and even much larger projectiles were of little use against stone walls. Hence, for the first seventy-five years of their use in Europe, the purpose of guns used in offensive warfare was mostly what Peter of Bruges had said to the town council in describing his gun – that it was an engine for "shooting *into* a walled town when it is besieged"[9] – rather than what, presumably at their insistence, he tried to do with it. Traditional engines such as traction trebuchets and springalds (roughly, giant tension- or torsion-powered crossbows) were also used for this same role alongside the cannon.

Each technology had advantages and disadvantages. Traction trebuchets, lever-action engines that used the force of a team of pullers hauling on ropes to launch

6 Text in L. Lacabane, "De la poudre à canon et de son introduction en France," *Bibliothèque de l'école des chartes*, 2ᵉ ser., I (1844), 45–46n. The total flight of the missile cannot be determined because there is no clear indication of how far from the wall the gun was initially sited, but the distance between the Morel gate (the target) and the plaza of St. Brice (where the victim was killed) is about 440 meters, and the distance from St. Brice to the inner wall is about 160m, based on Isaac van der Kloot's map of the 1709 siege (https://upload.wikimedia.org/wikipedia/commons/f/f1/Siege_of_Tournai_%28Doornik%29_in_1709_%28Isaac_van_der_Kloot%29.jpg) and the 1649 van Loon atlas (https://upload.wikimedia.org/wikipedia/commons/0/02/Tornacum_-_Tournai_-_Doornik_(Atlas_van_Loon).jpg). Since Peter was aiming for the wall, it is likely that he was firing at a relatively flat trajectory. The projectile must still have been on a rising trajectory when it passed the first wall, or it would not have made it over the second wall. Assuming the two walls were the same height, and recognizing that in a ballistic trajectory through air the projectile falls more steeply than it rises, the launch distance must therefore have been more than 160m from the outer wall, for a total horizontal flight of more than 600m.

7 For example, not accounting for air resistance, a missile launched at 10 degrees would need an initial velocity of 131 m/sec to travel 600 m. With the same initial velocity, fired at 45 degrees, the projectile would have an *in vacuo* range of over 1,750 m. These numbers and similar calculations reported in subsequent notes were calculated using the online tools at http://hyperphysics.phy-astr.gsu.edu/hbase/traj.html.

8 Assuming a total flight of 600 m and 10 degree elevation, the initial velocity would be over 130 m/sec, which is a close match for the median results achieved in modern testing with a gun of probably similar design, firing a lead ball and using dry-mixed fourteenth-century powder formulations (133–142 m/sec, though ranging up to 210 m/sec), though much higher than the maximum 87 m/sec achieved with an (unweighted) arrow. Medieval Gunpowder Research Group, "The Ho Experiments. Report Number 1: The Firing Trials," (Nykøbring. F., Denmark: Middelaldercentret, 2002), 10. A contemporary crossbow for battlefield use, by comparison, would have launched bolts with velocities somewhere around 39–50 m/sec. Andreas Bichler, "Testing Reconstructed Medieval Crossbows Featuring Composite Bows," www.historiavivens1300.at/biblio/beschuss/beschuss1-e.htm, accessed 2 September 2012; note also Stephen Grancsay, "Just How Good Was Armor?" *True* (April 1954), p. 90, for 42.16 m/sec from a 336 kg draw weight crannequin crossbow (with a light bolt).

9 Lacabane, "Poudre," 20n: "pour traire en une boine ville quand elle seroit assise."

their projectiles, did not require well-rounded stones, specially made quarrels, or gunpowder, so they were cheap to operate.[10] They could sustain a very high rate of fire and could be surprisingly accurate. Springalds may have been able to outrange bows and crossbows, and were definitely effective anti-personnel weapons even against armored combatants. Cannon, aside from the psychological value of their noise and smoke in frightening the enemy (which must have been short-lived as a major effect), were equally effective against armored men-at-arms, much cheaper to buy than springalds,[11] and easier to transport than either trebuchets or springalds. They also had the advantage of suffering far less deterioration over time than wooden siege engines.[12] On the other hand, they were at first terribly inaccurate and, relatively, very expensive to operate, because the cost of powder was much higher in the fourteenth century than in the fifteenth or later.[13]

10 For examples of the very rough rounding of 4.5 kg trebuchet stones, see Laure Barthet, "La prise de la barbacane de Montségur (Ariège) en février 1244: une introduction à l'archéologie de la poliorcétique," in Nicolas Prouteau, Emmanuel de Crouy-Chanel and Nicolas Faucherre, eds., *Artillerie et Fortification, 1200–1600* (Rennes: Presses Universitaires de Rennes, 2011), 44; note also Reynaud Beffeyte, *War Machines in the Middle Ages* (Rennes: Éditions Ouest-France, 2004), 28. The excavations of Burg Tannenberg (destroyed in 1399) found clearly distinguishable rough-hewn stones for trebuchets and smooth-rounded stones (for guns). J. von Hefner and J. W. Wolf, *Die Burg Tannenberg und ihre Ausgrabungen* (Frankfurt am Main: Verlag der S. Schmerber'schen Buchhandlung, 1850), 77. Stones that were not very carefully rounded could cause guns to burst. J. Bertin, "Le siège du château de Vellexon en 1409," *Bulletin de la société d'agriculture, sciences et arts du département de la Haute-Saône*, 3d Ser., 29 (1898), 120, 170.

11 Rogers, "The Military Revolutions of the Hundred Years' War," 64, for an example of a springald costing five times as much as a small cannon.

12 Edward I of England spent a small fortune on the giant trebuchets he had built at the Tower of London in 1278, but within 35 years they were unusable and valuable only for salvage. Michael Prestwich, "The Trebuchets of the Tower," in Gregory I. Halfond, ed., *The Medieval Way of War: Studies in Medieval Military History in Honor of Bernard S. Bachrach* (Farnham: Ashgate, 2015), 284–290. By contrast, the bronze bombard *Faule Mette*, cast in 1411, was used for the defense of Brunswick as late as 1550, and fired for ceremonial purposes as late as 1728. C. W. Sack, Emil Ferdinand Vogel, and Ferdinand Spehr, *Alterthümer der Stadt und des Landes Braunschweig*, 2d. ed. (Brunswick: Friedrich Otte, 1861), 77–82.

13 Inaccuracy: the experience of Peter of Bruges is reflected in modern tests with a replica of the Loshult gun: five shots all failed to hit a two-meter-square target at 200 paces. Medieval Gunpowder Research Group, "The Ho Experiments. Report Number 2," (Nykøbring. F., Denmark: Middelaldercentret, 2003), 4. Costs: Powder costs per kilogram were especially high early on – for example, powder purchased by the town of Arnhem in 1354 cost 13 *livres* for 11.7 kg, so each kilogram cost the equivalent of over 20 days' wages. Karl Jacobs, *Das Aufkommen der Feuerwaffen am Niederrheine bis zum Jahre 1400* (Bonn: Peter Hanstein, 1910), 33. During the fifteenth century the cost per kilogram fell dramatically, but the quantities of powder used increased even more dramatically. In France, for example, the price of gunpowder fell by around 80% from the late fourteenth century to the late fifteenth, due to a variety of factors including importation of mined saltpeter and economies of scale in production. See Bert S. Hall, *Weapons and Warfare in Renaissance Europe* (Baltimore: Johns Hopkins U.P., 1997), 58. However, this was more than offset by the need to purchase more and more gunpowder resulting from the use of increasingly powerful charges for increasingly large numbers of guns. For example, in 1346, for the Crécy-Calais campaign, Edward III of England made what was for the time an extremely large purchase of saltpeter

Although contemporary sources surprisingly do not emphasize the fact, it does seem from scattered references and modern testing that the biggest advantage of early guns, relative to mechanical artillery, was that they could impart much greater velocity to their projectiles, allowing for shots that carried much farther and still retained enough power to kill, even through armor. One recent analysis suggests springalds had maximum ranges similar to those of crossbows, at around 300 meters.[14] Both traction and counterweight trebuchets generally operated at shorter distances. In one fourteenth-century case, two large counterweight trebuchets were set up around 155 and 240 meters from their target, which is a good match for the range of distances achieved with modern replicas.[15] By contrast,

---

and sulphur, enough to make around 865 kg of gunpowder. French plans for the campaigns of 1406 and 1442, however, envisaged the provision of 2,940 and 9,800 kg of powder, respectively. By 1453, a well-informed Burgundian author reported that the Turks besieging Constantinople were using 490 kg of powder per day. By 1513, Henry VIII's siege train could use as much as 29,000 kg *per day.* T. F. Tout, "Firearms in England in the Fourteenth Century," *English Historical Review* 26 (1911), 690; Napoléon III and I. Favé, *Études sur le passé et l'avenir de l'artillerie*, 6 vols. (Paris: J. Dumaine, 1846–1871), 3:121; Hall, *Weapons and Warfare*, 62, and Philippe Contamine, "Les industries de guerre dans la France de la Renaissance: l'exemple de l'artillerie," *Revue historique* 271 (1984), 278; Jean de Waurin, *Croniques et anchiennes istories de la Grant Bretaigne*, vol. 5, ed. William and Edward L. C. P. Hardy (London: Rolls Series, 1891), 252; Michael Mallett, *Mercenaries and their Masters* (Totowa: Towman and Littlefield, 1974), 163; cf. also Philippe Contamine, *War in the Middle Ages* tr. Michael C. Jones (Oxford: Basil Blackwood Ltd., 1987), 48–49, and for 1567 Thomas F. Arnold, *The Renaissance at War* (London: Cassell, 2001), 33.

14 Peter Purton, *A History of the Late Medieval Siege* (Woodbridge: Boydell, 2010), 397, for J. Liebel's estimate of 180 m at fifteen degrees elevation, shooting from the top of a tower, which can be calculated to equate to about 300 meters on level ground and using a 45-degree angle. This is based on a torsion-model springald; springalds seem to usually have been tension-powered devices, and those might have had greater or lesser ranges. See the drawing of a springald in the 1315 charter of the town of Carlisle (https://upload.wikimedia.org/wikipedia/commons/4/48/Siege_of_Carlisle_1315.jpg); the drawing in Österreichisches Nationalbiliothek manuscript CPV 3069, fo. 22v; and the mention of an *archum* for a springald in David S. Bachrach, "English Artillery 1189–1307: The Implications of Terminology," in *English Historical Review* 121 (2006), 1414. Alain Salamagne, "L'Artillerie de la ville d'Arras en 1369," in Prouteau et al., *Artillerie et Fortification*, 52, gives a higher estimate of 500m based on a rather unclear statement by Marino Sanudo that he interprets as meaning that springalds outranged crossbows.

15 *Ville de Périgueux. Inventaire sommaire des archives communales antérieures à 1790*, ed. Michel Hardy (Périgueux: R. Delage & D. Joucla, 1894), 92; my thanks to Dr. Nicolas Savy for help identifying the locations. In 1320, the Dinanters besieging Bouvignes apparently had to set their trebuchet up near the bank of the Meuse River in order to throw house-smashing stones into the town, which, given the width of the river there, suggests a range of 180 m or less. Jean de Hocsem, *Chronique*, ed. Godefroid Kurth (Brussels: Commission royale d'histoire, 1927), 167. At the thirteenth-century sieges of Dryslwyn and Newcastle Emlyn, the operators of a large English trebuchet needed a protective shelter, suggesting they were operating at less than the range of the Welsh shortbow, which was probably not more than 200 m, if that much. John France, *Western Warfare in the Age of the Crusades, 1000–1300* (Ithaca: Cornell U.P., 1999), 123–24. For modern reconstructions, see the NOVA television show "Secrets of Lost Empires- Medieval Siege" [114 kg stones for 183 m]; Renaud Beffeyte, *L'art de la guerre au Moyen Âge* (Rennes: Éditions Ouest-France, 2005), 81; Beffeyte, *War Machines*, 13 [125 kg over 170m]; Peter Vemming Hansen, "Experimental Reconstruction of a Medieval Trébuchet," *Acta Archaeologica* 63 (1992), accessed

recent tests demonstrate that a small (9 kg) cannon of a very early style can send a lead ball over 1,000 meters.[16] An iron or stone ball, being lighter, would, with the same powder charge, have carried much farther still.[17] The results of this increase in range were demonstrated as early as the siege of Algeciras in 1342. Siege camps were usually positioned to be out of range of fire from the besieged fortification's walls, but in this case the defenders used cannon to shoot iron balls the size of apples at the besiegers, and some not only reached the camp, but passed right over and landed on its far side.[18]

The high velocity of the projectiles not only allowed for this long-range shooting, it also meant, at shorter ranges, very high impact energy compared to muscle-powered projectile weapons. Even an extraordinarily strong English longbow or a windlass-drawn composite-bowed crossbow would rarely impart more than 150 joules of energy to its missile, whereas a small fourteenth-century gun firing a 1.5 cm lead ball could easily launch its projectile with *ten times* that amount of punch, such that "no armor can withstand it," as Jean Buridan wrote as early as 1358.[19]

---

at http://web.archive.org/web/20070105190530/www.middelaldercentret.dk/acta.html; and (for shorter ranges) cf. Tanel Saimire, "Trebuchet – A Gravity-Operated Siege Engine. A Study in Experimental Archaeology," *Estonian Journal of Archaeology* 10 (2006), 75–78.

16 Medieval Gunpowder Research Group, "Report 2." New tests with various powder formulations, using both the *Feuerwerkbuch* loading method (which leaves an empty space in the powder chamber behind the tampon, and will be described more fully later) are much needed to improve our understanding of early artillery.

17 Ke=1/2 M v^2 so as mass falls, for a given amount of ke, v^2 increases in inverse proportion. Horizontal range increases in proportion with v^2 (though this does not account for air resistance). Hence, for example, cutting mass to 1/3 with unchanged ke allows for triple range. The specific gravities of lead, iron, and marble are around 11.35, 7.21, and approximately 2.56, so *in vacuo* ranges should be roughly 1.57 and 4.29 times as far for iron or marble vs. lead.

18 Köhler, *Entwickelung*, 3:1:223; José Antonio Conde, *Historia de la dominacion de los Arabes en España, sacada de varios manuscritos y memorias Arabigas* (Paris: Baudry, 1840), 604. Although the source is late, this is not the sort of story that mere anachronism would lead to, since a late writer transposing the warfare of his own time back to an earlier period would expect the camps to be farther back in proportion with the increased range of the defenders' weapons.

19 Kinetic energy for gun: Medieval Gunpowder Research Group, "Report 1," table 5, shot 2. Buridan: Hall, *Weapons and Warfare*, 45. The lethality of these early guns is also emphasized in Spanish sources. The description in the *Cronica de Alfonso XI* of iron balls able to pass entirely through an armored soldier (*pasaba un ome con todas sus armas*) at the siege of Algeciras in 1342 might perhaps be questioned, since the text dates to the early fifteenth century, but the poem *Alfonso el Onceno*, composed before 1348, also refers to cannon-shot killing men and beasts in the besiegers' camp. Francisco Javier López Martín, "La evolución de la artillería entre los siglos XIV y XVI, con especial atención a los manuscritos de Walter de Milemete y los primeros usos de la artillería en Europa," *Fortificações e Território na Península Ibérica e no Magreb (séculos VI a VXI)*, vol. 2, ed. Isabel Cristina F. Fernandes (Lisbon: Edições Colibri, 2013), 605–606. The validity of Buridan's statement (at least at short range) is, moreover, confirmed by the results of tests with the replica Loshult gun. Robert Douglas Smith, *Rewriting the History of Gunpowder* (Nykøbing Falster: Middelaldercentret, 2010), 117–18. My thanks to the author for sending me a copy of this publication. Arrows: Clifford J. Rogers, "The Battle of Agincourt," *The Hundred Years War (Part II): Different Vistas*, ed. L. J. Andrew Villalon and Donald J Kagay (Leiden: Brill, 2008), Appendix 1. Crossbow: Alan R. Williams, *The Knight and the Blast Furnace: A History of*

Growing Size and Power

Thanks to these advantages in range and penetrating power, between 1351 and 1375 cannon went from being something of a novelty to an absolute commonplace in siege operations. The largest ones also grew much bigger. As late as 1372, an English clerk could still describe a gun weighing 55 kg as "large,"[20] but just three years later an iron gun weighing around a tonne was forged at Cahors for the siege of St.-Sauveur-le-Vicomte.[21] From that point, cannon continued to get steadily bigger and more strongly built, so that they could handle stronger charges of powder and impart more energy to their projectiles. In 1394 a gunfounder in Frankfurt cast a bombard close to four times as massive as the largest St.-Sauveur gun. *Faule Mette*, cast in 1411, was more than twice as heavy as that, and *Luxembourg* (cast before 1447) was well more than twice the weight of *Faule Mette*.[22]

The destructive capacity of the weapons increased in close parallel with their weight. (Figure 5.3) Already by 1378 cannon had surpassed the strongest mechanical artillery in power.[23] A master engineer asked in 1297 to build the largest trebu-

---

the *Metallurgy of Armour in the Middle Ages & the Early Modern Period* (Leiden: Brill, 2003), 919–20, suggests 200J, but modern testing indicates much lower energies. (See note 8.) 150J may therefore be generous for the crossbow, but various indications suggest that windlass- or crannequin-drawn crossbows were comparable to longbow arrows in range and penetrating power, and therefore (since crossbow bolts and longbow arrows were comparable in mass) also in velocity and kinetic energy. Clifford J. Rogers, "The Longbow, the Infantry Revolution, and Technological Determinism," *The Journal of Medieval History* 37 (2011), 332 n. 73, and chapter three of this Variorum volume, p. 59 n. 80.

20 T. F. Tout, "Firearms in England in the Fourteenth Century," *English Historical Review* 26 (1911), 693: "pro ij gunnes grossis feri, precia pecia xl. s." Each would be a gun of around 55 kg given the normal pricing of 4 d./pound (454 g). Ibid., 682–3.

21 Peter J. Burkholder, "The Manufacture and Use of Cannons at the Siege of St-Sauveur-le-Vicomte, 1375," (MA Thesis, U. of Toronto, 1992); Napoléon III and Favé, *Études*, 4:p.j. xvii–xlii. There is also a reference to a 660 kg Pisan bombard as early as 1362 in the *Cronica di Pisa*, though that is obviously not as reliable a source as a financial account (and is not included in Fig. 3). In Muratori, *Rerum italicarum scriptores* XV, col. 1037.

22 The weights were respectively 3,995; 9,583; and at least 23,520 kg. Bernhard Rathgen, *Das Geschütz im Mittelalter* (Berlin: V.D.I. Verlag, 1928), 256–7; Sack, et al., *Alterthümer . . . Braunschweig*, 77; Joseph Garnier, *L'Artillerie des Ducs de Bourgogne, d'après les documents conservés aux archives de la Côte-d'Or* (Paris: Honoré Champion, 1895), 172.

23 Already in 1378 bombards could shoot 1,000 *Schritt* (note 29), so around four times as far as a trebuchet throwing large stones (note 15), indicating double the velocity and quadruple the energy for a given mass. Also in 1378 we have records of a "grooten donrebusse" of the count of Holland able to throw 400-pound (196 kg) stones, and a bombard made at Chalon firing 450-pound (221 kg) stones. However, it is likely that the bombards of 1378 with the thousand-Schritt range (which was presumably noted in the chronicle because it was an increase over previous experience) used the roughly 5% powder:ball ratio of the 1394 Frankfurt bombard and several 1397 *Centnerbüchse*, while the earlier big guns used only about half that proportional charge. (Notes 72, 71, 59). If so, that suggests the Chalon bombard of 1378 would have imparted about 1.5 times the kinetic energy of the 600-pound stones of the near-contemporary Périgueux trebuchet described later. Jacobs, *Aufkommen der Feuerwaffen*, 56; Napoléon III and Favé, *Études*, 3:103; J. R. Partington, *A History of Greek Fire and Gunpowder* (Cambridge: W. Heffer & Sons, 1960), 114, 122; Garnier, *Artillerie*, 12.

Highest Recorded Weight (European Guns)

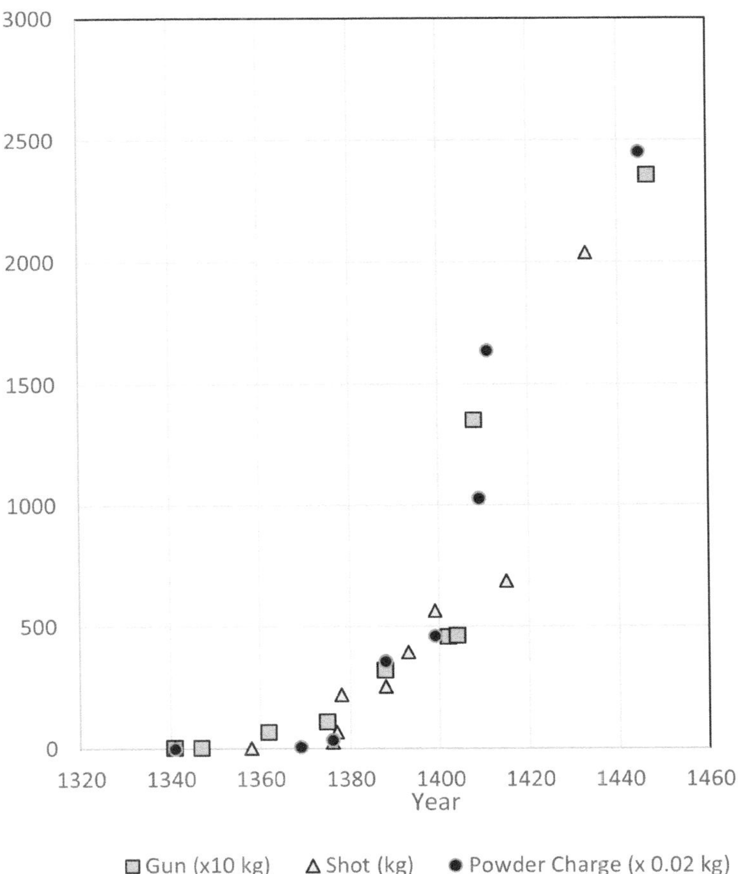

Figure 5.3 The Increasing Power of the Biggest Guns, 1340–1450.

chet possible constructed one that threw stones of 200 pounds.[24] A bigger weapon casting 400-pound stones had been built for the siege of Damietta in 1218–19, and this seems to have been a plateau level for the normal range of counterweight trebuchets: two hundred years later, French and Burgundian examples lobbed stones of

24 *Istore et croniques de Flandres*, ed. Kervyn de Lettenhove, 2 vols. (Brussels: F. Hayez, 1879–80), 1:213. The trebuchet stones excavated at Burg Tannenberg (from 1399) ranged from 24–36 cm, so up to a maximum of about 63 kg. Hefner, *Burg Tannenberg*, 77.

100, 300, and 400 pounds.[25] A few were even bigger: an engine owned by the town of Périgueux in 1398 launched 600-pound projectiles.[26] But in the period 1399 – c. 1415, as we know from both archival evidence and surviving weapons, bombards could throw gigantic stones of 700; 800; 1,100; even 1,700 pounds.[27]

25 Paul Chevedden, "The Hybrid Trebuchet: The Halfway Step to the Counterweight Trebuchet," in Donald J. Kagay and Theresa M. Vann, eds., *On the Social Origins of Medieval Institutions: Essays in Honor of Joseph O'Callaghan* (Leiden: Brill, 1998), 187; Contamine, *War*, 195; Ch. L. Grandmaison, ed., *Documents inédits pour servir à l'histoire des arts en Touraine* (Tours: Soc. Arch. de Touraine, 1870), 124. In between, for example, a French trebuchet at the siege of la Roche Derrien in 1347 cast stones of 300 pounds (147 kg?), while the engines of the count of Hainault at a siege of 1345 threw projectiles of "200 pounds [94 kg?] or more." *Grandes chroniques de France*, ed. Jules Viard, 9 vols. (Paris: SHF, 1920–53), 9:300; Hocsem, *Chronique*, 336.

The claim that two Venetian engines threw stones of three thousand pounds (*libbre tre mille*) (one tonne if the Venetian *libbra sottile* is meant) is not credible; wooden throwing arms simply cannot be made strong enough for that. *Istoria dell'assedio e della ricupera di Zara fata da'Veneziani nell'anno MCCCXLVI, scritta da autore contemporaneo*, in Jacopo Morelli, ed. *Monumenta Veneziani di varia Letteratura* (Venice: Carlo Palese, 1796), ix. The same goes for the stones of 2,700 pounds (*cantariorum . . . decem octo*) supposedly thrown by the Genoese engine *Troia* at Cyprus in 1373 (even if the *libra peso sottile* of 317 grams is meant). Giorgio Stella, *Annales Genuenses*, in *Rerum italicarum scriptores*, ed. Muratori, vol. 17 (Milan: n.p., 1730), 1104–05. Roland of Padua is a better source, but his claim that at the siege of Este in 1249 the Paduans and their allies had fourteen trebuchets casting stones of 1,200 pounds (408 kg) and more (*rotabant hedificia quedam lapides . . . ponderis librarum 1200 et ultra*) should be viewed with skepticism. *Rolandini Patavini chronica*, ed. Philippo Jaffé in MGH, *Scriptores*, ed. G. H. Pertz, 19 (Hannover: MGH, 1866), 90. It is not, however, beyond possibility. Guillaume Henri Dufour, *Mémoire sur l'artillerie des anciens et sur celle du moyen âge*, vol. 1 (Paris: A. Cherbuliez, 1840), 104, calculated that a 500 kg stone could be thrown by an engine with a 15,000 kg counterweight, and Edward I in 1278 had built a trebuchet with a lead counterweight of over 17,000 kg. By comparison, the Urquhart fixed-counterweight trebuchet built for the NOVA television show "Secrets of Lost Empires: Medieval Siege" was able to throw 250-pound stones for 200 yards with a counterweight of only eight tons (about 7,250 kg). "Secrets of Lost Empires: Medieval Siege"; Ed Levin, "The Highland Fling," *Timber Framing: Journal of the Timber Framers Guild* 50 (1998): 14–19.

26 *Petit Livre Noir* (Registre consulaire de Périgueux. 1360–1449. Archives départementales: BB. 13), fo. 47v, transcription by Jean Roux online at https://web.archive.org/web/20150405145020/ www.perigordoccitan.fr/Paleographie/Livre_Noir/LNtexte.htm, accessed 24 Jan. 2017.

27 The last-mentioned figure is for the still-extant *Pumhart von Steyr*, forged 1414–1424 (probably at the start of that period), now in the Heeresgeschictliches Museum in Vienna. It has a conical bore, to allow for stones of different sizes. Since it has a caliber at the mouth of 88 cm (tapering down to 76 cm), it could have handled a ball of at least 83 cm, which would weigh around 772 kg, depending on the density of the stone used. For the measurements, see Volker Schmidtchen, *Bombarden, Befestigungen, Büchsenmeister* (Düsseldorf: Droste Verlag, 1977), 32–4. In this article I have made calculations based on a specific gravity of 2.56 for stone, though in the majority of cases extant stone cannonballs that have been weighed have lower specific gravity, ranging from 2.02–2.39. Rathgen, *Geschütz*, 49, 203; a personal communication from Sascha Pries of the Kölnisches Stadtmuseum providing the diameter and weight of the ball associated with their "Steinmörser" indicates a s.g. of 2.34. These data, however, are all from Germany and mostly from the fourteenth century. English documents of 1428 and 1440 show a preference for cannonballs from Maidstone, indicating Kentish ragstone (s.g. ca. 2.7) and fifteenth-century Burgundian accounts describe bombard stones as marble (s.g. 2.4–2.7) or more rarely sandstone (commonly s.g. ca. 2.6). The National Archives, Kew, E101/51/27; Joseph Stevenson, (ed.)

Moreover, even before 1400 bombard stones had substantially greater velocity than trebuchet projectiles – and increasingly so over time, as shown by increases in range. In 1377 the Büchsenmeister Walter Judenkind of Arles contracted to forge an iron *Centnerbüchsen* (gun throwing hundred-pound stones) with a range of at least 300 *Schritt* (215 m) – about the same as a large trebuchet.[28] But three bombards cast in Frankfurt just one year later could shoot large projectiles 1,000 paces (715 m), indicating an initial velocity about 1.8 times that imparted by a large trebuchet.[29] Just one decade later, in 1388, a gunfounder in the same town offered to cast a cannon able to shoot 1,000 *Klaster* (1,715 meters).[30] That seems to have been exceptional for the time, but was quite normal by around 1405–10. A passage in the *Feuerwerkbuch* that probably dates to that period says that bombards should be able to shoot up to 2,500 *Schritt* (around 1,850

Letters and Papers Illustrative of the Wars of the English in France during the Reign of Henry the Sixth, King of England, 2 vols. (London: RS, 1861–4), 2:2:587; Garnier, *Artillerie*, 122, 127, 134, 173, 175, 87. Two 44 cm cannonballs recovered from the siege of Bolesławiec in 1396 have an average s.g. of 2.38. Piotr Strzyz, "Archaeological Traces of Use of Firearms During Sieges. Selected Examples from the Territory of Poland," *Fasisculi Archaeologiae Historicae* 28 (2015), 112. Granite and related stone (s.g. ca. 2.5) is the most common material for the stones at Olszytyn. Piotr Strzyz, Piotr Czubla, and Adam Mackiewicz, "Cannonballs from the Olsztyn Turret," *Fasisculi archaeologiae historicae* 28 (2015), 123–4. My thanks to Grzegorz Żabiński for pointing me to this article. For a Turkish bombard reportedly firing an 1,800-pound stone in 1453, see Waurin, *Croniques*, 5:252; for a gun at Belgrade with a 42" caliber (indicating a stone of over 2,000 kg) in 1433, see Bertrandon de la Broquière, *Le voyage d'Outremer* (Paris: Leroux, 1892), 214.

28 A pace (*Schritt*) of 2.5 feet varied in length from place to place, as did other traditional measurements of weight and distance. Throughout this article I have made use primarily of the following references to determine the appropriate metric equivalencies for traditional measures: Johann Leuchs, *Der Contorwissenschaft*, pt. 3, 4th ed. (Nuremberg: C. Leuchs und Comp., 1834); M. R. B. Gerhardt, *Allgemeiner Contorist*, pt. 1 (Berlin: Arnold Wever, 1791); P. Kelly, *The Universal Cambist*, 2nd ed., vol. 2 (London: Printed for the Author, 1821); Franz Mozhnik, *Lehrbuch des gesammten Rechnens für die vierte Classe der Hauptschulen in den k.k. Staaten* (Vienna: Verlage der k.k. Schulbücher Verschleiß-Administration, 1848). Note that traditional forms of measurement such as "miles" and "pounds" are sometimes retained instead of (rather than used alongside) metric linear measures, particularly when there is doubt about appropriate local usage.

29 Martin Crusius, *Annalium Svevicorum* (Frankfurt: Nicolai Bassaei, 1596), 291: three bronze bombards cast in 1378 that reportedly threw iron balls (*globum ferreum* – which is likely a misreading of an original document that described the *bombard* as iron, though the rest of the passage certainly seems to be based on a contemporary municipal fiscal account or gun-maker's contract) of 127, 70, and 50 pounds to a distance of a thousand paces (*ad spatium mille passuum*). The *passus* here is probably a translation of a German *Schritt* of 2.5 feet, rather than the proper Latin meaning of two *gradi* (five feet), because bombards of this period generally had a maximum powder charge of 5–8% of stone weight and used uncorned powder, whereas the later *Feuerwerkbuch*, which indicates a range of about a mile for bombards, also envisions a use of an 11% charge of corned powder 50% more powerful than uncorned powder. (See the following.) Initial velocity vs. trebuchet: 4x the range (taking average trebuchet range as 180 m) indicates double the initial velocity.

30 Rathgen, *Geschütz*, 249–51.

meters).[31] A slightly later text increases that to 3,000 *Schritt* (2,220 meters) or more.[32] Recorded shots of around 1,600 meters in 1430 and in 1452, and instances when large medieval bombards threw stones or 3,200 meters or more during the sixteenth century, show these generalizations are perfectly credible.[33] Guns manufactured at the very end of the Middle Ages were designed to have ranges as great 3,200 meters, roughly thirteen to sixteen times what a heavy trebuchet could manage.[34]

Thanks to their high velocity, already by the turn of the fifteenth century gunstones could hit with significantly more energy than even the largest projectiles from counterweight trebuchets. For example, the second-largest bombard used at

31　From a manuscript dated 1432 but incorporating various earlier strata of text: Gerhard W. Kramer, ed., *The Fireworkbook: Gunpowder in Medieval Germany*, tr. Klaus Leibnitz (London: Arms and Armour Society, 2001), 56, 58 (Freiburg MS 362, fo. 88), gives the range of a cannon-shot as 1,500 *Schritt* to (with stronger powder) 2,500 paces [1,850 meters or 1.15 English miles], from a gun with a powder charge of one ninth the weight of the stone, and a ball weighing a Venetian hundredweight (30 kg). This version of the text probably dates to c. 1405–10, since it recommends a 1:9 powder charge ratio (which matches a Burgundian gun cast in 1409: Garnier, *Artillerie*, 25–6).

32　Max Jähns, *Geschichte der Kriegswissenschaften, vornemlich in Deutschland*, pt. 1 (Munich: R. Oldenbourg, 1889), 1:389. This is from the Ambraser Sammlung codex 67, dating to 1408 or shortly thereafter. At this same time (1408), according to a private letter of the duke of Burgundy, the Flemings at Othée opened fire with bombards at a range of just three bowshots (perhaps 600–700 meters), but for battlefield use smoothbore guns were generally employed at only a fraction of their maximum range. Dom U. Plancher et al., *Histoire générale et particulière de Bourgogne, avec les preuves justificatives*, vol. 3 (Dijon, 1748), preuve justificative no. 260.

33　1430: *A Parisian Journal, 1405–1449*, tr. Janet Shirley (Oxford: Clarendon Press, 1968), 241 (from the Porte St. Denis of Paris to well beyond St. Lazare); Purton, *Late Medieval Siege*, 325 (Ottoman bombard cast in 1452 for the siege of Constantinople firing a test shot one mile). The Boxted bombard, which probably dates to the 1440s–50s, was used into the eighteenth century to throw balls into a hill "about a mile distant" for the amusement of fair-goers. Robert D. Smith and Ruth Rhynas Brown, *Bombards: Mons Meg and Her Sisters* (London: Royal Armouries, 1989), 52. In 1558, the medieval bombard Mons Meg fired a stone from Edinburgh castle to Wardie Muir, 3.2 km distant (though also about 75 meters lower). J. Hewitt, "Mons Meg, the Ancient Bombard Preserved at Edinburgh Castle," *Archaeological Journal* 10 (1853), 28. Judging by the style of gun-carriage shown in a c. 1600 stone-carving of the gun – reproduced on the back cover of Smith and Brown, *Bombards* – this was likely at a low angle (perhaps 15–20 degrees), not 45 degrees, suggesting a maximum potential range of three miles or more (albeit with Early Modern gunpowder). Even *Faule Mette*, cast in 1411, was recorded as firing from Brunswick to and over a besiegers' camp near Melverode, indicating a flight of approaching 5 km – though this was in 1550 and using a 70-pound charge rather than the c. 55-pound charge employed in earlier times. A 1717 shot at 45 degrees elevation with a 52-pound charge shot a 341 kg ball 660 *Ruthe* (around 3 km). C. W. Sack, Emil Ferdinand Vogel, and Ferdinand Spehr, *Alterthümer der Stadt und des Landes Braunschweig*, 2d. ed. (Brunswick: Friedrich Otte, 1861), 77–81.

34　In 1453 the Venentians employed a German gunfounder to cast bombards capable of throwing 400-pound stone balls with a two-mile range. Michael E. Mallett and J. R. Hale, *The Military Organisation of a Renaissance State: Venice c. 1400 to 1617* (Cambridge: Cambridge U.P., 1984), 83. In 1451 Stralsund purchased a bombard supposed to throw a stone a half of a *Meile*, so either somewhat under or somewhat over two English miles, depending on the *Meile* meant. Otto Piper, *Burgenkunde* (Munich: R. Piper, 1905), 372n.

the siege of Burg Tannenberg in 1399 already would likely have thrown its 54 cm stones with more than double the kinetic energy of the somewhat heavier stones cast by Périgueux's great trebuchet.[35]

These giant bombards could sometimes make a big difference in siege warfare. In 1405, the captain who held Warkworth castle for the rebellious earl of Northumberland at first refused to surrender the place to King Henry IV of England, because his garrison was well provided with men and supplies. Hearing the captain's final response, as Henry himself wrote shortly afterwards to his council, "We immediately directed Our cannon against the castle, and they did such service that within seven shots the captain and all the others of his company, crying for mercy, submitted themselves to Our mercy."[36] It made a big impression when a bombard knocked a hole in a fortification's wall with a single shot, as happened at a castle besieged by the bishop of Minden in 1408.[37] An English translator of Vegetius commented that same year that "great guns nowadays shoot stones of so great weight that no wall may withstand them, as has been well shown both in the north country and in the wars of Wales."[38]

Practical Capabilities in the Early Fifteenth Century

That last comment, however, was an exaggeration, more reflective of the psychological impact of the new bombards than their actual power. It has often been noted that Berwick town and castle were captured by Henry IV in 1405 using his bombards, but the rarely mentioned details are revealing. The town's walls were low, thin, and in such poor repair that they were on the verge of collapse from their own weight. The castle, however, was much stronger, and initially the stones from

---

35 As noted earlier (note 29) already in 1378 some bombards could shoot a thousand *Schritt*, so around 4x as far as a trebuchet throwing large stones (note 15), indicating double the velocity and quadruple the energy for a given mass, * 250 kg/293 kg for the difference in weight, = 3.41 ke. However, it is likely that the half-mile range was achieved by bombards with powder charges of around 8% of ball weight, rather than the 5% ratio that seems to have been more common in the 1370s–90s, both for *Centnerbüchsen* throwing hundred-pound stones and for larger bombards. (See notes 69–70.) So we can fairly conservatively estimate the power of the Tannenburg gun as 3.41 * 5/8 = 2.13 times that of the Périgueux trebuchet. The weight for the stone is calculated based on its diameter, for which see Hefner and Wolf, *Tannenberg*, 77.

36 *Proceedings and Ordinances of the Privy Council of England*, ed. N. H. Nicolas, 7 vols. (London: Record Commission, 1834–7), 1:275.

37 Minden: *Scriptorum rerum Brunsvicensia* (Hanover: Nicolas Förster, 1710), 2:203: "Vicit et recuperavit potenter *castrum Wedegonis*; Nam habuerunt ibi de *grotten Metten* & cum ea uno ictu jactaverunt intra muros unum magnum foramen & foderunt subtus terram, ita ut accepterint urnam de fonte, & aliqui de castro suspense sunt."

38 Malcolm Hebron, *The Medieval Siege: Theme and Image in Middle English Romance* (Oxford: Clarendon Press, 1997), 31. Similarly, Thomas Walsingham reports that Henry IV took to Berwick in 1405 several guns, one of which was so large that, "it was believed" (*ut creditur*), no wall could withstand it. Thomas Walsingham, *Annales Henrici Quarti, in Chronica et annales regnantibus Henrico Tertio, Edwardo Primo, Edwardo Segundo, Ricardo Segundo, et Henrico Quarto*, ed. Henry Thomas Riley (London: Rolls Series, 1868), 411.

the king's guns simply shattered against its solid walls. It was only when a lucky shot struck a window and passed through to kill several men that the demoralized garrison surrendered.[39]

A campaign conducted in 1411 by the young duke of Guienne (the heir of the king of France) against the rebellious duke of Berry provides another excellent case-study for delimiting what gunpowder artillery both could and could not do in this period.[40] Guienne's army aimed to capture the town of Étampes. To clear his lines of communication, he needed to capture the castle of Bretonnière. The captain of the place refused to surrender initially, but then fled with his garrison by night once he saw artillery emplaced for battery. The duke's army then pushed on to Étampes. The townsmen had no interest in facing a siege for the sake of the duke of Berry and surrendered while Guienne's army was still approaching. Berry's garrison, however, held out in the castle of the town. This fortification was old but strong, with large towers and substantial walls, and on high ground. Since its situation made it practically impossible to undermine, it was considered almost impregnable. Despite the presence of the artillery train that had successfully intimidated the captain of Bretonnière, the duke's initial attempts to assault the place, mounted over several days, seemed to bear out this conclusion. Each attack was thrown back, with serious losses to the besiegers. The situation changed, however, when the great bombards of Paris arrived.[41] "The wall was broken down in several places by the enormous, massive stones – as big as millstones – that these engines cast without ceasing," writes the chronicler of St.-Denys. "When the shot had at last broken a way through, the French consigned all the houses next to the walls to the voracious flames," compelling the defenders to retreat to the extremely strong twelfth-century donjon. It is not clear whether the cannon had opened a breach that French soldiers stormed through, or merely exposed flammable structures inside the walls to an attack with incendiaries, but either way the bombards had managed to do enough damage to solid masonry walls to make a major difference in the conduct of the siege.[42]

39 Town: Clifford J. Rogers, "The Military Revolutions of the Hundred Years' War," repr. with revisions in Clifford J. Rogers, ed. *The Military Revolution Debate* (Boulder: Westview, 1995), 88. Castle: *Eulogium historiarum*, ed. F. S. Haydon, 3 vols. (London: Rolls Series, 1858–63), 3:408. At Plaue in 1414 cannon destroyed walls so thick that a wagon could be driven along the ramparts – but they were only made of brick, not stone. Dorothea Goetz, *Die Anfänge der Artillerie* (Berlin: Militärverlag der DDR, 1985), 22.

40 This paragraph and the next are based on Louis Bellaguet, ed., *Chronique du religieux de St. Denys (1380–1422)*, 6 vols. (Paris: Crapelet, 1839–52), 4:572–577.

41 These are described in the *Chron. St. Denys* as "machinas jaculatorias" [throwing machines], and this has often been taken to mean trebuchets. It can, however, be taken as fairly certain that bombards are meant, because in the chronicler's account of the siege of Bourges the following year the author also uses the same noun to refer to what is unquestionably a bombard, named *Griete*, likewise throwing "immense stones," but specified as using a large charge of powder and making a thunderous noise. Ibid., 4:652.

42 The French translation that accompanies the Latin text of the *Chron. St. Denys* has the French entering by the breach, but the Latin does not actually say that. For bombardment at Harleur in

The next phase of the operation, however, showed the limits of the new bombards. Several assaults failed badly, and against the sixteen-foot-thick walls[43] of the donjon the gun-stones accomplished nothing but to inspire the "derision and mockery" of the ladies inside the stronghold, and the great displeasure of the leading besiegers. They saw the ineffectiveness of the missiles and worried that their efforts were being wasted and that the lengthening siege might end in failure. They even seriously considered breaking off the operation, though the failure of his first campaign would have brought "eternal shame" on the duke, and on the royal house of France. This disastrous result was avoided when a Parisian engineer organized the construction of a giant mantlet of extremely thick oak, which was emplaced against the walls. Under its shelter, teams of workers with steel picks worked away at the masonry. After five days, they had so undermined the base that they could threaten to bring down the whole structure by firing the props they had emplaced. The captain of the garrison "suppressed his stubborn pride" and surrendered to the duke.[44] Ancient techniques and technology, more than new ones, were the key to the success of the capture of the stronghold.

In this period the failure of the guns against the donjon of Étampes was still more the norm than artillery's success against the outer castle walls. The French offensives against English Aquitaine in 1404–1408 illustrate the situation well. In 1404, even the little towns of the Chalosse region in the Landes were still so confident of the superiority of the defense in siege warfare that when threatened by a substantial French army they declared they would not pay so much as fifty pennies to secure a truce. In 1405, the French required six months to capture the minor fortification of Chalais, and three months to take Mortagne in Saintonge.[45] In 1405–6, the town of Bourg was unsuccessfully besieged by a large army under the French king's brother, the duke of Orléans. The duke set up three bombards "in order to weaken the walls by the continuous blows of large stones," mounted frequent assaults, and rained a steady hail of arrows and bolts down on the defenders. But, from the end of October to mid-January, all his efforts failed. With his army suffering from miserable conditions and his finances in disarray, Orléans

---

1415 breaking down a barbican and making it "more liable to catch fire," see *Gesta Henrici Quinti: The Deeds of Henry V*, ed. and tr. F. Taylor and J. S. Roskell (Oxford: O.U.P., 1975), 46–7. For another very early example of the power of artillery at the turn of the century, see Denis Lalande, ed., *Le livre des fais du bon messire Jehan le Maingre, dit Bouciquaut, Mareschal de France et gouverneur de Jennes* (Geneva: Droz, 1985), 131, for a claim that Marshal Boucicaut's engines (including at least one huge trebuchet firing 600-pound stones) and cannon, firing constantly, were able to thoroughly demolish large sections of the walls of the "seemingly impregnable" stronghold of Montignac in 1398 – albeit over the course of two months.

43 The St. Denys chronicler (Michel Pintoin) says the walls, where they were eventually broken through, were ten feet (3 m) thick, but that would have been at the entrance, where internal guard-chambers left a void in the middle. The gunstones would have hit higher up where the wall was presumably solid.

44 *Chronique de St. Denys*, 4:576.

45 Guilhem Pépin, "The French Offensives of 1404–1407 against Anglo-Gascon Aquitaine," *Journal of Medieval Military History* 9 (2011): 1–40.

admitted failure and retired into France "in great dishonor."[46] Similarly, in Italy in 1405–6, the castle of Vicopisano was besieged for nine months by the Florentines, and although the interior of the place was completely demolished by bombardment (by both trebuchets and large bombards), the attackers' assaults on the walls were always repulsed "with shame and damage" until hunger forced the defenders to surrender.[47]

Up to this point, the spread of gunpowder artillery had actually probably favored defenders more than attackers. Either stone walls or earth-and-timber bulwarks were good protection against all but the largest cannon-shot. But during an assault the attackers had to leave their shelters and expose themselves to the lethal shot of guns, whereas defenders did not. Indeed, the besieging gunners manning large bombards were themselves vulnerable to accurate counter-fire from the smaller cannon of the defenders (which shot more accurately, and when fired from city walls had an advantageous falling trajectory). In 1419, for example, a Moroccan army with gunners on loan from Granada aimed to recapture Ceuta in North Africa from the Portuguese. They hoped

> to wear them down with that violent invention – the bombard – like the ones we [Portuguese had earlier] brought to conquer this place. The Moors had two, a lot for them, the mainstays of this diabolical offensive. They shot at the walls but with no success because we brought up our own, which inflicted real damage. Then, as the enemy was repairing, a wily engineer shot off his piece so well that he bested the enemy's main gun, slew the gunner, and then so adroitly suppressed the fire of the other weapons that the Moors lost all their guns.[48]

Thus, observed a German master-gunner of around 1410, the defender of a strong fortress provided with sufficient artillery ought to be able to repulse any attacks "until he is relieved, or the enemy is given a good thrashing and departs the siege."[49] As we have already seen, that is what happened at Bourg in 1406. Because of the great kinetic energy they could impart to their projectiles, which made them capable of lethal hits even against well-armored attackers, small to medium guns were very effective in defeating assaults. At Bourges in 1412, some of the Burgundian soldiers besieging the town tried to pursue a defeated Armagnac sally force back through the open town gates. The Armagnac defenders on

---

46 Ibid. and *Chron. St. Denys*, 3:452–61; Germaine Lefevre-Pontalis, "Petit chronique de Guyenne jusqu'à l'an 1442," *Bibliothèque de l'école des chartes* 47 (1886), 64 (quotation).

47 *L'Assiedo di Pisa (1405–1406). Scritti e documenti inediti*, ed. Giuseppe Odoardo Corazzini (Florence: Ulisse Diligenti, 1885), 66–7; Purton, *Late Medieval Siege*, 225–7.

48 Zurara, quoted in Weston Cook, "Warfare and Firearms in Fifteenth Century Morocco, 1400–1492," *War & Society* 11 (1993), 28.

49 Wilhelm Hassenstein, *Das Feuerwerkbuch von 1420. 600 Jahre Deutsche Pulverwaffen und Büchsenmeisterei* (Munich: Verlag der Deutschen Technik, 1941), 31.

the walls began to shoot so strongly with cannon that the Burgundians had to fall back. Although the besiegers were provided with "many large bombards for laying siege," some ten weeks later they still had accomplished nothing and decided to starve the defenders out with a distant blockade.[50]

Because guns were so effective in the defense of town walls, they became increasingly popular acquisitions for municipal governments and individual bourgeois. In 1406, the town of Mons, for example, possessed 66 guns.[51] In the same year Lille had an arsenal of at least 37 guns.[52] On top of that stock, in response to the siege of the nearby city of Arras in 1414, the latter town acquired an additional *63* cannon, almost all *veuglaires* or "fowlers," a recently developed type of longer-barreled gun shooting stone or lead shot. These weapons were breech-loaders provided with two or three beer-stein-shaped powder chambers each, and typically weighed around 50 kg per piece.[53] The large bombards of this era, firing stones each of which weighed as much as several *veuglaires*, were obviously far more expensive, and less useful (due to their slow rate of fire) in defeating an assault. However, as noted earlier, they did have occasional dramatic successes in offensive siege warfare. Such weapons were therefore sometimes acquired not only by territorial princes, but also by wealthy urban governments, both as status symbols and for their practical value.[54] In 1420, for example, the city of Bordeaux commissioned a great bombard to throw 346 kg stones.[55] The big guns could be used

50 Pierre de Fenin, *Mémoires de Pierre de Fenin*, ed. Mlle. Dupont (Paris: SHF, 1837), 27.

51 Michael Depreter, "La prince face à l'artillerie communale dans les Pays-Bas au début du XVe siècle: Flandre, Brabant, Hainault," in *Revue du Nord*, collection histoire (hors série) no. 35, *Autour d'Azincourt: Une société face à la guerre (v. 1370 – v. 1420)*, ed. Alain Marchandisse and Bertrand Schnerb (2017), 299, which also notes 95 cannon and 38 ribaudequins at Bruges in 1411, 50 more guns purchased by the same city in 1416, and a large iron bombard acquired the next year.

52 Alixandre de la Fons-Mélicocq, *De l'artillerie de la ville de Lille aux XIVe, XVe et XVIe siècles: archers, arbalétriers, canonniers* (Lille: Lefebvre-Ducrocq, 1854), 15. In 1412 there were four large cannon called bombards, four other large cannon, and 24 small iron guns, and the town councilors commissioned the casting of two more "portable" bronze guns, weighing 44 pounds combined, "considering that there are none of that style, and that all are of iron." Ibid.

53 Ibid., 16. Compare to the 11 guns (probably the town's first) purchased by Regensburg in 1379, which collectively weighed only 120 Pfund (67 kg). Köhler, *Entwickelung*, 3:1:247.

54 The inscription on the "large and noble" bombard *Redoubtée* (meaning "dreaded, feared") nicely reflects the typical motivations of townsmen in acquiring the big guns, pitching the purpose as defensive but implying also offensive use: "The year 1436/I was made, in order to spend my time/on guard and for the defense/against those who against Metz make offense/to punish them and bring them to justice. / Suitable I am for this profession; / and for those who would know my name/*Rebdoubtée* is what I'm called." Jean-François Huguenin, ed. *Les chroniques de la ville de Metz, 900–1552* (Metz: S. Lamort, 1838), 199.

55 Seven *quintaus* of 49.402 kg. *Archives Municipales de Bordeaux, IV: Registres de la Jurade. Délibérations de 1414 à 1416 et de 1420 à 1422* (Bordeaux: Imprimerie G. Gounouilhou, 1883), 426; for the weight of the quintal of Bordeaux, see Jean Cavignac, *Jean Pellet, commercant de gros, 1694–1772* (Paris: S.E.V.P.E.N., 1967), 35.

in enforcing a city's authority in its district, and towns that owned them gained opportunities to be useful to allies or lords, or to the king, by loaning them out.[56]

Quantitative Evolutions and the Artillery Revolution: More Guns

The increased breadth and pace of acquisition of cannon meant that the number of guns available for deployment at major sieges – both wall-breaking bombards and smaller supporting weapons – increased sharply. In the sieges of 1398–1406 mentioned earlier, even armies led by the French king's brother or by his marshal (one of the three highest-ranking military officers of France) had only two or three big guns.[57] In 1405, the wealthy count of Holland (who was also count of Zeeland and Hainault and duke of Bavaria-Straubing) had at least four big bombards (and around twenty fowlers) for the siege of Hagestein.[58] In 1407, King Juan II of Aragon had five large bombards for the unsuccessful siege of Setenil; his eldest son had only three for the prior siege of Zahara.[59] John the Fearless, duke of Burgundy, who may have had the best artillery train in the world at the time, had at first two and ultimately five at Vellexon in 1409–10.[60] In 1415, when

---

56 Although the theme of his article is how territorial princes worked to *reduce* their dependence on artillery loaned by the towns, the importance of such guns is clear in Depreter, "La prince face à l'artillerie," 285–304, especially 289–298. Middling nobles could also use their guns both for their own purposes and to aid their lords. Among the guns used by the duke of Burgundy's army for the siege of Vellexon in 1409–10, for example, were a large bombard borrowed from the lord of Villars, three small bombards belonging to the town of Besançon, and two medium bombards belonging to the lord of Montaigu. For the same duke's planned siege of Calais in 1406, he sent a messenger to ask three noblemen and two cities (Brussels and Louvain), for guns. See note 60; Bertrand Schnerb, "Un projet d'expédition contre Calais (1406)," in *Les champs relationnels en Europe du Nord et du Nord-Ouest des origines à la fin du Premier Empire: 1ère Colloque Historique de Calais*, ed. Stéphanie Curveiller et al. (Calais: N.P., 1994), 187, and see also Depreter, "La prince," 294.

57 As an additional example, a major Polish royal expedition in 1396 probably had just three bombards, of 26–50 cm caliber. Strzyz, "Archaeological Traces," 111.

58 M. J. Waale, *De Arkelse oorlog, 1401–1412. Een politieke, krijgskundige en economische analyse* (Den Haag: Koninklijke Bibliotheek, 1990), 114 (*grote bussen* and *vogelaars*). The defenders made use of two *bussen* and some *vogelaars* as well; this is by several years the earliest reference to *veuglaires* I have found, though the *Bochsen* cast in four parts [i.e., presumably, with three chambers] at Marienburg in 1403 may be *veuglaires* without the name. *Das Marienburger Tresslerbuch der Jahre 1399–1409*, ed. Erich Joachim (Königsberg: Thomas & Opperman, 1896), 217; Grzegorz Żabiński, "Technology of Manufacture of Firearms in the Teutonic Order's State in Prussia – Gun Barrels and Metal Projectiles," *Fasisculi archaeologiae historicae* 28 (2015), 102. Over the course of the siege of Hagestein, according to financial accounts, the besiegers' *grote bussen* used some 15,480 kg of powder, though one wonders if this actually covered the powder expended on the smaller guns also. Waale, *De Arkelse oorlog*, 254, 260.

59 Fernán Pérez de Guzmán, *Cronica del serenissimo rey don Juan el segundo deste nombre* (Valencia: Benito Monfort, 1779), cap. xli, xxxvi.

60 Initially two large bombards (one of bronze, one of iron) belonging to the marshal of Burgundy and three small bombards belonging to the town of Besançon; then the large bombard *Dijon*, a bronze

Henry V deployed nine large bombards of "unheard-of size" for "playing tennis" with the defenders of Harfleur, it was considered quite a remarkable show of force. But thirty-four years later Charles VII emplaced almost twice as many (sixteen) for the siege of the same town.[61] At Caen in 1450, King Charles had twenty-four great bombards, some of which were reportedly so large that a man could sit erect inside the barrel.[62] At Belgrade in 1456, Mehmet the Conqueror had twenty-eight newly cast bombards of roughly 55 cm caliber.[63] So the number of big bombards at a major siege increased by a factor of around ten over the course of a half-century.

This quantitative increase mattered in qualitative ways. What two bombards could do in two months, eight could do in two weeks, and twenty-four could do in five days. Indeed, the practical effect of larger numbers of big guns was even greater than that, because the large bombards fired so slowly that in between shots defenders could work to reinforce threatened sectors of walls, repair breaches, or build new defenses behind them.[64] Thus, slower progress in the work of demolition

---

gun weighing 833 kg; then an additional large bombard of iron (the bombard of "M. de Chalon" or of "M. d'Arlay" = belonging to the lord of Chalon Arlay); plus two more "medium" bombards ("à M. de Montagu"). On 17 December, the Burgundians commissioned a new large bronze bombard at Auxonne, designed to cast a stone of 157 kg and weighing 3,381 kg, including the bronze of the *Dijon*, which was melted down after bursting. The bombard of Chalon Arlay also burst and was rebuilt with a new chamber made of cast bronze, weighing 833 kg. This was a huge increase from the old chamber, which only weighed 147kg. The lords conducting the siege were clearly aware that larger powder charges made for more powerful shots, but also were dismayed by the frequency with which their guns were bursting, so they evidently were erring very much on the side of strengthening the guns. The new chamber was designed to support a 36-pound powder charge for a 320-pound stone (11.25%) – the ratio recommended in the *Feuerwerkbuch* – suggesting that the old chamber must have been intended for a much smaller charge of weaker powder, perhaps around 2.5% of "meal" powder. Bertin, "Le siège du château de Vellexon," 101, 90, 108.

Late in the siege two more bombards arrived; the large bombard the lord of Villars and another sent by the duchess of Austria (the duke of Burgundy's sister) from the castle of Maison-Vault. Villars' gun, however, burst at its first shot, so I have not counted it.

Another very large iron bombard was purchased for the siege, but at 3,773 kg it was so heavy that 31 horses and 32 oxen could only pull it around 5 kilometers per day, so that it did not arrive at the siege in time. It was designed (with a conical barrel able to accommodate stones of different sizes) for shot of 343–416.5 kg. Yet another large bombard, shooting 600-pound (294 kg) stones, was not brought because it burst during testing. All these details are from Garnier, *Artillerie*, 22–30; see also Bertin, "Siège."

61 1415: *Chron. St. Denys*, 5:536 (*inaudite*); *Ingulph's Chronicle of the Abbey of Croyland: With the Continuations by Peter of Blois and Anonymous Writers*, tr. Henry Thomas Riley (London: Bohn's, 1854), 365 (tennis). 1449: Jean Chartier, *Chronique de Charles VII*, vol. 2, ed. A. Vallet de Viriville (Paris: P. Jannet, 1858), 178.

62 Thomas Basin, *Histoire de Charles VII et Louis XI*, ed. J. Quicherat, vol. 1 (Paris: SHF, 1855), 240.

63 Gábor Ágoston "Firearms and Military Adaptation: The Ottomans and the European Military Revolution, 1450–1800," *Journal of World History* 25 (2014), 92; Purton, *Late Medieval Siege*, 331; *Magyar történelmi tár: a történelmi kútfők ismeretének előmozdítására*, vol. 9 (Pest: Eggenberger Ferdinánd M. Akad. Könyvárusnál, 1861), 66.

64 Smaller guns were used precisely to hinder these efforts, but they could not stop reinforcement of a wall before it was breached, or the construction of back-up retrenchments behind a weakened wall.

actually increased the total amount of demolition that needed to be done.[65] That logic would apply even if the guns themselves had not changed, but in fact – for reasons that will be laid out in the next two sections – the bombards of the 1430s–1440s were much more powerful than those of the period up to 1420, and could be fired significantly more rapidly.

Quantitative Evolutions and the Artillery Revolution: More Shots Per Day

Up to around 1400, bombard barrels were very short, like early-modern mortars. Powder chambers were of smaller diameter and relatively long. Working from the mouth of the weapon, the gunner had to ladle and pack in enough powder to fill 3/5 of the chamber, and then plug the opening tightly with a wooden tampon that filled another fifth of the space, leaving an equal volume of empty air between the powder and the wood. The ball was then loaded in, wedged in place with triangular pieces of soft wood, and covered with some mixture of mud, loam, and hay, which then had to be allowed to dry. All of this served to keep the ball in place as a plug for the powder chamber until the pressure of the gas generated by the burning powder built to a high level, which was necessary because all the explosive force of the small charges that were then used had to be applied in a single spitting push, since the shortness of the barrel allowed for relatively little application of pressure after the ball started to move. Because of this complicated procedure (which of course was especially difficult and time-consuming when it involved manipulating balls of 300 or 700 kg), rates of fire were very slow, typically 5–6 shots per day.[66] This was a poor showing in comparison to trebuchets – of which even the largest could be fired as many as 89 times in 24 hours, though the practical norm was more like 10–18 shots per day.[67]

---

65 To cite one example of many: at Melun in 1420, for some eighteen weeks, "as fast as their walls were broken down by the engines of their adversaries, [the defenders] strengthened them with hogsheads full of earth, straw, wool," and other suitable materials. John de Waurin, *A Collection of the Chronicles and Ancient Histories of Great Britain, Now Called England (1422–31)* tr. William and Edward L. C. P. Hardy (London: Rolls Series, 1887), 311.

66 Johann Ottsen, "Über den derzeitigen Stand unserer Kentnisse von den Anfängen der Pulverwaffen," *Beiträge zur Geschichte der Technik und Industrie* 13 (1923), 8 (40 shots in 7 days at Tannenberg in 1399); Pérez de Guzmán, *Cronica Juan el Segundo*, cap. xli, xliij (8 shots per gun, firing all day and part of the night, at Setenil in 1407); Wenceslai Hagecii, *Böhmische Chronica*, vol. 2 (Cadan: J. S. Zluticensem, 1596), fo. 114v; Thomas Fudge, ed. and tr., *The Crusade against Heretics in Bohemia, 1418–1437: Sources and Documents for the Hussite Crusades* (Aldershot: Ashgate, 2002), 155–6 (largest guns only six or seven times per day, at Carlstein in 1422). Note also: *Chroniques de la ville de Metz*, ed. J. F. Hugenin (Metz: S. Lamort, 1838), 200 (3 shots in a day, at different targets, 1437).

67 *Petit livre Noire* of Périgueux, fo. 47v: "hi tiret en I jorn XLV peyras, en Iᵃ nuech XLIIII." For an average of only 18 per day at Karlstein in 1422 see Rathgen, *Geschütz*, 369. In the secondary literature I have often read of the siege of Edinburgh in 1296 as an example of a rate of fire of around 18 stones per day, based on 158 stones thrown by three engines in three days. However, although the figure

But the situation changed as gunners improved their art and as changes in gun design and loading technique favored a higher rate of fire. Gun barrels grew steadily longer, up to around 5 calibers by 1440s. Longer barrels allowed gunners to dispense with the wedges holding the ball in place, and to do without the loam seal at the mouth used in earlier guns.[68] By around 1410 gunners were also using "loading irons" that enabled them to consistently and rapidly set in place the right amount of powder to load a gun, without needing to measure the distance filled.[69] The rate of fire increased correspondingly. It is difficult to say how fast fifteenth-century gunners normally served their weapons, but at Luxembourg in 1443 the highest rate of fire was 12 shots per day.[70] During a particularly intense period of firing at the siege of Rheinfelden two years later, four large guns managed 74 shots in 8 hours: two per gun per hour.[71] Four Milanese and Genoese bombards in 1464 each managed ten shots in just a very few hours, so around two to three per hour per gun.[72] When *Mons Meg* – forged in 1449 – was put into use in 1571, its crew fired four times per hour.[73] But as with the trebuchet, the practical norm

---

of 158 stones in three days and nights is given by the Lanercost chronicle, the number of engines is not. It was probably well more than three, since the same source describes 'immense machines for casting stones having been set *all around* [*circumquaque*]" the castle, and Hemingbugh's chronicle says the English had "plentiful" [*copiosas*] large machines. Hence this suggests an average of at *most* 10 shots per 24 hours per machine, assuming "plentiful" and "all around" imply at least five engines. Joseph Stevenson, ed., *Chronicon de Lanercost, MCCI-MCCCXLVI* (Edinburgh: The Bannatyne Club, 1839), 178; Walter of Hemingburgh, *Chronicon domini Walteri de Hemingburgh* ed. H. C. Hamilton (London: English Historical Society, 1849), 2:105. For 10–12 heavy stones thrown per day in the 1380s, see Jean Froissart, *Chronicles*, tr. Thomas Johnes, vol 2 (London: Routledge, 1874), 254. During a large siege by the count of Hainault in 1345, 10 trebuchets ("mangonalibus") threw 1,100 large stones over seven weeks – an average of less than three per day per machine. Jean de Hocsem, *La chronique de Jean de Hocsem*, ed. Godefroid Kurth (Brussels: Commission Royale d'Histoire, 1927), 336. We get almost the same low figure for the siege of Ypres in 1383, where two trebuchets together threw 400 or 450 stones over 8 weeks: 3.5–4/machine/day. Kervyn de Lettenhove, ed., *Istore et croniques de Flandres*, vol. 2 (Brussels: F. Hayez, 1880), 296, 311–12. This last example is often mistakenly cited in secondary literature as indicative of the rate of fire of *cannon* at the time, but the chronicle is clearly referring to trebuchets, not guns.

68 This was clearly expressed in a 1445 manuscript of the *Feuerwerkbuch* quoted by Max Jähns: "Diewyle die Buchssen vor dem pulversak als kurtz waren, wenn der stein dar in geladen wart, dass er ein wenig für die Buchs gieng, zu den zyten und zu denselben Buchssen, was bedurfft, das man den stein verbisset. Aber zu den Buchssen, die man yetzunt hat, die langen Ror haben vor dem pulversack, so die Buchs eingeladen wirt mit pulver und mit stein, da bedarff der stein nichts denn usschoppens." Jähns, *Geschichte*, 397.

69 Quoted in Jähns, *Geschichte*, 387.

70 Ralph Lange, "L'Artillerie de Luxembourg devient bourguignonne," in *E' wor emol e kanonéier. L'artillerie au Luxembourg*, ed. François Reinert (Luxembourg: Musée Dräi Eechelen, 2019), 10.

71 "Am mendag, als ich und Schaltenbrant in der forstat uff wacht giengend, von 8 zaltend wir 74 höptchütz bisz früge uff 4." Ed. A. Gessler, "Basler Geschütznamen," *Basler Zeitschrift für Geschichte und Altertumskunde* 14 (1915), 87.

72 Luca Beltrami "Le bombarde milanese a Genova nel 1464," *Archivo storico lombardo*, 2nd ser., 5 (1888), 802.

73 I.e. ten times in two and a half hours (not 24 times, as I have read in a number of modern works), according to an eye-witness. Richard Bannatyne, *Journal of the Transactions in Scotland during*

seems to have been much lower than the theoretical maximum, perhaps never passing 12–19 shots per day – still a large increase over the standard of the turn of the century.[74]

Quantitative Evolutions and the Artillery Revolution: More Kinetic Energy Per Shot

The first great bombards fired slowly in the other sense as well: the muzzle velocity of their projectiles, though higher than that of mechanical artillery, was quite low in relation to cannon of the mid-fifteen century. Compared to later guns, the earliest bombards used very small amounts of powder relative to the weight of their shot, so the amount of potential energy available to be converted to kinetic energy was low, and so was the velocity of the projectile. When the first bombards throwing large stones were built in the 1370s, they had very short barrels (just long enough to hold the ball, or a bit longer) and small powder chambers designed to hold about 2.5% of their projectiles' weight in dry-mixed powder.[75] A generation later the design was similar, though powder chambers were now

---

the Contest between the Adherents of Queen Mary and those of Her Son (Edinburgh: Constable, 1806), 153. For a similar rate of 140 shots in a day from two guns in 1484, see Prescott, *History of the Reigns of Ferdinand and Isabella*, 1:268.

74 At Rheinfelden in 1445 over two days the average was 18.75 shots for each of four guns (including the period of more intense fire just noted). Gessler, "Basler Geschütznamen," 87. For 14 shots/day in 1474: ibid., 89, and Vaughan, *Charles the Bold*, 295. Note also: *Chronique de St. Denys*, 652 (*Griete* at Bourges in 1412: 12 shots in a day); Napoléon III and Favé, *Études*, 2:96) (guns with 1,200-pound stones once per two hours by the Turks in 1453).

75 For a record-based example of a 2.5% powder ratio for a 1376–7 gun firing a 60-lb. stone with a 1.5 lb. charge, see Contamine, *War in the Middle Ages*, 141. The Nürnberger bombard *Kriemhilt* or *Chriemhilde* is often given in secondary literature as firing a 560-pound stone with a 14-pound charge (2.5%), but actually the source does not give the weight of the stone. *Chroniken der deutschen Städte. Nürnberg*, vol. 1 (Leipzig: S. Hirzel, 1862), Beilage IV.B., p. 177. The Bodiam Mortar, which can be dated c. 1405 based on its very close similarity to an illustration in Conrad Kyeser's *Bellifortis* (see Schmidtchen, *Bombarden*, 15), has been estimated to have used a 2.2% charge, apparently assuming corned powder loaded up to the base of the ball. Brig. Gen. J. H. Lefroy, *Official Catalogue of the Museum of Artillery in the Rotunda, Woolich* (London: HMSO, 1864), no. 1. It would have been correspondingly lower if used with a 3/5 chamber fill. The Amsterdam bombard, which may date to the 1370s, would have used a similarly small charge, though it is difficult to estimate due to the uncertainty of loading method and powder type that early. In any case, it has less than one third the ratio of powder chamber volume:stone volume of the otherwise fairly similar in form (though more sophisticated in manufacture) *Centnerbüchse* of the Kölnisches Museum, which can be estimated to have held a powder charge of about 4.8%, based on a specific gravity of .88 and a 3/5 load. I have determined to use that figure (which falls between the .80 used by Angelucci and the .94 used by Rathgen) for specific gravity because it is the number that would make the charge weight ratio match the volume ratios given in the early fifteenth century text of HS 1481a. of the Germansiches Nationalmuseum [hereafter GNM] in Nürnberg, fos. 23v-24 (printed in Köhler, *Entwickelung*, 3:1:291–2n), assuming a specific gravity for stone of 2.56. My thanks to Sascha Pries of the city museum for providing me with measurements of the Köln gun.

constructed more solidly and powder charges had increased to about 5% of ball weight.[76]

From then until the 1430s, gun design changed in ways that allowed bombards to fire more rapidly in both senses. The most obvious change was that (as already noted) barrels grew proportionally longer. At the turn of the century, barrels were still commonly about 1–1.5 calibers long: that is, with an internal length 1–1.5 times their bore width. One text of the start of the fifteenth century says the barrel should be 1.5 times the height of the stone (so just under 1.5 calibers). A manuscript probably written shortly after 1408, but incorporating an earlier text, implies a norm of barrels 1.6 calibers long.[77] *Faule Mette*, a bronze bombard cast in 1411, had a barrel 2.5 calibers long.[78] A manuscript of about the same date seems to say that bombards with barrels 2.8 calibers long shoot the farthest.[79] A manuscript drawing from the 1420s (Figure 5.4) shows a bombard designed based on a barrel length of 3 calibers.[80] The contract for the casting of the bombard *Bourgogne* specified a barrel the length of four stones (just slightly below four calibers) in 1430, and when a new barrel was made for the gun 1437–8 the same ratio was retained.[81] Large late-medieval bombards like the two "Michelettes" and *Mons Meg* had barrels of from 5 to 5.6 calibers.[82]

76 Financial records show that the bombard of Frankfurt at the siege of Tannenberg in 1399 used 7 Centner and 33 pounds pounds of powder to fire forty stone balls weighing 3.5 Centner each (5.2%). Rathgen, *Geschütz*, 6; Schmidtchen, *Bombarden*, 15. In 1397 an Imperial campaign against a robber baron included three contingents each bringing a *Centnerbüchsen* along with 20 hundred-pound stones and a hundred pounds of powder, indicating a 5% powder:ball ratio. Karl Heinrich Lang and Max von Freyberg eds., *Regesta sive rerum boicarum*, vol. 11 (Munich: Royal Press, 1847), 108.

77 The former text, from GNM HS 1481a fo. 23v-24, is transcribed with minor differences but substantively accurately in Kohler, *Entwickelung*, 292–93n and (a little less well) in Jähns, *Geschichte*, 1:390–91. The latter text, quoted ibid., 389, calls for filling a gun to a maximum of 1/3 with powder, to avoid risk of rupturing the gun. (What seems to be the oldest extant *Feuerwerkbuch* text, possibly dating to 1400, also calls for loading the gun 1/3 with powder. Bucharest, Arhivele Naţionale ale României – Centrale, Colecţia Manuscrise, Nr. 2286, fo. 21v.) One third of total length would equate to 3/5 of the powder chamber length given an overall design of 3 (powder) + 1 (space) +1 (tampon) + 4 (barrel). Given that the ratio of powder chamber bore (= Klotz thickness and diameter) to barrel caliber was 2/5 (per Jähns, *Geschichte*, 391), a barrel 4 Klotz long = 8/5 or 1.6 calibers long.

78 Sack, et al., *Alterthümer . . . Braunschweig*, 76–82. However, because it had a conical barrel, the barrel length was 3 rather than 2.5 times the length of a ball that would fit at the base of the barrel.

79 I.e. 7 Klotz, the Klotz (equal to the powder chamber diameter) being 2/5 calibers at that time. Jähns, *Geschichte*, 388–92. Read literally the text, as given by Jähns, is actually describing a whole *gun* (Büchse) 7 Klotz long. But that would be too short to allow for a standard 5-Klotz powder chamber, a necessary thick base (which should be ½ caliber thick, according to GNM HSS 1481a, fo. 23v and 24347, fo. 121v) behind the touchhole, and a barrel even 1 caliber long.

80 Hassentstein, *Feuerwerkbuch*, 101; Köhler, *Entwickelung*, 3:1:293.

81 Garnier, *Artillerie*, 70; Monique Sommé, "Les mesures dans l'artillerie bourguignonne au XVe siècle," *Le médiéviste et la métrologie historique: Table ronde Université Charles de Gaulle, Lille III, 9 Mai 1989*, 1st part (Lille: Université Charles de Gaulle, 1989), 49.

82 Brown and Smith, *Bombards*. The *Mons Meg* group of bombards have barrels of from 5.1 (the Basel bombard) to 5.6 (*Mons Meg*) calibers.

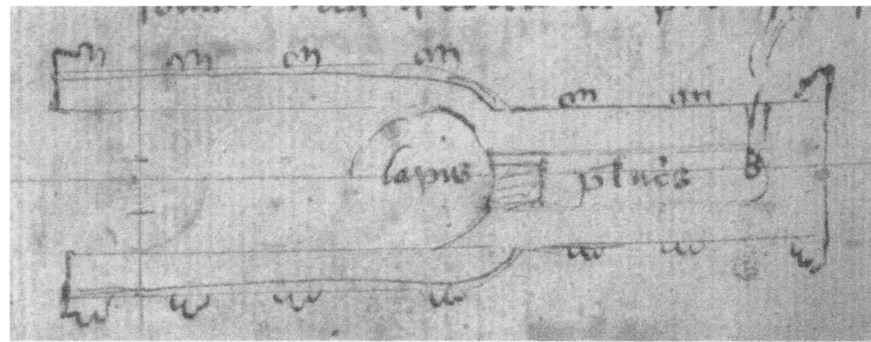

Figure 5.4 Sketch of a Bombard, ca. 1422. With one exception, the proportions in this sketch of a bombard are the same called for in a text of about 15 years earlier. The length of the powder chamber is five times its own diameter and twice as long as the gunstone's diameter, and the gun's base is as thick as the stone's radius. The exception is that the barrel is now the length of 3 stones instead of 1.5 stones. Germanisches Nationalmuseum, Handschrift 24347, fo. 121v. Public domain image courtesy of the GNM.

These longer barrels meant that the ball remained subject to the impulsion of exploding powder for a longer period of time, making them more suitable for larger charges of powder that could give balls more velocity and kinetic energy. *Faule Mette* (1411) used a powder charge of around 7–8% of the weight of its balls – about a 50% increase from what was typical at the turn of the century, but already on the low side for its own day.[83] A text of about that same date or a little earlier calls for a 10% powder load.[84] Burgundian bombards of 1409 already were designed for even larger powder charges: around 1:9 (11.1%), as also recommended in the *Feuerwerkbuch*.[85] The more than fourfold increase in proportional powder charge since the first substantial stone-throwing cannon required greatly increasing the quantities of metal used in manufacturing guns, to make them strong enough to contain and direct more powerful explosions. Events at the siege of Vellexon in 1409 illustrated this dramatically. After an old bombard burst in operation, a new powder chamber was made to go with the old barrel. The new

83 *Faule Mette:* Sack, et al., *Alterthümer . . . Braunschweig*, 76–81.
84 GNM HS 1481a, fos. 23v-24; printed version in Köhler, *Entwickelung*, 3:1:291–2n. Note that the same physical proportions, implying the same weight proportions, are still shown in GNM HS 24347, ca. 1422: see Fig. 4 of this article.
85 Sack, et al., *Alterthümer . . . Braunschweig*, 76–82. *Feuerwerkbuch*, ed. Kramer, 25 (with this part of the text probably dating to around 1410, rather than to c. 1380 as Kramer thinks, since records show powder loads of only 5% in 1379–99 [note 72], and around 11% only around 1409). For three guns with powder:ball ratios of 10.42% to 12.15% in 1409, see Garnier, *Artillerie*, 25–6.

piece was made with almost *seven times* the weight of bronze used for the part it replaced.[86]

In 1430, Burgundian authorities commissioned a bombard that was to fire a 600-pound (294 kg) stone using 70 pounds (34.3 kg) of powder – so still using a powder-to-ball ratio of 11.6%, similar to the one for the bombard commissioned in 1409 and still aligning closely with the *Feuerwerkbuch*'s recommendation for a 1:9 powder-to-ball ratio.[87] However, there was not enough metal available to make the new gun as planned, so its barrel was redesigned for a 400-pound (196 kg) stone, allowing for a charge of 17.5% of the ball weight. It is quite possible this accident led to an appreciation of the value of larger powder charges, since in 1446 another Burgundian bombard was designed from the start to fire a 350-pound (172 kg) stone with 72 pounds (35.3 kg) of powder (20.6%), and the year before that *Bourgogne* was re-built to use a 100-pound (49 kg) charge for a stone of about 174 kg – a charge of about 28%.[88] Thus, the amount of powder per pound of gunstone, for the largest guns, had increased around ten-fold over 75 years, and almost six-fold in just 50 years. The increasing powder charges made guns substantially more powerful even though smaller stones were being used. Even if both had used the same formulation of gunpowder, for example, a 174 kg stone thrown from the remade *Bourgogne* (1445) would have hit with about five times the kinetic energy of the 190 kg stones from the large bombard cast at Frankfurt in 1394, and more than triple the punch of the much larger stone of the *Pumhart von Steyr* (c. 1415).[89]

But the guns of the 1430s–1440s were not only using much more powder per pound of projectile than had been the case a century or a half-century earlier, they were also using powder made in a different way, so that it was in practice significantly stronger. In the mid-fourteenth century, before the first substantial stone-throwers came into use, the ingredients of gunpowder were normally finely powdered and mixed together dry. Probably in the 1370s, many

---

86 Garnier, *Artillerie*, 26, 265; Bertin, "Siege," 100.

87 Burgundian: Garnier, *Artillerie*, 70. In 1428, three English bombards (two firing 16" and one firing 18" stones) were tested using 80 pounds of powder, indicating a similar charge of around 11.8%. The National Archives, Kew, E101/51/27 (Receipts and other documents relating to the account of John Parker, master of the ordnance under the earl of Salisbury); Dan Spencer, "The Provision of Artillery for the 1428 Expedition to France," *Journal of Medieval Military History* 13 (2015): 179–92.

88 Garnier, *Artillerie*, 70, 112, 109. For the Burgundian pound for artillery (490 grams), see Sommé, "Mésures," 46.

89 If the same formulation of powder was used in each case and the gun conveyed its potential energy into kinetic energy with equal efficiency, the muzzle energy would increase in proportion with the powder charge, so that the 1446 bombard would impart 4.95 times the muzzle energy of the 1394 Frankfurt gun, and 3.15 the muzzle energy of the Pumhart (assuming a 3/5 load of 0.88 s.g. powder). However, the later gun would have a substantially bigger advantage in *impact* energy, because its smaller stone would lose less energy to air resistance. Moreover, because of its longer barrel, the 1446 bombard would have done a better job capturing the potential energy in the powder and transferring it to the shot, for a higher muzzle energy from the same powder charge.

master gunners had learned the "secret" of making wet-mixed powder. When the main ingredients were moistened with brandy or other liquids during the manufacturing process, the sulfur and saltpeter were absorbed into the fibrous structure of the charcoal. Modern tests using the 3/5 loading method have not yet been performed, but such tests as have been done do seem to confirm that wet-mixed powder is significantly more powerful than dry-mixed.[90] The earliest large stone-throwers probably already used this type of powder. The next stage was the development of "corned" powder, in which the gunpowder was formed into larger or smaller balls or grains. The interstices between the powder grains provided the air needed for an explosion to progress by the more rapid inter-grain burn rate, rather than the slower intra-grain rate. The *Feuerwerkbuch* manuscripts consistently say that such powder was 50% more powerful than the older style of meal powder.[91]

Thus, relative to a major siege of around 1400, one conducted around 1450 might involve something on the order of 8 times as many great bombards, each delivering 7.5 times as much kinetic energy per shot, and shooting 3 times as frequently, for a total of around *one hundred and eighty times* as much destructive power deliverable per day.[92] This offers an excellent example of how the multi-

90 The earliest text on the subject says that wet-mixed powder with a 4:1 saltpeter:sulfur ratio, reinforced with small quantities of salammoniac and camphor, was three times as powerful as earlier formulations. Österreichische Nationalbibliothek Wien, Codex 3069, fo. 2 (1411); cf. Rainer Leng, ed. *Anleitung Schiesspulver zu bereiten, Büchsen zu laden und zu beschiessen: Eine kriegtechnische Bilderhandscrhift im cgm 600 der Bayerischen Staatsbibliothek München* (Wiesbaden, Reichert Verlag 2000), 72–3. That ratio happens to almost exactly match the results of modern testing with a replica fourteenth-century gun (not employing the tampon loading method), which found that two shots with dry-mixed "Marcus Graecus" (6/1/2 Saltpeter/Sulphur/Charcoal) powder averaged muzzle velocities of 101 m/sec, whereas two shots using the same formulation but wet-mixed resulted in average muzzle velocities of 176 m/sec – i.e. triple (3.03 times) the kinetic energy. However, this may be due to an improper loading for the first dry-mixed powder shot; discarding that as an outlier, the wet-mixed powder was about 76% more powerful than the dry-mixed. Medieval Gunpowder Research Group, "Report 1," table 5. This is another area where more testing needs to be done, using the loading methods and powder recipes specified in the same text.

91 The various German manuscripts of the *Feuerwerkbuch* all agree that two pounds of powder wet-mixed and rolled into balls (*Knollenpulver*) does more than three pounds of "meal" powders, which latter are called either (depending on the manuscript) *gestossen* [meaning either "stamped" or "rammed"] or "sieved" powders. E.g. Royal Armouries *Feuerwerkbuch* (MS I-34), fo. 4v: "knollen pulfer zway pfunt mer tun dann gestossen pulfer drey pfunt"; Staatsbibliothek zu Berlin, Ms. germ. qu. 1187, fo. 15v; Kramer, *Feuerwerkbuch*, p. 25; Hassenstein, *Feuerwerkbuch*, 45; Heidelberg Universitätsbibliothek *Feuerwerkbuch*, Cod. Pal. germ. 502, fo. 4. Cf. the French translation of c. 1430 in Napoléon III and Favé, *Études*, 3:146, which clarifies the meaning of *Knollen* and makes the ratio 3:1 rather than 3:2: "La question est que si les pouldres ainsi meslées et faites par boulles et pelottes sont meilleurs . . . les dictes pouldres accumulées en monceaux en façon de boulles et pelottes vallent et font mieux et plus une seule livre que trois d'autres pouldres criblées."

92 Reckoning 3 bombards firing 5 times per day with 5% charges of wet-mixed "meal" powder vs. 24 bombards firing 15 times per day using 25% charges of 50% more powerful corned powder.

plicative interaction of several more linear or "evolutionary" developments can cause exponential or "revolutionary" change.

The Effects of the Artillery Revolution:
Siege Warfare and Strategy

Fortification design responded in a variety of ways to the increasing power of gunpowder artillery, but the strength of defenses rose nowhere near as rapidly as the strength of artillery siege trains, and so for the first time in history it became possible to rapidly breach strong masonry fortresses by means of bombardment. This changed the dynamic of siege warfare dramatically, with tremendous consequences for the art of war and indeed for the course of European and World history.

Although, as we have seen, the development of efficient wall-breaking artillery resulted from the interaction of several strands of evolutionary change, the real tipping point seems to have come in the mid- to late 1420s and the 1430s, at least in France. Before that time, as noted earlier, guns aided defenders as much as attackers, and only relatively weak walls like those of Berwick in 1405 could be quickly breached. Major sieges – including sieges of well-built but small strongholds – continued to last, typically, from two to six months, and to end by starvation rather than assault. Long sieges were then still the expectation as well as the reality. In 1400, the indenture between the king of England and the castellan of Roxburgh forbade the castellan, if besieged, to surrender until three months had passed – the same standard that had been applied in the case of Berwick back in 1356. In 1409, Christine de Pizan's essay on modern siege warfare, which called for besiegers to deploy numerous heavy guns, advised defenders to provide themselves with six months' worth of food supplies, which shows how little she expected the attackers' bombards to actually accomplish.[93] Those expectations not only reflected but also shaped reality, since defenders who expected to be able to sustain long sieges would lay in food supplies accordingly, and since garrison commanders and soldiers would be highly motivated to hold out for a respectable amount of time.

Long sieges meant that there was a huge amount of "strategic inertia": The normal advantage of the defensive in warfare was magnified, and it was far easier to retain a region than to capture it. This encouraged commanders on the offensive to avoid efforts at gradual conquest and to rely instead on devastating expeditions (*chevauchées*) that bypassed strong fortifications and took the war to the relatively vulnerable agricultural population. For those on the strategic defensive, however, the tendency was to prefer a Fabian or "Vegetian" style of defense, which relied

---

93 Michael Prestwich, *Armies and Warfare in the Middle Ages* (New Haven: Yale U.P., 1999), 241; Christine de Pizan, *The Book of Deeds of Arms and of Chivalry*, tr. Sumner and Charity C. Willard (University Park: Penn. State U. P., 1999), 110–11.

on stone walls to provide a secure "reservoir of lordship."[94] Small powers (little states or sub-sovereign territorial lords) could use the leverage of their fortresses effectively to resist strong neighbors or their own superior lords. All this tended to eternalize war and to make it especially damaging to civilians and to the European economy.[95]

By 1425, however, it took only eight days for a large Burgundian siege train to leave the walls of Braine-le-Comte so "ripped apart and broken down" that the place would soon have been taken by assault if the garrison had not surrendered.[96] Stronger places could still hold out longer, but by 1449–1451 even the formidably defended town of Harfleur was captured after just seventeen days, and the walls of Blaye "were completely thrown down in many places" after only five days.[97] The full extent of the Artillery Revolution's impact on the conduct of war really became clear in those years, when Charles VII of France was able to overrun first Normandy and then Aquitaine each within a single year. "Never," wrote the Berry Herald, "was so large a country conquered in such a short time."[98] At the other end of the continent, Mehmet the Conqueror used many massive bombards in his successful siege of Constantinople in 1453. His siege of Belgrade in 1456 did fail, but only because of the extraordinarily vigorous defense mounted by Jan Hunyadi. Although the city was then "probably one of the strongest fortresses in Europe," Mehmet's numerous great bombards were so effective that, according to an eyewitness, "within the space of ten days almost all the walls of the citadel were leveled to the ground." A few days later, Belgrade's defenses were so thoroughly demolished that, according to Hunyadi himself, the place "could be called a field, rather than a fortress": *non castrum, sed campum.*[99]

94 On the Vegetian defensive strategy see John Gillingham, "Richard I and the Science of War in the Middle Ages," in John Gillingham & J. C. Holt, eds., *War and Government in the Middle Ages* (Boydell Press, Woodbridge 1984); on its limits, see Clifford J. Rogers, "The Vegetian 'Science of Warfare' in the Middle Ages," *Journal of Medieval Military History* 1 (2003): 1–20.

95 On the destructiveness of warfare in this period, see Clifford J. Rogers, "By Fire and Sword: *Bellum Hostile* and 'Civilians' in the Hundred Years War," in *Civilians in the Path of War*, ed. Mark Grimsley and Clifford J. Rogers (Lincoln: University of Nebraska Press, 2002): 33–78.

96 Jean de Waurin, *Recueil des croniques et anchiennes istories de la Grant Bretaigne, a present nomme Engleterre*, 5 vols. (London: Rolls Series, 1864–1891), 5:165 ("deschirez et desrompus"); similarly at Jargeau in 1428, ibid., 5:241.

97 Rogers, "Military Revolutions," 66–7, and Clifford J. Rogers, "The Artillery and Artillery Fortress Revolutions Revisited," in Nicolas Prouteau, Emmanuel de Crouy-Chanel and Nicolas Faucherre, eds., *Artillerie et Fortification, 1200–1600* (Rennes: Presses Universitaires de Rennes, 2011), 77–78, or this volume, Chapter Four, p. 92 for these and other examples.

98 Rogers, "Military Revolutions," 67, 89n. 110.

99 The number of large bombards is given in various places as 12, 22, or 28. Gábor Ágoston, *Guns for the Sultan: Military Power and the Weapons Industry in the Ottoman Empire* (Cambridge: Cambridge U.P., 2005), 67; Kenneth M. Setton, *The Papacy and the Levant, 1204–1571*, vol. 2 (Philadelphia: American Philosophical Society, 1978), 174–181, quotations pp. 176, 177–8. Hunyadi, in *Monumenta Hungariae historica*, 33 (Budapest: Kiadja a Magyrar Tudományos Akadémia, 1907), 208: "In tantum enim castrum per ictus bombardarum destruxit, quod ipsum castrum non castrum sed campum dicere possimus, quod usque ad terram murus castri est destructus."

Although the new artillery contributed greatly to the rapid conquests of the mid-century, there was of course much more to the story. A siege train, no matter how powerful, is of little use if you cannot bring it to bear on the enemy's fortresses, and it is difficult to do that if your army cannot hold its own in open battle. For most of the Hundred Years War, the French had failed signally to convert their superior resources into tactically superior field armies, but by the 1440s English field forces had grown weaker, while their enemies had experienced a large improvement in their qualitative effectiveness. The latter development was largely the result of a sustained and energetic program of military reform that began to be implemented in 1444–45, after the French monarchy broke noble resistance to the increased taxes necessary to sustain a substantial body of soldiers during a period of extended truce with England. Charles VII broke up the private mercenary bands that had formed the bulk of his army to that point, then commissioned a carefully selected sub-set of their leaders to head twenty new regular or "ordinance" companies, each composed of 100 veteran men-at-arms (armored cavalry), 100 light cavalry, 200 mounted archers, and 200 support personnel. All these soldiers, who were enlisted for long-term, year-round service, were selected on the basis of their strength, their fighting prowess, and the quality of their mounts, armor, and arms. It was the ordinance companies that provided the hard edge of the French armies that conquered Normandy and Aquitaine in 1450–51 and that – along with the impressive French artillery train – made France's army by far the most powerful in Christendom.[100]

Conclusion

Over the course of their first century of use European gunpowder weapons had come a very long way. By the end of the fourteenth century, and even more so by 1410, big bombards far surpassed the largest trebuchets in range and hitting power. They then continued to see a rapid growth in the strength of powder charge they could handle – more than doubling again the punch of the largest guns, even though the stones they used fell to "only" a few hundred pounds. Meanwhile the number of wall-breaking bombards employed rose steadily, so that by the 1440s there had been a true quantum leap in the total amount of destructive power a king or great noble could bring to bear against a besieged stronghold. The multiplicative interaction of the effects of more guns shooting more rapidly with more power meant that it was possible for a royal siege train of 1450 to do as much damage to a fortification in one day of firing as could have been accomplished in a half-year of bombardment in 1400. This completely changed the balance between offense

---

100 Paul Solon, "Charles VII and the *Compagnies d'Ordonnance*, 1445–61: A Study in Medieval Reform" (Ph.D. Dissertation, Brown University, 1970); idem, "Valois Military Administration on the Norman Frontier, 1445–61: A Study in Medieval Reform," *Speculum* 51 (1976): 91–111; Philippe Contamine, *Guerre, état et société à la fin du moyen âge. Etudes sur les armées des rois de France, 1337–1494* (Paris: Mouton, 1972), 277–530.

and defense in siege warfare and therefore in warfare as a whole, in ways that are important for explaining the rise of the expansive empires of the second half of the fifteenth century, including France, Spain, and the Ottomans. The consequent military revolution would not have been possible using mechanical artillery, which had already reached more or less its maximum potential by the end of the thirteenth century. Thus, despite Lynn White's suggestion to the contrary, there is no need to look to non-rational factors to explain why medieval rulers and artillerists turned to gunpowder artillery first to supplement the trebuchet and then – well before 1426 – to replace it as the principal technology for bombarding fortified positions.

6

TACTICS AND THE FACE OF BATTLE, 1350–1750[1]

Prologue: High Medieval Tactics

Although revisionist historians have lately argued otherwise, in the High Middle Ages (c. 1000–1300) cavalry was the dominant arm in battle.[2] Infantry was valued, and was present at most battles, but its role was almost purely defensive, and it usually takes offensive action to win a true victory. As Jim Bradbury observed in concluding his study of Anglo-Norman battles (in which dismounted knights played a substantial role), the biggest threat on the battlefield was a cavalry charge, but the strongest defense against a head-on mounted attack was a steady line of spear-armed footmen supported by archery.[3] In other words, cavalry and footmen could work together like sword and shield. A swordsman surely values a sturdy shield, and even knows it can sometimes be used to strike important blows, but would sooner dispense with it than try fighting without his blade.

Battles in this period tended to follow one of a few basic patterns. Often the opposing armies would face off, infantry-line to infantry-line, each with its cavalry to the flanks or rear. Then the mounted forces would charge each other at a trot, while the footmen either held their ground or advanced to contact very slowly, so that they could keep their order. Since the former closed much more quickly, and since mounted combat was typically decided more rapidly than a contest between shield-walls, the issue of the battle as a whole was likely to be determined by the cavalry fight. At least some of the victorious horsemen could be expected to come in on the flank and rear of the enemy footmen while the latter were still engaged by infantry in front; when attacked from all directions, infantry had little chance of escape, and less of victory.

---

1 I owe thanks for their comments and assistance to John F. Guilmartin, Jr., Michael de Jong, Geoffrey Parker, John Stapleton, David Trim, and the conference participants, and for financial assistance to the United States Military Academy. This version is substantially expanded from the originally published text.
2 For the two opposing viewpoints, see the articles by J. F. Verbruggen, "The Role of the Cavalry in Medieval Warfare," *Journal of Medieval Military History* 3 (2005) and Bernard S. Bachrach, "Verbruggen's 'Cavalry' and the Lyon-Thesis," *Journal of Medieval Military History* 4 (2006).
3 Jim Bradbury, "Battles in England and Normandy, 1066–1154," in Matthew Strickland (ed.), *Anglo-Norman Warfare* (Woodbridge: Boydell, 1992), 193.

DOI: 10.4324/9781003399971-8

The side that was weaker in cavalry therefore often avoided battle, but if constrained to fight might try dismounting most of its knights and fighting primarily on the defensive. In this scenario, the men on foot would aim to occupy a position where the terrain protected their flanks, then to hold it against the attacks of the enemy infantry and cavalry until the latter had been worn down and disordered to the point where they were vulnerable to the strike of a cavalry reserve, or at least until they simply gave up attacking.

Such tactics sometimes (though not often) brought victory to the footmen, but even then the force with superior cavalry was typically able to retire without heavy loss or great strategic disadvantage.

Over the course of the thirteenth century, increasing urbanization, expanding wealth, and the growing power of royal governments all contributed to the rising importance of infantry. At the start of the fourteenth century this trend reached a tipping point. Already there were normally several times more infantrymen in an army than there were cavalrymen, and while the men-at-arms remained qualitatively superior soldiers, their margin of advantage declined as the footsoldiers (especially urban militiamen) improved their equipment and strengthened their cohesion and morale through the continued development of more martial civic cultures. In the last decades of the thirteenth century the bowmen of Edward I of England and the Catalan *almugavers* proved invaluable, though they did not yet surpass the horsemen in importance. In the thirteenth century, James the Conqueror boasted that the townsmen loyal to him knew "as much of war" as some Aragonese knights who threatened rebellion, and the armies he used to win his sobriquet included far more infantry than cavalry, but he nonetheless clearly viewed his armored horse as the most important arm of his forces.[4] In one battle, for example, Ramon Muntaner went with 1,000 foot, 220 armored horse, and 30 light horse to fight a large Muslim army composed almost entirely of infantry. The badly outnumbered Catalan footmen were routed by the opposing infantry, but then the small body of horse charged into their mass and won a complete victory.[5]

Late Medieval Tactics: the Infantry Revolution

In the fourteenth century, this balance was reversed.[6] Cavalry remained very important in battle, and even more so in the broader conduct of war, but it was men

4 James came to the throne because his father lost the battle of Muret and was killed there, even though he had an army with a large contingent of footsoldiers and Simon de Montfort, who beat him, had almost none. Throughout his memoir, there is evidence for the principal importance he and his nobles placed on armored horsemen: James I of Aragon *The Book of Deeds of James I of Aragon: A Translation of the Medieval Catalan* Llibre dels Fets, tr. Damian Smith and Helena Buffery (Aldershot: Ashgate, 2003), 70–78, 83–5 ("a horse is worth more than twenty Saracens"), 106–8, 124, 217, 376.

5 Ramon Muntaner, *The Chronicle of Muntaner*, trans. Henrietta Margaret Goodenough, online reprint (Cambridge, Ontario: In parentheses Publications, 2000), 510–11; cf. 117–9.

6 Clifford J. Rogers, "The Military Revolutions of the Hundred Years' War," in idem, ed., *The Military Revolution Debate* (Boulder: Westview, 1995), 56–64.

fighting on foot who won the battles of Courtrai, Bannockburn, Morgarten, Dupplin Moor and Halidon Hill, Laupen, Staveren, Crécy, Vottem, Canturino, Nájera, Cascina, Aljubarrota, etc. The early victories in this series were won through a combination of favorable terrain and enemy errors, but their outcome also owed a great deal to the willingness of the infantry commanders to undertake offensive actions on the battlefield. Yet it remained true that the defense had a huge advantage in infantry combat, especially when fighting from a prepared position. The key to victory was to keep a formation in good order, and it was extremely difficult for attackers to avoid disrupting their array as they advanced, especially if they had to march through effective missile fire, or if the defenders were fighting behind trenches, marshes, hedges, stake-barriers, or other obstacles – as they commonly did.[7] John Chandos's observation of 1368 that "it goes ill with the side that attacks first"[8] was echoed by Jean de Bueil in the late fifteenth century: "All those who advance on foot break their own array and put themselves out of breath, and ordinarily they suffer defeat."[9] Thus, strategy often involved the use of devastation, maneuvers, or sieges designed to push the enemy into accepting the burden of the tactical initiative. But it required skillful generalship or serious mistakes by the opposing commander to lure an enemy into attacking a ready army on its chosen ground; often when two armies faced off, there was a substantial delay while each waited for the enemy to attack.

Sometimes such stand-offs did not lead to battles at all, but usually political, strategic, or operational considerations pushed one side into taking action to initiate the fight. The side with greater missile power might use its shot to provoke or force the other side into attacking. If the side with superior firepower also had the advantage in cavalry, it could use the threat or the action of the horsemen to keep the enemy foot pinned in place, then wear it down with shot until it was sufficiently shaken to be broken by a shock attack. Rather than undertaking a risky offensive or standing and suffering prolonged attrition, the opposing commander could of course choose to retreat from the field, but that too could lead easily to disaster. Another option was to launch a pinning attack in front, while sending an

---

7 It is not commonly appreciated that even the Swiss very commonly used earthworks called *Letzi*; Swiss battles were often won by lightly defending such a work, then counterattacking with a strong force against an enemy disordered by the process of storming it. This was the case for example at Näfels 1388, Vögelinsegg in 1403 and Stoss in 1405. Albert Lynn Winkler, "The Swiss and War: The Impact of Society on the Swiss Military in the Fourteenth and Fifteenth Centuries," (Ph.D. Diss., Brigham Young University, 1982), 48.

8 Jean Cuvelier, *La chanson de Bertrand du Guesclin*, ed. J.-C. Faucon, 3 vols. (Toulouse: Editions universitaires du sud, 1990–91), ll. 5875–78: "Laissiez-nous . . . tenons nos conroiz sans nous adesfouquier; / Car on voit bien souvent, je le di sans cuidier, / Qu'il meschiet à celui qui assault le premier."

9 Jean de Bueil, *Le Jouvencel*, ed. Léon Lecestre, 2 vols. (Paris: SHF, 1887–89), 1:189; see also 1:153 and discussion in Clifford J. Rogers, "The Offensive/Defensive in Medieval Strategy," in *From Crécy to Mohács: Warfare in the Late Middle Ages (1346–1526). Acta of the XXIInd Colloquium of the International Commission of Military History (Vienna, 1996)* (Vienna: Heeresgeschichtliches Museum/Militärhistorisches Institut, 1997), 158–171.

enveloping force to win the battle by attacking the enemy's flank or rear. Cavalry was of course the best arm for such a maneuver, and this is one reason it remained important on the battlefield despite the Infantry Revolution.

The Early Modern Period: Pike Squares

When they fought Charles the Bold of Burgundy or Emperor Maximilian, of course, the Swiss did not have superior firepower or superior cavalry, yet they still won, and in so doing added great impetus to the spread of a new model for European infantry tactics, one founded on the pike square, supplemented and eventually surpassed in importance by attached formations of handgunners.

Until the advent of the Swiss, medieval infantry (and cavalry) tactics were usually essentially "linear," meaning that the normal infantry formation was much longer than it was deep. Like the ancient Greeks, medieval footmen typically fought in four to eight ranks. The width of the array then depended on the strength of the army. The Swiss, by contrast, might draw up their formations in squares forty or more men deep. Although these blocks are conventionally referred to as "pike squares," they usually contained a core of men with shorter weapons such as halberds or bills, belted with pikemen, in varying proportion. In the late fifteenth century, Philipp von Seldeneck called for an infantry force of 10,000 men to be drawn up in a square 100 men deep and 100 across. In this sixteenth century, this would have been termed a "just square of men"; other common arrangements were the "square of ground" which was square in area but only three men deep for every seven in front, since files were more tightly closed than ranks, and the "broad square," at least twice as broad as deep. In the sixteenth century, squares as much as 60 or 75 ranks deep were sometimes seen.[10]

Square formations of this sort, like any other arrangement of soldiers, had advantages and disadvantages. On the negative side, they were particularly vulnerable to artillery, since a single cannonball could kill many men.[11] Like all very deep close-order infantry formations, they were also vulnerable to the greatest disaster that could strike a pre-modern army: if they became disordered, they could become compressed to the point that the soldiers could not fight effectively, and even to the point that they crushed and smothered each other to death, as happened for example to the unusually deep formations of the Scots at Dupplin Moor, the Flemings at Roosebeke, and the French at Agincourt. It took the Swiss "daily exercise" (and a culture in which boys and youths engaged in war-games

---

10 David Eltis, *The Military Revolution in Sixteenth-Century Europe* (London: Tauris, 1995), 52–3; Philip von Seldeneck, *Kriegsbuch*, ed. Kurt Neubauer (Ph.D. Diss., Heidelberg, 1963), 91; Robert Barret, *The Theoretike and Practicke of Moderne Warres* (London: William Posonby, 1598), 52, 75–7, 94–5.

11 E.g. At the battle of Calven in 1499, one ball killed seven men, including four brothers. Winkler, 140.

that prepared them for the task)[12] to develop the ability to maneuver and fight in formation without fatally compromising the integrity of their array. Also, because of their greatly reduced frontage, squares were much harder to anchor on terrain obstacles such as woods or villages. But then, one of the main purposes of having a square formation was to make the sides of the block inherently almost as strong as the front,[13] so that the danger of an exposed flank was much reduced. And on the offense, the narrower front was an advantage. Any little hedge or ditch posed a significant problem for a block of footmen trying to advance in formation,[14] and it is much easier to find an unobstructed approach to the enemy forty yards wide than to find one three hundred or a thousand yards across.

When a pike-square lumbered forward, its mass gave it tremendous momentum. If the attacking footmen were well armored, they could usually drive through even heavy rains of arrows or bolts and come to grips with missile-firing infantry, which would then almost inevitably break. A thinner formation of footmen armed with pole-arms likewise had little hope of resisting the onslaught of such a powerful mass. Of course, a thinner line would also, for a given number of men, be wider, which theoretically created a possibility of enveloping the square on the flanks and defeating it there even as it broke through the line in one spot, but this possibility was largely illusory. As already noted, the square was not terribly vulnerable to flank attacks. More importantly, the leaders of the deep mass could slam it directly against the principal banners of the opposing force, usually found in the center of the line. Banners were the principal means of tactical command and control, and once they went down, those who fought under them usually broke; even if they continued fighting, they were much less able to do so effectively. "The standard," explained a fifteenth-century banner-bearer, "is like a torch set in room to give light to all men; if by some accident it is put out, all remain in darkness and unseeing and are beaten."[15]

In the Italian Wars at the turn of the sixteenth century, experience showed that a Swiss attack could be halted, at least temporarily, in various ways – by cavalry attacks against the column's flank (as at Marignano), by the use of entrenchments, sunken roads, or water barriers (as for example at Bicocca and Cerignola), or by an opposing block of other pikemen. But none of these methods would serve, alone, to prevent the advance from resuming, or to truly defeat the attackers. To secure a genuine victory over these fearsome warriors (or others using their tactics), it

---

12 Winkler, 123n: "man auf obgennantem Tag sich reathen und mit Vollmacht der Obrigkeiten eine Ordnung machen soll, wie die Sache sürderhin geregelt werden solle mit Hauptleuten und Fähnloin, damit der krieg teglich geübt . . ." William Garrard, *The Arte of Warre* (London: Roger Warde, 1591), 9, 89, also calls for "daily practise" for armored pike, including engaging in pushing-matches, disengaging on command, and "to couch and crosse, for defence of horse."

13 Barret, 76–7.

14 de Bueil, 2:63.

15 Guttiere Diaz de Gamez, quoted in Clifford J. Rogers, *Soldiers' Lives through History: The Middle Ages* (Westport: Greenwood, 2007), 198. Note also Christine de Pizan, *Livre des fais d'armes*, MS Bodleian 824, fo. 37: "au regart de la baniere se gouverne l'ost."

131

was necessary to hammer the square with shot while it advanced and while it was checked. But because of new developments in armor technology, the old missile weapons (crossbows and longbows) were not really up to this job. Handguns and field artillery *were*, and their growing importance in the first half of the fifteenth century was the other half of the tactical transformation that marks the shift from medieval to early-modern on the battlefield.

The Rise of the Handgunner

There is a striking degree of continuity in tactics from Ancient Greece through the early Renaissance, deriving from the essentially constant capabilities of the human body and the limited possibilities for variation in hand-held weapons of steel and wood. That fact lies at the core of one sixteenth-century observer's comment that tactics have been "from the first creation of the world until now the very same, the disposition of the people only varying in the difference of weapons, engines, and instruments, which have been invented."[16] But the gun was new and fundamentally different, and came increasingly to replace the bow and crossbow, and to reduce even the cavalryman's reliance on the ancient weapons of cold steel. "The arms of our grandfathers," commented Jean de Saulx-Tavannes, "were the lance, the axe, the mace and the sword. The last we still use, but the rest are considered of little value partly because of armor of proof, which they neither pierce nor penetrate easily, and partly because of the invention of better pistols."[17] This change in weapons naturally demanded changes in fighting methods, both individual and collective: "at this day we are constrained to varie our order [from classical forms]," wrote William Garrard in 1591, "considering our armes be varied, which do now fetch and wound much more and further off, and are more pearcing then those of antient time."[18] Robert Barret, a few years later, had little patience for his compatriots who still thought the English bill and bow could win great victories, as they had done through the medieval period, and even into the early sixteenth century: "Time altereth the order of warre, with many new inventions daily"; "then was then, and now is now; the wars are much altered since the fierie weapons first came up."[19]

The importance of the shift to gunpowder weapons has long been recognized – and was well appreciated even in the sixteenth century[20] – but it is important to

---

16 Edward Hoby, quoted H. J. Webb, *Elizbethan Military Science* (Madison: U. of Wisconsin Press, 1965), 16.
17 Jean de Saulx-Tavannes, *Mémoires de Gaspard et Guillaume de Saulx-Tavannes*, in Michaud et Poujolat (eds.), *Mémoirs pour servir à l'histoire de France*, VIII (Paris: Guyot, 1854), 191–2.
18 Garrard, 64.
19 Barret, 2.
20 In addition to the last three citations: Barnabie Riche, *The Fruites of Long Experience* (London: Jeffrey Chorlton, 1604), 68: certain classical "forms and proportions" have been outdated by "the fury of shot." John Cruso, *Militarie Instructions for the Cavallrie* (Cambridge: Alsop, 1632), A3v, views the changes made in response to "later inventions of fire weapons" as the only thing "in these modern warres, which is not borrowed from antiquitie." But cf. Smythe, *2v.

note that that change was itself not simply caused by the "push" of an inherently superior weapon,[21] but rather (or also) by demand pull on two fronts. First, there was demand for weapons systems that would allow exploitation of increases in population and revenue that did not comprise a corresponding increase in the martial classes. Pikes and guns all required training to use well – particularly *group* training (drill) – but they did not require soldiers who were raised and trained to their use from early youth, as the longbow and (to an extent) the lance did. Indeed, a mere fortnight's drill was considered sufficient to make a recruit into a pikeman or arquebusier "mete to serve."[22] Humphrey Barwick meant to emphasize the difficulty of mastering matchlocks when he wrote that "the armed pikes and halberds, launces and speares are better to be made perfect in six daies then the fiery weapons are in 60 daies" – but even sixty days' training was a relatively small investment of time compared to the years of practice that could profitably be applied to perfect the older martial skills (and the muscular development) of the man-at-arms or the archer. Indeed, it was proverbially recognized that only someone who had trained from boyhood could serve respectably as a knight or a longbowman.[23]

The other factor creating a demand-pull that encouraged the adoption of firearms was – as Tavannes indicated – the problem posed to older weapons by new metallurgical developments. Around the turn of the fifteenth century, Milanese smiths developed new techniques that allowed for a radical improvement in the practical effectiveness of the steel used in plate armor. The key was quenching the armor pieces rather than allowing them to air-cool. The technique was very difficult to master, but made a tremendous difference in the practical protective value of the harness. Top-quality armor was, however, very expensive, and metallurgical analysis of surviving pieces shows that for a long time after the new methods were developed, there were plenty of mediocre harnesses being produced, and even more remained in service on the battlefield.[24]

21 Eltis, 102; note also Smythe, A2v.

22 Thomas Audley, "A Treatise on the Art of War," ed. W. St. P. Bunbury, in *Journal of the Society for Army Historical Research* VI (1927), 69; Michael Roberts, *Gustavus Adolphus*, vol. 2 (London: Longman's, 1958), 239, and cf. Webb, 35.

23 Barwick, 24 (and cf. 11, where he claimed he shot as well with an arquebus after five months as with a bow after many years). Knight: Marc Bloch, *Feudal Society*, tr. L. A. Manyon, vol. 2 (University of Chicago Press, 1961), 293–4; cf. Cruso, 30. The Dutch believed a lancer could be trained in three years (J. P. Puype, "Victory at Nieuwpoort," in Marco van der Hoeven, ed., *Exercise of Arms* [Leiden: Brill, 1997], 82); by contrast, Tavannes believed only three months' training was needed for a pistoleer who would not be required to duel on horseback. Tavannes, 194. Smythe "would not have any [longbowmen] in the army who had not been trained in archery from boyhood." Matthew Strickland and Robert Hardy, *The Great Warbow: A History of the Military Archer* (New York: Sutton, 2005), 405; Christine de Pizan, *Livre des fais d'armes*, MS Bodleian 824, fo. 27v.

24 In 1590, for example, Sir Henry Lee tested two breastplates, one of German and one of English metal, with a pistol. The former "more than a littell dent of the pellet nothinge perced," but the latter was shot "clean through." Bert Hall, *Weapons and Warfare in Renaissance Europe* (Baltimore: Johns Hopkins University Press, 1997), 147; A. R. Williams, *The Knight and the Blast Furnace: A History of the Metallurgy of Armour in the Middle Ages & the Early Modern Period* (Leiden: Brill, 2003), passim.

Even mediocre plate armor was very tough to overcome with weapons powered by human muscle. Normal bows and light crossbows were practically unable to do it; in fact, the very definition of a quality suit of plate was that it had been tested with an arrow or a bolt and found to be "proof" against it.[25] Still, heavy crossbows and arrows loosed from the extraordinarily powerful English warbow *could* penetrate mild-steel plate – thin pieces like gauntlets, greaves, or visors fairly easily, thick breastplates or bascinet-tops only at close range and with an ideal angle of impact. Properly quenched high-carbon steel, with a Vickers hardness of around 350 (as opposed to around 150 for air-cooled medium-carbon steel) was another matter. It could successfully resist a weapon striking with *double* the kinetic energy required to defeat an old-style harness.[26] It was nearly impossible for even the strongest longbow or windlass-drawn crossbow to cause serious injury through a breastplate of that sort, and only a lucky shot from a strong bowman would even be able to cause a limb-wound worth mentioning.[27] At Flodden in 1513, for example, English bowmen found that the Scottish pikemen were so well armored that arrows "did them no harm."[28]

Handguns, on the other hand, were not limited by the energy-producing capability of the human body. Modern tests have demonstrated that the very strongest archers (who had "hands and arms of iron" and "bodies stronger than other men's")[29] could put a formidable 130–150 joules of kinetic energy behind their armor-piercing shafts. A heavy steel-bowed crossbow produced somewhat more energy initially, perhaps up to around 200J.[30] But a well-charged 1.5-oz. musket ball could leave its barrel with around 3100 joules; a 1-oz. arquebus-ball with around 2700J; even a cavalryman's pistol could deliver over 1000J.[31] It is not simple to interpret the implications of these numbers for practical effectiveness, because ballistics is a complicated subject. For example, round bullets in their flight rapidly lose energy to friction with the air, unlike streamlined arrows or modern bullets, and hardened steel arrowheads are far better suited for penetrating armor than are soft lead spheres.[32] Hence, a pistol ball at 200 yards could easily

25 Williams, *Knight*, 924.

26 Ibid., 62, 693–4.

27 Clifford Rogers, "The Battle of Agincourt," *New Perspectives on the Hundred Years War*, vol. 2, ed. L. J. Andrew Villalon and Donald J. Kagay (Leiden: Brill, 2008), Appendix I.

28 Strickland and Hardy, 397–9.

29 Dominic Mancini, *The Usurpation of Richard the Third*, tr. C. A. J. Armstrong (London: Sutton, 1989), 99.

30 Arrows: Rogers, "Agincourt," appendix I. Crossbow: Williams, *Knight*, 919–920, but cf. Stephen Grancasy, "Just How Good Was Armor?" *True* (April 1954), 90 (31J/740lb).

31 P. Kalaus, "Schiessversuche mit historischen Feuerwaffen," in *Von alten Handfeuerwaffen* (Graz: Landesmuseum Joanneum, 1989): (musket) 52, 56, 59; (28g ball), 57; (18g ball, 988J), 53; (pistols) 55, 61; note also Grancasy, 91.

32 Round bullets lose kinetic energy about three times as fast as modern bullets. Hall, 137. On bullets' penetration, see Williams, *Knight*; also Hall, 145–6 and Kalaus, passim. The effects of projectile shape and hardness were such that, according to the calculations of A. R. Williams, it would take a bullet 450J of energy to defeat a 1 mm steel plate which could be effectively penetrated by an arrow with only 55J.

be stopped by even the cheapest vambrace, whereas a strong archer's war-shaft would still punch through the armor like a cobbler's awl through leather.[33] But if fired from just outside the reach of an enemy's lance, a pistol stood a fair chance of killing a well-armored sixteenth-century man-at-arms, which an arrow or bolt did not.[34] A musket ball could penetrate a high-quality corselet at 200 yards, defeat an average one at 400 yards, and ruin an unarmored horse or man even at 600 yards,[35] a distance far outside the range at which even the best bowman could return fire.[36]

33 Bullets, which are shot at a basically flat trajectory, lose velocity fairly steadily and fairly rapidly; at 200 yards, a pistol bullet would be down to something like 225J, about half what it would need to penetrate even 1 mm of mild steel with a direct hit. The dynamics for arrows are more complex. They are more streamlined and do not lose velocity as quickly, and in long-range shooting they actually regain velocity as they lose altitude on the descending half of their flight. Hence (depending on the strength of the bow and the weight of the arrow) an arrow might hit with just as much punch at 200 yards as it does at 70 yards, but less at 160 yards and more at 100! In Mark Stretton's experiments with an extremely powerful longbow, he found that his arrows retained around 2/3 of their penetrative capacity at 200 yards, leaving them well more than necessary to penetrate a 1 mm steel plate (or a 1.9 mm munition-armor plate of essentially wrought iron). Mark Stretton, "Medieval Arrowheads. Practical Tests, Part Two," *The Glade* 108 (2005), 55; see also Strickland and Hardy, 411; Williams, *Knight*, 928–9, 942; Hall, 137, 1, 146–7. Note, however, Barwick's belief that even at 120 yards even the best archer could do no harm against a man armored in pistol-proof. Barwick, 20v.
34 Tavannes, 191–2, 336; François de la Noue, *Politike and Militarie Discourses* (London: Thomas Orwin, 1587), 199.
35 Barwick, 10v: "musketes are weapons of great force, and at this day, bothe with leaders and followers, much feared: for fewe or no Armours, will or can defend the force thereof, being neerehand, which is as well a terror to the best armed, as to the meanest: it will kill the armed of proofe at ten skore yardes, the common armours at twenty score, and the unarmed at thirty skore, being well used in bullet and tried powder." Williams, based on experimental results and calculations, finds these statements "probably valid." *Knight*, 947–8; Roger Williams, *A Brief Discourse of Warre* (London: Thomas Orwin 1590), 40–41, also fairly closely agreed: "the musket spoils horse or man thirty score off, if the powder be anything good and the bearer of any judgment. If armed men give the charge, few or none carry arms of the proof of the musket, being delivered within ten or twelve score." (Modernized.) See also Barret, 2–3; Blaise de Monluc, *Military Memoirs*, ed. Ian Roy (Hamden, CT: Archon, 1972), 186; Smythe, 14–14v.
       Hall, however, claims that at around 100 meters or more "the greatest effect that could be expected [even for a heavy musket] would be to unhorse a man just by striking him," apparently because he thinks that the musket could not penetrate armor at that range. But a musket ball would have about double the intial ke of an arquebus ball (double the mass, the same velocity); hence at 100 yards it would still have roughly the same ke as an arquebus had at short range, and at short range one of the Austrian tests with a 27g musket-ball (i.e. by sixteenth-century standards an arquebus shot) penetrated a 3 mm modern steel sheet – quite thick – and still retained enough energy not only to create a 25 cubic centimeter wound, but also to send splinters from the armor 8 cm into the target. Hall, 146. See also Eltis, 13–15.
36 Strickland and Hardy, 410–411; cf. also Barret, 2–3, who argued based on "daily experience," which "hath made it most manifest in these our later warres" that bows were unable to do much damage to armored men at 200 yards, whereas at that distance even a caliver (arquebus) would penetrate most armor, and still had power to injure naked flesh at 333 or 400 yards, ranges which few archers could reach at all. Cf. Monluc, *Memoirs*, 129, whose experiences on the receiving end of mid-sixteenth-century English archery more or less bear out Barret's conclusions about the near-invulnerability of well-armed men to arrow fire; Barwick, 20v.

Moreover, the wounds inflicted by early-modern bullets were generally much worse than those caused by arrows. A bodkin arrowhead that fully penetrated into its target might create a wound cavity of around thirty cubic centimeters; by contrast, modern testing suggests that at close range an arquebus ball might blast a hole in a human body three times that size at one hundred meters, or eight times at nine meters.[37] It was common for men wearing good fifteenth-century armor to suffer multiple wounds from arrows or crossbows and still to be able to fight. By contrast, Bayard was killed by a ball from a musket that struck him in the side, apparently penetrated his armor, and continued on to break his spine.[38]

To be sure, firearms had their disadvantages. Even a skilled handgunner with a well-made weapon and a tight-fitting ball could not hope to match the long-range accuracy of a good bowman. In Tudor times, the *minimum* target range allowed for longbow practice by full-grown men was 220 yards, aiming at 18" round targets! When the married men of Calais challenged the bachelors of the town to a shooting-match in 1478, the distance between the targets was set at 260 yards.[39] By contrast, arquebusiers in 1560 were provided with butts twenty *feet* wide by sixteen feet high, with target circles 54" in diameter, for shooting at just 120 yards.[40] Barwick's claim that he could hit a man standing at 120 yards

---

37 Barwick, 14v-15: "now the weakest of us are able to give greater wounds, then the greatest and strongest archer." Arquebus (28 g ball into soap-block): Kalaus, 77, cf. 57; Hall, 146. Arrow: based on 1.5 cm diameter head and 25 cm penetration; cf. B. Karger et al., "Experimental Arrow Wounds: Ballistics and Traumatology," *Journal of Trauma* 45 (1998), 497–99.

    Of course, not all bullet-wounds were lethal; Monluc survived seven, for example. But it was much more common for a single ball to kill an armored man outright, or disable him, than for an arrow to do so. At Neville's Cross, David II of Scotland was wounded with two arrows lodged solidly in his skull, but was nonetheless able to put up a stout fight against his captors. Clifford J. Rogers, "The Efficacy of the Medieval Longbow: A Reply to Kelly DeVries," *War in History* 5 no. 2 (1998), 238; Andrew of Wyntoun, *The Original Chronicle of Andrew of Wyntoun*, ed. F. J. Amours (Edinburgh: Blackwood, 1908), 6:184. At Pontevedra in 1397, Pero Niño had an arrow knit his camail to his neck, yet he "felt the wound little or not at all," and kept fighting to great effect. Guttiere Diez de Games, *Chronica de Don Pedro Niño*, ed. Eugenio de Llaguno Amirola (Madrid: Antonio de Sancha, 1782), 40. Similarly, Joan of Arc once "was wounded by an arrow which penetrated her flesh between her neck and shoulder for a depth of six inches," but kept fighting (Strickland and Hardy, 286). In Ireland in 1230, one warrior "had five arrows in his body when a horseman attacked him, and he had nothing but an axe; yet he kept the horseman at a distance, parrying his spear with the axe." *Annals of Connacht* (via CELT Corpus of Electronic Texts, https://celt.ucc.ie), 1230.6.

38 Guyard de Berville, *Histoire du Pierre de Terrail, dit Chevalier Bayard* (Lyon: Yvernault et Cabin, 1809), 379. Similarly, Monluc was shot with an arquebus ball in the thigh. Presumably after penetrating his armor, it passed through his body and entered his stomach, so that he died a few days later. Blaise de Monluc, *Commentaires et lettres*, ed. Alphonse de Ruble, 5 vols. (Paris: SHF, 1864–1872), 1:71, 75, 203–4. Barwick (3) is pithy, as ususal: "whereas Manuell the Emperour of Constantinople, had in his armour or target the number of 30. arrowes sticking: one harquebuze or Musket shot would have dispatched the matter," for (2b v) – "now by reason of the force of weapons, neither horse nor man is able to beare armours sufficient to defend their bodies from death."

39 Strickland and Hardy, 381; see also Smythe, 27v; Fourquevaux, 1:12.

40 Soar, 194–5.

with an arquebus was a boast,[41] and modern tests give reason tc doubt even that claim. Due to the Magnus effect, the unpredictable spin of a smooth-bore musket's ball causes it to curve unpredictably from its target line.[42] The inaccuracy of the weapon would have been seriously worsened by the normal battlefield practice of using balls significantly smaller than the gun-bore.[43] Of course, there was no great need for precision on an early-modern battlefield, where the target would normally be an enemy formation dozens of yards wide. But even against such a target, effective range was quite limited. Digges considered that a "trained shot" ought to be able to deliver his ball between foot-height and head-height at 133 or 166 yards, but he admitted that "our common shot, if they discharge not within an hundred pace [83 yards], they will waste their powder, and do little or no hurt at all to their Enimies."[44]

An even bigger disadvantage was the time-consuming complexity of the reloading process, well described by Bert Hall:

> The soldier first had to dismount and secure his match; then to blow any sparks from his firing pan; then to prime the pan with special fine gunpowder, remembering to shake any excess from the pan and to tamp the pan with his finger; then to recharge his piece with regular gunpowder and to reload it with wadding and shot, drawing out his ramrod and tamping the powder and ball with just the right amount of pressure. He was then ready to cock his mechanism, blow his match to life, fix it in the matchlock's jaws, present his piece, and give fire.[45]

In de Gheyn's precisely scripted exercise of arms, this all required 42 motions, not counting the movements of coming forward to the front of a formation to deliver fire, or retiring to its rear to reload.[46] Hence, some sixteenth-century cap-

---

41 Webb, 94; cf. Barwick, 17v–18.

42 Hall, 141–2. As a spinning ball flies, it creates a band of rotating air arounc it. Unless the spin is perpendicular to the flight (as with a rifle), this air is accelerated by the "windstream" on one side of the ball, and decelerated by it on the other, creating a pressure differential that sucks the ball towards the faster-moving (lower pressure) air.

    A test of 1742 found a set of five balls fired from a musket on a fixed block had a horizontal spread of around 1 foot at 50 ft, nearly 2 ft at 100 ft, but nearly 8 ft at 100 yds. Hall, 138–40, 142–3; Strickland and Hardy, 31.

43 Smythe, 17v–18.

44 Leonard and Thomas Digges, *Stratioticos* (London: Henrie Bynneman, 1579), 108, 122. Smythe, in the context of a tract in favor of bows over handguns, was even more pessimistic: he opined that "arquebuses (considering their inaccuracy) are [not] to be used . . . in the field, above three or four scores [50–66 yards] at the farthest." Smythe, 5v (modernized); cf. ibid., 14v (muskets 133–166 yards), 15v (17–42 yards), 17, 28v; Monluc, *Memoirs*, 105 (83–100 yds).

45 Hall, 149.

46 An excellent sense of the requirements for reloading can be gained from the words of command used in the process: "1. Shoulder your Peece and matche. 2. Unshoulder your Peece. 3. And with the Right hand hold it up. 4. In the left hand take your Peece. 5. In the right hand take your

tains considered that arquebusiers could not be expected to fire more than ten shots in an hour.[47] True, Humphrey Barwick suggested a much higher rate of forty shots per hour – for an individual of unusual skill, on a practice ground. But for a common arquebusier acting in formation under combat conditions, one shot per two or three minutes is probably realistic.[48] The larger-caliber musket, due to its great weight, required a rest. The need to manage that as well as the match and the weapon itself made its rate of fire even slower.[49] An archer, by contrast, could certainly loose as many as ten arrows a minute.[50] But then, it was better to deliver one ball that even against armored men "carries death and fear with it," than to loose numerous arrows or bolts that collectively posed "little hazard of life."[51]

---

matche. 6. Hold wel your match and blow it of. 7. Cock your match. 8. Try your match. 9. Blow your match, and open your pann. 10. Present your Peece. 11. Give fire. 12. Take downe your Peece, and in the left hand hold it well. 13. Uncock your matche. 14. And joyne it againe betwixt your fingers. 15. Blow out your pan. 16. Proyme your pan. 17. Shut your pan. 18. Shake of your pan. 19. Blow of your pan. 20. Turne about your peece. 21. And to your left side let it sink. 22. Open your charges. 23. Charge your peece. 24. Your skowring-stick draw out. 25. Your skowring-stick take shorter. 26. Ramme your pouder. 27. Your skowring-stick draw out your peece. 28. And take it shorter. 29. Put up your skowring sticke. 30. With the left hand bring forward your Peece. 31. And with the right hand hold it up. 32. Shoulder your peece. 33. Hold your Peece well upon your shoulder, and marche to the place of garde. 34. Unshoulder your peece. And in the left hand let it sinke. 36. Hold your Peece well. 37. With the left hand hold your Peece. 38. In the right hand take your match. 39. Blow of your match. 40. Cock your match. 41. Try your match. 42. Garde your pan and stand readie." De Gheyn, *Exercise of Arms.*
47 Garrard, 6.
48 Barwick, 4v; Gordon R. Mork, "Flint and Steel: A Study in Military Technology and Tactics in 17th-century Europe," *Smithsonian Journal of History* 2 (1967), 27; Digges, 124, claimed it would take over 15 minutes for 10,000 handgunners in formation each to deliver one shot. Cf. Eltis, 15 (end 17th c., 1 musket shot/minute considered good). 20–50% misfires would reduce the effective rate further. Frank Tallett, *War and Society in Early Modern Europe* (London: Routledge, 1992), 23; Smythe 21–21v.
49 Williams, *Discourse*, 41 (probably overstated; cf. de Gheyn.)
50 Strickland and Hardy, 31; Rogers, "Agincourt," n. 132; note also Matthew Sutcliffe, *The Practice, Proceedings, and Law of Armes* (London: Christopher Barker, 1593), 190. However, Smythe, though an archery advocate, only claimed that bowmen could loose four or five shots to a gunner's one. Smythe, 20v.
51 Tavannes, 192; Barret, 3–4. De Saulx opined that bullets killed a hundred times more men than did cold steel. Claude Gaier, "L'opinion des chefs de guerre français du XVIème siècle sur les progrès de l'art militaire," *Revue internationale d'histoire militaire* 29 (1970), 733. Barwick made a similar claim: "where as there hath beene one slaine with arrowes, there hath been a hundred slaine with manual wepons of fire, since the use of the same hath beene practised and rightly knowen" (18v); the handgun "is the onely weapon, that hath beene the greatest cause of the deaths of suche numbers, as have beene of late dayes overthrowne, in great incounters, yea and in small skirmishes." (8v) Hence, the musket in particular was "bothe with leaders and followers, much feared . . . [and] as well a terror to the best armed [i.e. armored], as to the meanest [i.e. worst.]" (10v) He specifically contrasts this with older missile weapons (14): "to my knowledge, I never saw any [one] slaine out right with an arrowe, and but with Quarels few, but with harquebuze and pistoll shot, I have been at severall times, where 20000. hath beene slaine outright, besides manie wounded and maimed." As a result, armored soldiers could charge, close with, and disperse archers with little

Hence – and also for many other reasons, ranging from cultural and dietary changes in England to the accelerating urbanization and monetization of the European economy to improved guns to the declining price of saltpeter and the rising price of yew staves[52] – firearms became, by the mid-sixteenth century, the predominant shot weapon for European soldiers.[53] Moreover, the importance of missile-firing troops relative to close-combat infantry rose dramatically in most armies, the exception being the ones in which the shot already played a key role, as in England or Bohemia. But even in those places, it was something new in the sixteenth century to be able to say that "many Tymes it hath bene sene that Battail hathe been [won] by shott onlie, without pushe or strocke stricken."[54]

Pike and Shot

As the handgun replaced the crossbow and longbow, the sword, mace, and axe – weapons which had played important parts in hand-to-hand combat literally for millennia – were largely replaced in infantry formations by the pike (supplemented by shorter weapons like the halberd). Why did this occur? For many reasons:

- Plate armor was generally too strong to be penetrated by one-handed edged weapons such as axes and swords, and distributed the force of impacts well enough to make maces largely ineffectual. (Swords were retained because they were useful for defense as well as attack, being well suited to parrying, and because although they could not well hack or punch through armor, they could be precisely thrust through joints or eye-slits, and could also – like other short weapons – be used to beat an enemy down. And they remained excellent weapons for use against unarmored or nearly-so soldiers, such as handgunners.[55])

---

fear, since resolute soldiers do not much "feare any woundes, so that life may be in little or no daunger." See also 15v.

52 Hall, 58; Strickland and Hardy, ch. 20.

53 Although rarely noted in the modern literature, the technical improvements of the weapons themselves (particularly lengthening barrels) contributed significantly to this development. Williams, *Knight*, 854–5 (noting mislabeling of columns, p. 854); Barwick, (B 1) noted in the 1590s that when he had first served in the field in 1548, the arquebuses used in England were "but mean stuffe in comparison of those that are now in use"; see also Tavannes, 191–2.

54 Audley, 68. Similarly Barret, 75: "And againe it is rarely seene in our dayes, that men come often to hand-blowes, as in old time they did: For now in this age, the shot so employeth and busieth the field (being well backed with a resolute stand of pikes) that the most valiantest and skilfullest therein do commonly import the victorie, or the best, at the least wise, before men come to many hand blows." Barwick, B 0v: in the days of the longbow and crossbow, "the enemies did then commonly joyne both with long and short weapons . . . but in these daies where the weapons of fire hath beene rightlye used, it half beene scarselye seene that either Pike or Halbert hath come to joyne at any time before one partie did turne their faces, by reason of the tarrible force of the great and small shot." See also Riche, 68; Bernardino de Mendoza, *Theoretique and Practise of Warre*, tr. Edward Hoby (London: N.P., 1598), 109, but cf. Audley in Strickland and Hardy, 403.

55 Tavannes, 191–2.

Weapons that put the full force of both arms behind a single point did better against sheet steel. Jean de Waurin, late in the fifteenth century, wrote of Flemish pikes that "there is no armor so good that they cannot penetrate or break it."[56]

- Pikes were the best weapons for defense against cavalry. The Swiss learned at Arbedo in 1422 what Barret affirmed much later: "Against horse . . . farre better is the Pike, then either Bill, or Halbard."[57] A pike outreached a cavalryman's lance, and "any [charging] horse struck in the breast by a pike must invariably die."[58] Moreover, the great length of the weapon allowed pikemen to put not one or two but five or six rows of pike-points between the mass of a speeding horse and the relatively frail flesh of the front-line footman. Even if a skilled rider could get a war-trained mount to make a suicidal charge into the serried front of a pike phalanx, so long as the men stood their ground the horse would find it "as hard to be pierced . . . as an angry porcupine or hedgehog with the end of a bare finger."[59] This defensive strength was particularly important because by the late fifteenth century, heavy cavalry had experienced something of a resurgence on the battlefield, thanks to the improvements in armor already discussed and also because of the reduced volume of shot the horses had to pass through, due to the rising predominance of slow-firing firearms.[60]

- Footmen too, unless they were also arrayed and armed in the same way, would find attacking a resolute stand of pikes as profitless as assailing "a wall of bronze."[61] "It may be taken as axiomatic," wrote Montecuccoli, "that no battalion of pikemen can ever be ruptured in a head-on attack."[62]

- The impressive successes of the Swiss (and to an extent the Flemings),[63] particularly from their defeats of Charles the Bold through the early Italian Wars, naturally called forth emulation. The reasons for the Swiss victories were much broader than their use of the pike, but their tactics and weapons could easily be copied (unlike the social underpinnings of their exceptional cohesion).

---

56 Jean de Waurin, *Anchiennes Cronicques*, ed. E. Dupont, 3 vols. (Paris: Société de l'Histoire de France, 1858–63), 3:74; for halberds and bills, A. W. Boardman, *The Medieval Soldier in the Wars of the Roses* (Stroud: Sutton, 1998), 175; C. W. C. Oman, *A History of the Art of War in the Middle Ages*, vol. 2 (London: Methuen, 1924), 251n.

57 Winkler, 49–53; note also 31–2; Barret, 4; Raimondo Montecuccoli, *Sulle Battaglia*, tr. in Thomas M. Barker, *The Military Intellectual and Battle* (Albany, NY: State University of New York Press, 1975), 88.

58 Waurin, 3:74.

59 Garrard, 229, modernized. Barret, 4: "Against horse . . . farre better is the Pike, then either Bill, or Halbard."

60 Malcolm Vale, *War and Chivalry* (London: Ducksworth, 1981).

61 Barret, 47 (modernized). Hence (p. 4): "for the plaine field, neither blacke bill, Halbard, nor Patizan [is] comparable to the Pike." However, it should be noted that at Stoke Field and Flodden English billmen did very well against pikes.

62 Montecuccoli, *Sulle Battaglia*, 104.

63 Particularly at Guinegate in 1479.

- The pike, unlike any other hand-to-hand weapon, allowed a formation to protect not only itself, but a significant number of shot troops attached to it. When threatened by an enemy advance (particularly by enemy cavalry), arquebusiers could kneel, roll, or dive under the pike-hedge. A short file of gunners could even stand and deliver ready shots while squeezed between two stacks of pikes presented by the men behind them, though reloading in such conditions would be very difficult.[64] On the battlefields of the late fifteenth through the seventeenth centuries, shot troops badly needed that protection. At Agincourt, the English archers had been able to defend themselves largely by their own arrows, with just a little help from a rapidly set up hedge of stakes.[65] Fifty years later, that was not really practical, thanks to the improvements in armor already mentioned, and also to the declining number of really good bowmen.[66] Less fearsome archers than England's had never really been expected to play that role. Handgunners, with their much slower rate of fire, could not hope to stand their ground against a determined attack by cavalry or by armored pikemen over open ground.[67] This simple fact rose in significance in tandem with the proportion of shot in the army, and that figure generally climbed in most nations over the sixteenth century, until around half to two thirds of all infantrymen were armed with firearms.

One solution to this tactical problem was a long, thin pike-wall with the shot stationed either behind the line or in alternating files. This tactic was employed, for example, by the Burgundians in the late fifteenth century.[68] It was better suited, however, to bowmen (who often arced their fire, and who could provide enough volume of shot to help defend the thin pike-line from a denser attack column) than to fire-shot. In particular, such formations proved unable to halt the steamroller charge of a deep Swiss-style square.

Another response to the vulnerability of handgunners (and other missile troops) that enjoyed substantial success in the fifteenth century was the use of

---

64 Cf. Barret, 48, 96–8; Garrard, 106; or kneeling under the pikes, Smythe, 16v.

65 Rogers, "Agincourt."

66 Barret, with reference to bows: "What souldier is he, that commeth against a weapon wherein there is little hazard of life, which will not more resolutely charge, then against a weapon, whose execution he knowth to be present death?" Barret, 3; see also Barwick, B 0v, 20v, 21. In 1545, with a force of 120 mixed arquebusiers, halberds, and pikemen, Monluc successfully charged and put to flight 200–300 English archers (with a few Italian arquebusiers). He noted that their arrows "would do no execution" unless fired from short range, by contrast with arquebuses, which were fired "at a great distance." Monluc, *Memoirs*, 129–30; Smythe, 26v; but cf. Smythe, 31v.

67 Barret, 69; Webb, 118–9; La Noue (203) believed that 700–800 gendarmes could defeat 18,000 arquebusiers; Smythe, 19–19v; Garrard, 117, opined that arquebusiers "ought never to be sent about any enterprise" without the "fellowship" of "armed pikes and corselets." Barret, 69, remarked that "any troupe of shot, though never so brave & expert, being in open field, having no stand of pikes, or such other weapons, nor hedge, ditch, trench, or rampier, to relieve and succour them, could not long endure the force of horse, especially Launciers."

68 E.g. Richard Vaughan, *Valois Burgundy* (Hamden, CT: Archon, 1975), 128, 125.

wagon-fortresses. These, however, were normally only practical for an army on the defense. The Hussites scored very impressive successes by employing the *Wagenburg*, but their opponents lacked effective field artillery. An army which did have a sufficiency of mobile guns could concentrate them against one segment of the "wall," and then storm the "breach," enjoying the classic benefit of a massed attack against the divided forces of a static perimeter.[69]

Deep blocks of pike could resist other pike, had enough manpower to pay the cost of closing with and defeating unprotected shot, and were especially effective at defending themselves (and attached shot) against cavalry attacks, whether from front or flank. In some ways a square of men thirty or more ranks deep was very wasteful (only the first six or seven ranks could use their weapons to the front),[70] but the men further back could assist by providing weight to a push of pike, and replacements for the fallen when charging shot.[71] This is one principal line of explanation for the universal adoption in the sixteenth century of infantry formations composed basically of a core of pike encircled or flanked (or both) by shot.

It is crucial to realize, however, that the purpose of this pairing of pike and shot was as much for the protection of the former as for the safety of the latter. Pike blocks were very vulnerable to fire if they could not simply overrun the enemy shot, which they might not be able to do if the latter were protected by formidable entrenchments or shielded by obstacles like streams or sunken roads, or even if the gunners simply had room to maneuver away from the more cumbersome pike block. Hence, thought Barret, "by every man's judgment" an unsupported stand of pikemen, "though never so well armed," could be defeated by an equal number of arquebusiers. Moreover, enemy lancers through repeated charges (or sometimes even by mere threat of charges) could force a pike column to halt and form for defense, thereby preventing it from closing with its target.[72] Finally, pikemen on their own had no really effective response to fire delivered by cavalry squadrons. As will be discussed in more detail later, pistoleers could ride up by ranks to just a few yards beyond the pike-points, discharge their pieces, then wheel and retire. The damage a squadron could do in an iteration of this "caracole" technique was not negligible, and against a pike-block unsupported by shot it was almost entirely *one-sided*, and could be continued indefinitely.[73] Thus, pike and shot existed in a kind of tactical symbiosis. On the one hand, shot needed to be able to "retyre safely" to the protection of pikemen "whensoever they shall happen to be charged with Lances," or by pike. On the other, the shot were needed "to . . . succour their Pykes, whensoever any attempt shalbe made by Argoletiers [dragoons] or Pistols to breake their array."[74] "The one without the other," wrote Barret in 1591, "is weakened the better halfe of their strength. Therefore of necessitie

<hr>

69 Hall, 130.
70 Montecuccoli, *Sulle Battaglia*, 88; cf Barret, 75.
71 Eltis, 54.
72 Barret, 69; Sutcliffe, 181, 186, 189; but cf. la Noue, 199.
73 Barriffe, quoted Mork, 27.
74 Digges, 110, 144.

(according to the course of the warres in these dayes) the one is to be coupled & matched with the other, in such convenient proportion, that the advantage of the one may helpe the disadvantage of the other."[75]

The mutual protection of pike and shot did not extend a long distance. As already noted, an arquebus volley was relatively ineffectual past 100 yards (especially under battlefield conditions, when the weapons would often not be optimally loaded).[76] In the medieval tactical style it had been common to build a central battle of heavy infantry and have it flanked by two wings of shot; the central battle might have around 250 or 300 yards of frontage, or more, and the wings would stretch further out. But if arquebusiers on the far flanks of the formation were 250 yards away from the edge of the pike-block in the center, their fire could not reach far enough with any accuracy to provide much assistance even to the pikemen on the very edge of their mass, much less those in the center. Even those gunners immediately adjacent to the heavy foot would have been hard-pressed to do much against pistoleers making a caracole against the center of the line. [Figure 6.1]

Moreover, horsemen charging the shot at the far flank would be able to overrun them long before they could reach the shelter of the supporting pike.

The basic solution to this problem was to attach sleeves of shot to multiple smaller battalions of pike, rather than one single great block. But this array posed another problem. Placing shot units in the front line made it virtually impossible to prevent an enemy force of pike from breaking the line where the gunners stood.

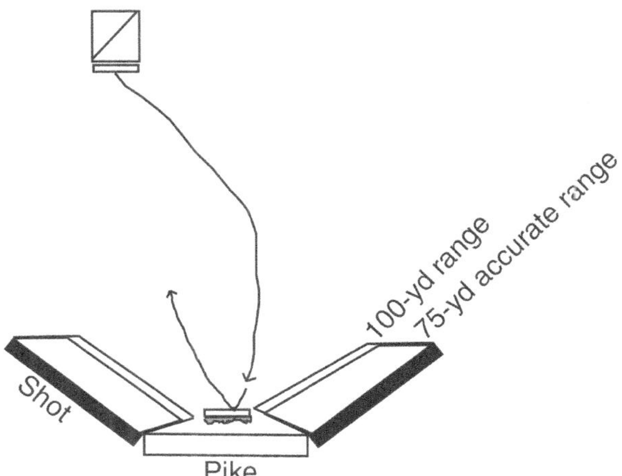

Figure 6.1 A Caracole.

75 Barret, 69; Sutcliffe, 182–6; Audley, 66.

76 Proper loading made a huge difference in effectiveness. Smythe, 18v: "the first volley, carefully loaded out of harm's way, had more effect than the next four put together." Eltis, 15; Tavannes, 191; la Noue, 202; Williams, *Discourse*, 38–9.

In late-medieval infantry tactics, the line was always contiguous, and if broken at any point was likely to collapse into rout.

A more segmented line, where each unit was physically separated from the ones to either side of it, had greater capacity to resist such a catastrophic rupture, especially if the soldiers knew that their side had a reserve line, with units that could be individually sent forward to respond to breakthroughs in the first line. Already by the time of the Wars of Religion, French infantry was often arrayed in a formation similar to that used by the Romans' manipular legions, with battalions of pike arranged in two or three staggered lines, in which each formation was separated from the next by a space equal to its own frontage, and where each such void space was faced by a unit in the next line back.[77] This was at that time more the exception than the rule; the Spanish and their many emulators preferred more solid fronts in single lines (with smaller reserves behind). But even those who argued that the use of many small battalions separated by intervals made for a weaker main battle-line recognized that a Roman-style checkerboard arrangement was suitable for the advanced shot companies of the forlorn hope, and tactics of this sort became more common over time. There were many variations on the theme, but the general trend in the early modern period was increasingly to view the battle-line not as a single entity, or as the combination of just three elements (left, center, and right), but rather as a larger team of cooperating but essentially *tactically independent* troops, battalions, or brigades. Whether these smaller units were separated by broad intervals or only by narrow gaps, the battle-front was no longer a solid, contiguous front, a difference that would have been striking to any medieval commander viewing an early-modern army's deployment. The implications for tactics, and for the nature of officership, were enormous.

Intervals Between Units: Infantry

In the 1590s, Barret described the advantage of dividing battalions of pike from one another by broad intervals (equal to the frontage of the units) as follows:

> The voide spaces may serve for the troupes of shot to sallie out [to] skirmish with the enemy, and to retire againe, and also for ye battallions of the second front, to march up and pass betwixt them, [as] the battallions of the first front having encountered the enemy, and feeling themselves distressed, are warily and orderly to retire with their faces and weapon point bent upon the enemie [and similarly, eventually, with a third line, advancing on the flanks]. By which order it should seeme, fortune [would have] to abandon them thrice before that they should be quighte vanquished.[78]

77 John A. Lynn, "Tactical Evolution in the French Army, 1560–1660," *French Historical Studies* 14 (1985), 178–9.
78 Barret, 85.

In this passage we see identified the two main advantages of tactics based on independent small units: the ability to employ maneuver, including rotating fresh troops in to replace worn-down forces, and the possibility of suffering a partial defeat without losing the ability to withdraw gracefully from combat. When a single battalion collapsed, the presence of two friendly formations immediately behind and to each side of it made it practically impossible for the victorious unit opposing it to pursue, and the interval to its rear allowed it to escape without disordering the troops behind it. Moreover, the separation of the units made it much more likely that panic could be contained to that one element, without infecting its neighbors. They would of course be disadvantaged, but if necessary could withdraw in good order as their supports came up. As Montecuccoli put it, in praise of such a deployment, "If a smaller troop is overturned, the disorder is not so great; nor is it as difficult to repair the breach which is not as wide."[79]

A unit operating as an independent battalion in this manner was more likely to break than one that was part of a contiguous front, since its flanks were more vulnerable and since the individual soldiers had better prospects of escape (for the reasons just noted), thus reducing their motivation to stand fast. But, to repeat, the consequences of its breaking would be much less.

This meant that the role of chance in determining the outcome of the battle was reduced. Moreover, the switch from a few large, linked combat elements to many smaller, independent ones gave the wing- or army-commander much more of a role in shaping the battle. He had, as Michael Roberts put it, more "tactical small change,"[80] allowing him more opportunities to intervene effectively over the course of the combat. He had a whole hand of cards to play, rather than staking his fate on one roll of the dice. For Montecuccoli, this was a point of crucial importance:

> the science of being a general consists . . . [in] understanding how to make [troops] fight in an opportune fashion, i.e., successively and not simultaneously. The military leader must bear in mind that that his troops will not obey him well after they have been called into combat; all the entreaties in the world will not make them stand firm once they have begun to flee. Conversely, troops in good order will execute a command properly, will affront the enemy, and will force him to desist from pursuit so that their fugitive comrades will have both time and space to pull themselves together.[81]

Another advantage of dividing an army into small units – especially if they had empty space on each side of them – was their greater ability to deal with difficult ground when on the tactical offensive. A battalion could swing to the side and

---

79 Montecuccoli, *Sulle Battaglia*, 90.
80 Michael Roberts, "Gustav Adolf and the Art of War," *Essays in Swedish History* (London: Weidenfeld and Nicholson, 1967), 60.
81 Ibid., 84.

pass around a farmhouse or bush in its path; by contrast, "large bodies of men are inconvenient because they require a very level battlefield; without the latter they cannot maintain good order."[82]

Nonetheless, Barret in 1598 recommended to his contemporaries that they continue to draw up their main forces in one or a few large squares – as they usually did – rather than many smaller units. He recognized the superiority of the latter arrangement in theory, for soldiers who were "perfect and ready," but he thought that in practice it would not work out well, since the soldiers of his day were not as disciplined as the ancient Romans. It is "dangerous to frame many batallions," he concluded, "except men be very skilfull and well practised therein, by reason of the difficultie in seconding one another."[83] Hence, it was better to rely on the square, which "common reason and experience hath made most men confesse and agree . . . is the most assured, most strong, and most apt to bee reduced into any other forme."[84]

What Barret had in mind here was the ad-hoc dividing of a large army into separate fighting battalions. But even as he wrote, armies were coming more and more to be organized with permanent battalions of around 500 men (pike and shot) that drilled together more frequently, and that in battle fought in intervaled array. Erik XIV of Sweden and Maurice of Orange had led the way in this change, and Gustavus Adolphus dramatically demonstrated the potential value of the new structures.[85] True, they required a larger complement of officers – but officers were, as Montecuccoli wrote, worth at least what they cost, expensive though they were.[86] Officers had always been important for maintaining morale and control, on the battlefield and off, but as units became more capable of independent action at a smaller level – and not just independent action, but complicated combined-arms action, requiring coordination at least of pike and shot, and perhaps of fully and light-armed pike, musketeers and arquebusiers, and halbardiers or billmen – they became to a much greater extent decision-makers, as well as retaining their importance for supervision and inspiration.

As already noted, the use of more numerous independent units compartmentalized by intervals served to increase greatly the chance that a breach in a battle line could be repaired without causing a broader collapse. This tended to make battles last longer, a tendency already inherent in the use of slow-firing shot and deep, tough blocks of pike. This in turn had an unintended consequence of great import for

---

82 Ibid., 90

83 Barret, 75, 82: multiple small units ("battles") could be used "with great reason and advantage, were men experte and skilfull. For as it is venturous to set ones fortune upon the brunt of one sole battell, so it is dangerous to frame many batallions, except men be very skilfull and well practised therein, by reason of the difficultie in seconding one another." Cf. Sutcliffe, 175.

84 Barret, 45; see also 76–7, 94–5. Bérault Stuart, in his 1508 military treatise, gives no hint of the idea of intervals, and advises against intermixing cavalry and infantry. *Traité sur l'art de la Guerre*, ed. Élie de Comminges (La Haye: Martinus Nijhoff, 1976), 7.

85 Roberts, *Gustavus*, 192–4, 182–185, 219–221, 247–51.

86 Montecuccoli, *Mémoires* (Paris: Muzier, 1712), 37.

soldiers and commanders alike: there was a marked tendency for battlefield casualties to be more evenly distributed between winners and losers than in the Middle Ages. In the earlier period, much of the fighting was done by well-armored combatants who were very tough to kill, so long as they stayed in their formations. When one side broke, the slaughter could be horrendous, but it was nearly all on one side. In an early-modern battle involving many units, it was typical for each side to suffer localized, partial defeats before the overall outcome of the combat was decided, and so to suffer significant losses. Even if this did not occur, the attritional nature of exchanges of arquebus fire meant many killed on both sides before the balance of victory tipped in one direction or the other.[87] A main impetus for the creation of battle arrays with mutually supporting lines was to ensure that a battle would not be "lost in the twinkling of an eye at the first joining";[88] thus, these tactics extended the period of mutual loss before the one-sided pursuit began. The depth of the pike-square had a similar effect. When pikemen fighting in the Swiss style clashed, it typically meant a relatively large number of killed on both sides, a phenomenon already observable in battles in which handguns played little role.[89]

The combined effect of these factors meant a major change. There are plenty of medieval battles in which the winner's losses amounted only to dozens of killed – or even less. The ratio of victors:defeated killed might be as low as 1:50 or even 1:100.[90] In the early modern period, the victor's dead were almost certain to number in the hundreds, if not the thousands, and the ratio of killed was much more likely to be on the order of 1:2 to 1:6. Indeed, an army that went on the offensive, in the somewhat unlikely event that it won a battle, might even suffer *more* killed than the "loser" of the engagement.[91] "Many of these sanguinary triumphs," as Montecuccoli rightly observed, "are simply defeats in disguise."[92]

For the strategist, the main significance of this trend was that it further reduced the already limited ability of a battlefield victory to bring decisive strategic or even operational results. Although that is an important effect, it is beyond the scope of this essay. But this was also of great significance to the individual soldier. The risks he faced in battle were significantly changed, and so therefore were the demands placed on his courage. Before, an armored soldier could go into combat with a reasonably high confidence that he would survive, at least if his side won the day. But in the new military era, that was unrealistic. Humphrey Barwick recognized how things had changed: "now by reason of the force of weapons, neither horse nor man is able to beare armours sufficient to defend their bodies from death, wheras in

---

87 Montecuccoli, *Sulle Battaglia*, 162: "during the fray, there are just as many victims on one side as there are on another."
88 Webb, 47; cf. Mendoza, 107.
89 Winkler, 66–8, 45, 42–3, 92, 96; Barret, 85.
90 Rogers, *Soldiers' Lives*, 214–6, 229–30.
91 E.g. Nieuwpoort: about 3–4,000 Spanish vs. about 2,500 English and Dutch: Puype, 106, 106n. Ravenna, approximately 4,000 vs. 9,000: Oman, 147–8. Marignano, around 5,000–6,000 on each side: ibid., 170.
92 Montecuccoli, *Sulle Battaglia*, 162.

the former times . . . woundes was the worst to have been doubted [i.e. feared]."[93] The soldier's odds of survival were worse, and the moral impact of this fact was redoubled by the fact that death was not only more likely, but also more random. An enemy's sword or lance could be parried; even an arrow or javelin could be spotted on its way in and dodged or blocked. But there was nothing the soldier could do to escape or deflect a musket-ball or cannonball. When engaged in hand-to-hand combat, a soldier's attention is typically absorbed by the immediate needs of physical action in response to visible threats. But, as Montecuccoli observed in another context, soldiers can "become greatly agitated [when] they recognize that the question of their escape or destruction will be decided by the *virtù* of others."[94] This situation – the requirement for a soldier to overcome the particular sort of fear tinted by the knowledge that one's fate was not in one's own hands – was the norm on the early modern battlefield, where "every moment you felt cannon balls and arquebus shot whistling past your ears, and every hour you saw the ground sown with the minute pieces of the shattered bodies of your companions."[95]

Tactical formations and drills are designed – whether through the conscious decisions of military leaders, or through the gradual molding of the abstract forces of survival-of-the-fittest mechanisms – to support soldiers in facing moral challenges, as well as to facilitate their application of physical force to the enemy. The deeper formations and dance-like[96] movement-in-unison characteristic of early-modern combat were developed in part in response to the moral challenges of the new risk-environment. This was explicitly recognized by some early-modern observers, but particularly in the context of cavalry rather than infantry formations. This essay has so far focused on how new formations and weapons affected infantry combat, but horsemen too had their way of fighting transformed and transformed again in this period, in similar ways and for similar reasons. And the cavalry of the sixteenth century remained of very great importance on the battlefield despite (indeed, partly because of) the rise of firearms. A good case could be made – though most contemporary observers held the opposite opinion – that the cavalry in the sixteenth century returned to its pre-fourteenth-century position as the most important arm in determining victory or defeat.[97] In early modern battles, if one side had clearly greater strength in infantry, and the other had a strong advantage in horsemen, the latter was more likely to win the fight overall, as at the battles of Ravenna, Ceresole, and St. Quentin.[98]

---

93 Barwick 2b v.; note also 10v; Tavannes, 191.

94 *Sulle Battaglia*, 154.

95 Collado, quoted Eltis, 31.

96 Digges, 225.

97 Compare Eltis, 24; Mark Fissell, *English Warfare, 1511–1642* (London: Routledge, 2001), 233–5; Audley, 131; and Montecuccoli, *Sulle Battaglia*, 151, with Mendoza, 54; Thomas Styward, *Pathwaie to Martiall Discipline* (London: Milo Jennings, 1582), 157; Garrard, 139.

98 Oman, 241–2. Also arguably Agnadello (1590) Ibid., 93. The reverse was still true at Guinegate in 1479, where the French cavalry won its fight, but the French infantry were defeated and lost the battle as a whole.

Cavalry Tactics

The sources are usually laconic when it comes to the details of medieval cavalry tactics, but it seems that horsemen usually fought in thin linear formations just one or two ranks deep. They sometimes employed open order (with each horse separated by an empty space roughly equal to its own width), sometimes closed order (stirrup to stirrup). When two opposing formations were both in open order, they would normally pass through each other when charging, each man attacking his opposite number in passing. When in closed order, if neither side shied away from contact then both would at the last moment slow to a walk, since otherwise the result would be an ugly-pile up. In this case the units would fight it out with hand-strokes until one or the other was penetrated and broke.[99]

If attacking infantry, cavalry sometimes used a tactic similar to the early-modern maneuver usually known as the caracole,[100] riding up in open order at a gallop, throwing javelins (which would carry farther and hit harder because the velocity of the projectile would be enhanced by the speed of the horse), and then wheeling back before coming into range of the footmen's weapons. Alternately, the cavalry might try simply to ride down the infantry. In this case the horsemen would normally advance in close order at the trot, in a small unit, while other units waited to follow up on the first unit's attack if it proved successful, or to charge any footmen foolish enough to break formation in pursuit if it did not. If the charging horsemen detected wavering in their target (as they likely would unless the footsoldiers were of good quality), they would charge home and try to shatter the defending formation, for "upon any smal disorder, they carrie with them victory."[101] If there was no sign of such an opportunity, they might try charging in anyway, a maneuver which required excellent horsemanship and very well-trained horses, and was very dangerous for themselves and even more so for their horses. Such a shock action could potentially smash a hole in the enemy line for the next troop to exploit, though particularly against footmen armed with pike it was more likely to be forced to "retire with losse and disadvantage."[102]

By 1480s, cavalry was sometimes being formed in much deeper formations analogous to pike squares.[103] It may be that at first these columnar arrays were not for fighting in, but rather deployed with the intent that individual ranks would step off from the front, charge, then wheel and retire to the rear of the formation as the next rank charged

---

99 Rogers, *Soldiers' Lives*, 191–6.
100 Though that word in the early modern sources often refers to various other maneuvers.
101 Garrard, 228–9.
102 Barret, 76, 139; Sutcliffe, 182–3, 163; Styward, 157. On the possibility of cavalry shock against infantry formations, see Rogers, *Soldiers' Lives*, 183–4.
103 Seldeneck, 107–9.

in turn.[104] As is explicitly recognized in the early modern treatises, only one or at most two or three ranks of lancers could effectively be employed simultaneously.[105]

The introduction of the wheel-lock pistol was the main influence on the drastic changes in cavalry tactics that took place in the sixteenth century. The pistol had three basic advantages over the lance which, between the 1560s and the 1590s, caused it to become the principal arm of the horseman. First was its penetrative capacity. De la Noue considered it a "miracle" if anyone in full sixteenth-century plate armor was killed by a lance, and although this is somewhat of an exaggeration (a charging horse put a great deal of kinetic energy and a massive amount of momentum behind a lance-point) it is true that we read of far fewer prominent individuals killed by cold steel than by bullets or cannonballs.[106] Tavannes, indeed, considered that firearms had increased the risk of death for military leaders a hundredfold.[107] Second, although a pistol was effective only at very short ranges, even if the horseman reserved fire until "he could see the white in the eyes of the enemy," he would have the advantage in range over a lancer.[108] Third, and perhaps most important, the pistol could be used effectively by horsemen attacking in a deep column,[109] who had many of the same advantages relative to thinner formations as did pike-squares. When a column of pistoleers passed through a line of horse, the *Reiters* could fire their pistols to the side, in such a way that they enfiladed the enemy line. And, if a mêlée ensued, their pistols and secondary weapons (swords) were generally preferable to lances, which were unwieldy and so not very effective unless employed in the charge.[110] Because of these three advantages, experience proved that pistol-armed German *Reiters*, and those who emulated them with sufficient skill, had a clear-cut advantage over lancers[111] – even aside from the facts that the pistoleers took less training and did not require such expensive mounts.[112] As Montecuccoli explained it (paraphrasing the much earlier de la Noue), it might be true that

104 As much later described in Cruso, 97.

105 Cruso, 30; Montecuccoli, *Sulle Battaglia*, 106.

106 La Noue, 201, 199; similarly Tavannes, 192; but cf. Williams, *Discourse*, 38–9.

107 Tavannes, 336, 191–2; note also Barwick, 21v.

108 Puype, 85.

109 Eltis, 63; Monluc, *Memoirs*, 3:484–5; la Noue, 201–2.

110 Montecuccoli, *Sulle Battaglia*, 91; Tavannes, 192; De Vere, quoted Puype, 86n.

111 Ronald S. Love, "'All the King's Horsemen:' The Equestrian Army of Henri IV, 1585–1598," *Sixteenth Century Journal*, 22 (1991), 514–20, 531–2. This superiority was of course dependent on circumstances; Spanish lancers at Mook in 1574 ripped up a formation of pistoleers when they caught them reloading, and some commentators thought them generally superior: Puype, 85; Mendoza, 51, 54–5; cf. Williams, *Discourse*, 37–9.

112 Puype, 82: "Maurits abolished the lancers in 1597 and changed them into 'pistoleers', or cuirassiers armed with pistols, since the training of a lancer took at least three years and because the stallions they used as mounts were too expensive." Cruso, 30: Pistoleers were introduced "when the Lanciers proved hard to be gotten, first, by reason of their horses, which must be very good, and exceedingly well exercised . . . and principally, because of the scarcitie of such as were practised and exercised to use the lance, it being a thing of much labour and industry to learn." Tavannes, 191.

only the first rank of a column could use its weapons to full advantage, but in arraying a troop of horse

> one does not pay attention merely to the fact that each reiter fires one pistol shot and delivers a thrust of his sword. Rather, one should take care that the formation be able to rout whatever it confronts. This is the sole objective of whoever commits himself to battle, and it is accomplished much more readily when cavalry is disposed in [deep] squadrons [rather than in line].
>
> To be sure, one might argue in reply that squadrons cannot overturn more than fifteen or sixteen horses of a troop which is fighting [in a single line]. While this is true, the breach will have been effected in the sector of the line where the flag is located and where the captain and better soldiers are posted. Even if that portion of the [line] which has not yet clicked its hoofs were to rush the flanks of the assailant, the squadrons that supported the latter would block the maneuver. . . . If three or four troops thus arrayed [in line] were to attack successively, a sole opposing squadron could overturn them as easily as a bowling ball can be used to knock down one row of pins after another.[113]

Montecuccoli recognized another key advantage for deep squadrons, a moral rather than physical one:

> One may add that if several ranks are formed into a squadron, it is probable that the troopers who are stationed in front will be choice material and that those of the next row will second them in valor . . . As for the remainder, whom we may regard as considerably more hesitant, they will be protected by those posted in front . . . The valor of those in front will cause them to move ahead, and the assurance of security should make the others follow. Conversely, when a troop is formed [in line], even though the brave soldiers (who are normally in the minority) proceed resolutely to the fray, the others (who have no desire to take the bit into their mouth) remain behind. And so, over a distance of 200 paces, one sees this long rank thin out and dissolve.[114]

When formations of fully armored lancers clashed, most soldiers derived from their armor the "assurance of security" that emboldened them to stick with their comrades in the assault. But, as we have already remarked in our discussion of infantry combat, in the age of firearms "neither horse nor man [was] able to beare

---

113 Montecuccoli, *Sulle Battaglia*, 91–2; la Noue, 186–7.
114 Montecuccoli, *Sulle Battaglia*, 91–2 (paraphrasing la Noue); see also 110 and la Noue 198–203, 188: less brave horsemen may feign a nosebleed or a broken stirrup or lost horseshoe, so that after 200 paces of advance "we shall see glasse windowes in that long file, & great breaches wil appeare therein . . . and many times among an hundred horse, scarce 25 dc enter in" to melee.

armours sufficient to defend their bodies from death"; no longer could one's own skill shield against the random strike of a lead ball; no longer were "woundes . . . the worst to have been doubted."[115] Hence, a new premium was put on the use of a tactical formation that did not rely on every man's courage being sufficient to bear the risk of "present death,"[116] but rather helped to ameliorate the problem that "large pistols have made [armor] useless, and made the melee so perilous that each person is eager to leave the fight."[117]

Moreover, the pistol gave the horseman an effective way to attack a block of pike, provided that the latter had not been properly provided with attached shot, or after the latter had been detached or driven off. This was done through the caracole, the firearmed equivalent of the old javelin tactics, with ranks of horsemen riding to just outside the striking-range of pikemen, shooting their pistols, and retiring to reload. Although this tactic has been much disparaged by modern historians, it was by no means scorned by the footmen who had to face it. As already noted earlier, Digges reckoned that one key role of arquebusiers was "to . . . succour their Pykes, whensoever any attempt shalbe made by Argoletiers or Pistols to breake their array."[118] The English cavalry at Turnhout in 1597 demonstrated what could be accomplished by horsemen using the caracole to create an opportunity for shock action: "we charged their pikes," reported Sir Francis Vere, "not breaking through them at the first push, as was anciently used by the men-at-arms with their barded horses: but as the long pistols, delivered at hand [i.e. fired at very short range], had made the ranks thin, so thereupon, the rest of the horse got within them."[119] Even if faced with a pike-square too determined and well-armored to be shaken by this technique, the cavalrymen could use it to gradually wear down their enemies while simultaneously pinning them in place, or at least severely hindering their movement, through the constant threat of a charge with cold steel.

When forming up a wing of horsemen arrayed in deep squadrons, the commander essentially had no choice but to make use of intervals at least equal in width to the frontages of the units. In addition to the general advantages of having the line broken into individual units already discussed earlier, the intervals were necessary to allow the squadrons to employ the fire-and-retire technique of the caracole, and also allowed the squadrons of the second line the space they needed to protect those of the first line from threats to their flanks. Moreover, the empty spaces allowed room for other sorts of troops (particularly artillerymen or handgunners) to maneuver to aid the cavaliers, and vice versa.

115 Barwick, 2b v.
116 Barret, 3.
117 Tavannes, 191; see also la Noue, 202; Mendoza, 107. See also Barwick, 21v, for the names of prominent, well-armored noblemen killed with pistols at St. Quentin.
118 Digges, 110.
119 Quoted Fissel, *English Warfare*, fig. 7.3; here modernized.

Combined Arms

The caracole was especially effective if employed in combination with field artillery: an immobile pike-square was a big target, and each cannonball had the possibility of opening a lane down which the riders could charge into the midst of the pike. If they got past the hedge of spearpoints, they would be able to slaughter the footmen like wolves inside a flock of sheep. Much the same could be said of the combination of *Reiters* and musketeers, whether used against a pike square or against an all-cavalry force. When all three arms were employed together, the results could be even better. To quote the insightful veteran commander Raimondo Montecuccoli again:

> Musketry and cavalry should be combined. The former makes the latter bolder. If the enemy is stronger in horsed troops, the musketry can establish an equilibrium; if he is weaker in this arm, the footmen will be able to rout him. It is incredible how much damage can be wreaked, how great a gap can be torn when musketry fires properly and when lances and cuirasses [*Reiters*] charge suddenly from behind this hail of bullets. It is quite certain that when a squadron sees nine or ten men felled from its front rank, the remainder thinks only of its own fate. Another favorable factor with regard to such musketry is that small cannon may be placed in its midst: this makes for a great slaughter of enemy horses.[120]

Or, he might have added, of enemy pikemen. To successfully coordinate horsemen, pike, musketeers, and cannon, however, was by no means easy. At Breitenfeld, Gustavus Adolphus illustrated just how much could be achieved by these means, but it remained profoundly difficult to accomplish, whether at the small-unit level or for an army commander.

Complexity

At Agincourt, the English had only two types of soldiers: archers and dismounted men-at-arms. Although the French had some other troops on the field, the only ones who played a significant role were men-at-arms (mostly dismounted) and a relatively small number of *gros valets* who were in tactical terms nearly equivalent to dismounted men-at-arms. The English army was divided into only five main units (two wings of archers and three nearly joined battles of men-at-arms) all arrayed in line, with the wings angled forward. The French, similarly, had two wings of cavalry, a body of crossbowmen and other shot, two lines of footmen, and a mounted reserve composed of *gros valets* and a few men-at-arms.

By contrast, the infantry of an army of the sixteenth century would typically include around six types of footmen (armored pikemen; part- or un-armored pikemen,

---

120 Montecuccoli, *Sulle Battaglia*, 106.

halbardiers or billmen; swordsmen with shields; arquebusiers; musketeers; possibly archers or grenadiers or men with musketoons or calivers)[121] and up to six types of horsemen (lancers or gendarmes, cuirassiers with pistols and swords, demi-cuirassed carabineers, dragoons with pikes, dragoons with muskets, and perhaps demi-lances or stradiots or other eastern-European light cavalry). In addition, field artillery, which only began to be of real significance by the mid-fifteenth century, had by the mid-sixteenth begun to play an integral and normal role as a major combat arm.[122]

All these men would be organized into far more independent tactical units: dozens of squadrons of *Reiters*, companies of lancers, battalions of foot, and units of shot detached to support the cavalry. Rather than five or six or nine the total number might be as many as around a hundred and fifty, as for example in the battle-array of the prince of Orange before Rees in 1614.[123] As noted earlier, each of these small units was truly independent: it was formed up with a substantial interval between it and the unit to its side, it could be committed to action and maneuver independently, and it could even break without automatically causing a disaster for its side. Moreover, even the individual battalions of foot were really composed of at least three distinct, articulated sections: a central body of pikemen and two wings or "sleeves" of shot. And even these were not simple, uniform bodies. The pike-block had to be formed with the fully armored men around the perimeter, the light-armed pike closer to the center, and the ensign, musicians, and halbardiers or billmen in the center. The complexity of the machine, with all its permutations and combinations, had increased a hundredfold. As a result, to quote Garrard, "there nowadays be over-many" rules and reckonings of military discipline.[124] The growth of tactical complexity had been stimulated by but was not solely due to the introduction of firearms.

Epilogue: From Complexity to Simplicity, 1670–1740

The combined-arms tactics based on intervaled lines and independently acting small units that the Swedes employed with such finesse in the 1630s were the norm by the later seventeenth century – but before the middle of the eighteenth century they had largely vanished, to be replaced with the simpler modes of "linear tactics." This change was not the result of a decline in the technical proficiency of European troops: on the contrary, mid-eighteenth-century units, now marching to a cadenced pace, were capable of executing tactical deployments with a precision that would have amazed Gustavus Adolphus.

---

121 E.g. Garrard, 57, figures that within a band there will be "targets of proof" (men with heavy shields), musketeers, arquebusiers, armed pike, light-armed pike, and halberds or bills.

122 Cf. Lynn, "Evolution," 184–6; Sutcliffe, 191; Mendoza, 107–9; Roberts, *Gustavus*, 228–234. By Breitenfeld (1632), many Imperial regiments were "completely shattered by bombardment before they were able to enter action." Montecuccoli, *Sulle Battaglia*, 140.

123 Cruso, fig. 12, par. 4, cap. 8.

124 Garrard, 57.

Probably the single most important factor in driving tactical change over the turn of the eighteenth century was the near-universal re-arming of European infantry with bayonet-equipped flintlocks (or "fusils").[125] Eliminating the slowmatch and the fork (and using paper cartridges) simplified the manual of arms substantially, allowing a trained soldier to fire two to three times as quickly, with fewer misfires.[126] Increased firepower, in combination with the bayonet, allowed fusiliers to defend their own fronts effectively: "if the infantry knows its own strength," commented Marshal Puységur, "cavalry will not break it."[127] The same troops could also now push forward and close with the enemy, as formerly pikemen had done. The proportion of pike had already been in decline, falling to under a third of the infantry by the 1650s, and a quarter by around 1680.[128] In 1687 Vauban noted that musketeers used their fire in every engagement, whereas pikemen might cross weapons with their adversaries in only three fights out of twenty.[129] The adoption of an effective socket bayonet finally convinced the conservatives that the pike had become obsolete.[130]

The rearming of the pikemen meant a given number of soldiers now had a third to a half more shooters, each firing two or three times as fast, thus in total putting three or four times the amount of lead in the air. If cavalry charged the line, "the Fire of one Platoon, given in due time, [was] sufficient to break any Squadron"; the bayonets behind the curtain of fire served mainly to ensure that the individual horsemen into which the squadron fragmented could not in turn break the battalion.[131]

When, on the other hand, a battalion of fusiliers advanced to drive off an enemy, it was a "receiv'd Maxim, that those who preserve their fire longest, will be sure to Conquer."[132] Typically, the side that fired first might give its discharge at around 60 paces. A volley even at that distance, however, was "less dreadful, and fewer lives are lost by it, than is generally imagined."[133] Marshal Saxe argued that not even the most efficacious blast of short-range fire could suffice to prevent the

125 Jeremy Black, *European Warfare, 1494–1660* (London: Routledge 2002), 39; Jacques François de Chastenet de Puységur, *Art de la guerre* (Paris: Charles-Antoine Jombert, 1743), 57.

126 Puységur, 71 and 67; John Lynn, *Giant of the Grand Siècle* (Cambridge: Cambridge U.P., 1997), 460; Brent Nosworthy, *The Anatomy of Victory: Battle Tactics 1689–1763* (New York: Hippocrene Books, 1990), 39–41.

127 Puységur, 70–71; cf. Humphrey Bland, *A Treatise of Military Discipline*, 2d. ed. (London: Samuel Buckley, 1727), 91.

128 Puységur, 3, 36, 56; Lynn, *Giant*, 476.

129 Quoted Lynn, *Giant*, 457.

130 Though cf. Maurice, Comte de Saxe, *Reveries, or Memoirs upon the Art of War* (London: J. Nourse, 1757), 32–3.

131 Bland, 91, 80; Puységur, 70–71.

132 Bland, 91; Maurice, Comte de Saxe, *Mes rêveries*, vol. 1 (Paris: Desaint et Saillant, 1757), 134; Charles Sévin, marquis de Quincy, *Maximes et instructions sur l'art militaire*, vol. VII, part 2 of *Histoire militaire du règne de Louis le Grand* (Paris: Denis Mariette, 1726), 67; Nosworthy, 104 (but cf. 108).

133 Saxe, *Reveries*, 20; cf. *Mes rêveries*, 40; but also cf. Christopher Duffy, *The Military Experience in the Age of Reason* (New York: Routledge & Kegan Paul, 1987), 204–6.

unit on the receiving end from stepping through the smoke, giving its own volley at much closer range, and then charging home with the bayonet. "It is by this method only," he concluded, "that numbers are to be destroyed, and victories obtained."[134] Saxe's emphasis on the decisiveness of the final charge led him to write of "the error of shooting," and to argue for a return to deeper formations including pikemen, as more suited to shock action – but his opinion did not prevail.[135] First, though it could be very effective to march through the enemy's fire without shooting back, it was very difficult to accomplish: in practice – whether the officers wished it or not – many advances devolved into firefights at relatively long range.[136] Second, armies often defended their fronts with entrenchments or natural obstacles. In such circumstances charges were often repulsed, in which case the casualties inflicted by the defenders' firepower (both during the attack and during the retreat) were of primary importance.[137] Third, when bayonet charges *were* successful, it was more by psychological than physical force. When the defending line delivered a close-range volley, the soldiers naturally expected their fire to have a devastating effect, and hoped that it would stop the attackers in their tracks. If not, the disappointment of these expectations naturally shook the defenders' morale, and the subsequent bayonet charge often, even usually, caused them to break and run before a single bayonet was bloodied.[138] But in this case what really led to panic among the defenders was seeing "their enemy advancing upon them through the smoke *with his fire reserved*;"[139] "it [is] certain, that when Troops see others Advance, and *going to pour in their Fire upon them, when theirs is gone*, they will immediately give way, or at least it happens seldom otherwise."[140] Thus the real *cause* of the rout might be more the fear of suffering a blast at spitting distance than unwillingness to exchange blows with the bayonet.

The importance of the infantryman's increased firepower put a premium on formations that maximized the number of soldiers positioned where they could shoot effectively. This favored thinner formations – 3–4 ranks deep – in solid, contiguous lines. Moreover, the increased speed of reloading made the countermarch unnecessary. This eliminated the need for open lanes between files and allowed infantrymen to form up much more densely, with one fusilier per 2' of battalion frontage, instead of one musketeer per 5'.[141] The empty space required to gain the maneuver-

---

134 Saxe, *Reveries*, 20; cf. *Mes rêveries*, 40.

135 Saxe, *Mes rêveries*, 38 ("l'abus de la tirerie"); cf. *Reveries*, 19 ("insignificancy of small-arms"); see also Duffy, 205–14.

136 Duffy, 207.

137 Saxe, *Reveries*, 71; *Mes rêveries*, 134–5.

138 Saxe, *Reveries*, 19, 71; *Mes rêveries*, 37–8, 134; Nosworthy, 111.

139 Saxe, *Reveries*, 19, emphasis added; the italicized text is not in the French edition (*Mes*, I, 37–8). Saxe certainly did, nonetheless, consider the retained potential of the charging soldiers to deliver shots at range close enough to burn the target's coat with the muzzle-flash ["à brûle-purpoint"] to be a key part of the threat they posed. *Mes rêveries*, 40.

140 Bland, 134, emphasis added; similarly, Nosworthy, 104 (Catinat) and Quincy, 67.

141 Puységur, 62–4; 36.

ability advantages of an intervaled line came to seem too costly in lost firepower to be justified, especially since new drills had been developed that allowed for battalions to maneuver effectively without needing empty space to turn through.[142]

The main reason given at the time for eliminating intervals between battalions, however, was that a unit not linked to friendly units on each side was vulnerable to enemy attacks on the flank, particularly by cavalry. To prevent the gaps that could create such vulnerabilities, "the whole Line must act like one Battalion, both in Advancing, Attacking, and Pursuing the Enemy together."[143] Such uniform action also much simplified command and control, obviating the need to send orders separately to each battalion.[144] Moreover, with the highly disciplined troops of the eighteenth century, the buffer against panic that intervals created was considered less necessary – especially since allowing a victorious battalion to push its advantage would expose its own flanks to attack, particularly by the units of cavalry typically held ready for just that purpose.[145]

With intervals in the infantry line minimized or eliminated, the prospects for combined-arms cooperation at the battalion level were much reduced. Moreover, the practical experience of battle in the later seventeenth century showed that the outcome of the engagement as a whole was often decided by a clash of saber-charging horsemen on the armies' flanks.[146] For both these reasons, it became the norm (though not the universal practice) to push most of the horsemen to the wings of the army, thus reinstating an overall formation that had been common in the middle ages and antiquity, with two central lines of contiguous units of footmen, flanked by bodies of horse and backed up by a third line or reserve.[147]

Meanwhile, for reasons largely extraneous to tactical developments, armies were becoming much larger. Bigger armies spread more thinly[148] could make it difficult to find suitable battlefields: "It is sometimes impossible," observed Marshal de Saxe, "to find a piece of ground in a whole province, sufficient to contain a hundred thousand men, in order of battle."[149] It took hours to form armies for battle, the duration of the deployment being roughly proportional to the length of the

142 Puységur, 147–8.

143 Puységur, 3; Bland, 134–5.

144 Bland, 139.

145 Bland, 134–5.

146 D'Aurignac wrote in 1663 that "it is the cavalry that ordinarily wins battles.' Quoted Lynn, *Giant*, 489. Note also Henri I, Duc de Rohan, *La parfaict capitaine* (Paris: Jean Houze, 1636), 263; Quincy, 56.

147 *The Art of War, in Four Parts* (London: J. Morphew, 1707), 234; Saxe, *Reveries*, 18, 63; Jean Charles, Chevalier de Folard, *L'esprit de Chevalier Folard* (ed. Frederick II of Prussia) (Leipzig: N.P., 1761), 88; Frederick II of Prussia, *Instructions to His Generals* in *Roots of Strategy*, ed. T. R. Phillips (Harrisburg, PA: Stackpole, 1985), 342; Quincy, 56.

148 John Lynn, "Recalculating French Amy Growth in the *Grand Siècle*, 1610–1715," in Rogers, *Military Revolution Debate*, 117–148; Puységur, 57–8, though cf. 52; Saxe, *Reveries*, 18–19; Montecuccoli, *Sulle*, 90.

149 Saxe, *Reveries*, 76.

formation. The near-elimination of intervals between battalions helped bring the armies back closer into proportion with the landscape and the length of the day.[150]

Since the battle-line was basically just a set of march columns rotated 90 degrees, armies often formed for battle by having the lead units turn to march parallel to the enemy's front, with each unit facing left or right once in position. This could involve a several-mile march and most of a day, which meant that battles (other than meeting engagements) normally took place only when both sides were willing.[151] An army not desiring battle could position itself on broken ground and entrench, thus deterring an attack, or it could simply move off while the enemy conducted his slow preparations for offensive action. When general engagements did take place, they often began only in the afternoon, and lasted until near dusk. For this reason, and because linear formations were ill-suited to rapid movement, and also because desertion was a huge problem for most eighteenth-century armies (making it impractical to allow infantrymen to chase scattering fugitives) an effective pursuit became very difficult to effect. Since the firepower of the flintlock and the increasing number and effectiveness of artillery pieces deployed on eighteenth-century battlefields only increased the advantage of the defense and the costliness of the attrition phase of battle, the problem of Pyrrhic victories grew significantly worse. Frederick the Great, tellingly, noted that the superior discipline and training of his Prussian infantry would have enabled him to conquer "the entire universe, were it not for the fact *that their victories were as fatal to them as to their enemies.*"[152] His "oblique order" attacks aimed to fix, or at least ameliorate, that problem, by holding the majority of his army back from the meat-grinder while the advanced wing won the battle. But with that observation, we pass beyond the chronological scope of this essay.

Looking over the period as a whole, there is no question that, truly, tactics were "much altered since the fierie weapons first came up."[153] However, the reasons for changes in tactics and the soldier's experience of battle, and thus the art of command, were, as we have seen, broader and more complex than simply the invention of firearms.

150 Puységur, 147–8, 123. Frederick the Great in 1747 prescribed that normally intervals for his infantry battalions should be only large enough to allow for the artillery, and that squadrons should be separated only by four paces; but on the other hand, he assumed that non-Prussian cavalry would operate with larger intervals, and that for some tactical purposes attack should be made with battalions in "checkerboard" (i.e. with full intervals) – which Quincy still took as the norm in 1726. *Instructions*, 383–5, 395; Quincy, 57, 105.

151 Nosworthy, 69–77.

152 Frederick II of Prussia, *Les principes généraux de la guerre* in *Oeuvres de Frédéric le Grand*, vol. 28, ed. J. D. E. Preuss (Berlin: Imprimerie Royale, 1856), 8 (emphasis added); cf. Puységur, 6–7.

153 Barret, 2.

Part 2

STRATEGY

7

GIRALDUS CAMBRENSIS, EDWARD I, AND THE CONQUEST OF WALES

In 1063, Harold Godwinson, the heir to the English throne, launched a major invasion of Wales. His forces spread fire and slaughter through the rough Welsh terrain, killing so many men, Gerald of Wales tells us, that he "left not one that pisseth against a wall."[1] Such large-scale campaigns of devastation were typical of medieval warfare, and the result was also typical, at least for the Early and High Middle Ages: the numerous Welsh "princes," who independently governed their own mini-states, "submitted" to the English and acknowledged their over-lordship in a loose way. Harold and his men then went home with their booty (mostly cattle, no doubt), confident they had both weakened the Welsh and taught them a lesson, so that they would make little trouble for years to come. Harold did not, so far as our limited sources indicate, annex any territory, or depose and replace any Welsh ruler, or hold and garrison outposts within Wales.

Having succeeded by the standards of the day, Harold found himself on the defensive in England three years later against two foreign attackers: Norway's Harald Hardrada, then William of Normandy. Against the Norwegians, who fought in the same style as the English, he won a decisive battle, but the Normans defeated him at Hastings. The result was the Norman Conquest of England, an extremely thorough occupation and domination vastly different from the loose subordination Harold had imposed on Wales. A French-speaking aristocracy of knights, barons, and counts supplanted the thegns and ealdormen who had been the principal land-holders and political elite of the Anglo-Saxon realm. French replaced English as the language of the royal court and of law-courts, and as the second language (after Latin) for writing history or literary works. Although William retained some elements of the old system, broad aspects of Norman military and political orga-nization arrived with the conquerors. For more than two centuries thereafter, the

---

1 Gerald of Wales, *Description of Wales*, in *The Journey through Wales and The Description of Wales*, tr. Lewis Thorpe (London, 1978), p. 266. Further citations to the *Description* are to this translation unless noted otherwise.

DOI: 10.4324/9781003399971-10

dominant arm of Anglo-Norman armies was the armored cavalry rather than heavy infantry.[2]

Throughout that period, the Anglo-Norman kings of England fought intermittently with the Welsh, but the patterns of these conflicts resembled Harold's expeditions rather than William the Conqueror's conquest of England. This is in some ways surprising, for in this period the Normans showed an astounding capacity for conquest, with Norman warlords seizing control over substantial portions of Ireland, the southern half of the Italian peninsula, the island of Sicily, and even far-flung Antioch. Anglo-Norman "Marcher" barons, whose English estates bordered on Wales, did make their own conquests within Wales, mainly along the southern coast, where they built towns and brought in English and Flemish settlers to populate the conquests. Yet, William and his successors, with the huge resources of England, Normandy, Maine, and later Aquitaine at their disposal, failed to push beyond the negligible barrier of Offa's Dyke to bring the 200,000 to 300,000 inhabitants of Wales under their direct rule. Not until the wars of 1277 and 1282–1283 did Edward I achieve just that, with consequences for the Welsh comparable to the effect of 1066 on the English.[3]

A principal reason for this is simply that the Welsh, for all their relative poverty and small numbers, were not easy to defeat. William managed to impose fealty and tribute on the Welsh princes by invading the country in 1081, but his son William Rufus accomplished nothing by his costly invasions of 1095 and 1097.[4] Henry I, by spending liberally on his invasions of 1114 and 1121, extracted hostages and submission from the Welsh generally, and imposed heavy indemnities on the princes of Gwynedd and Powys, but left the native rulers in possession of their lands.[5] In the troubled reign of King Stephen, the Welsh largely threw off

2 This assertion is not one that all scholars would accept, because of a fundamental disagreement about the importance of cavalry in medieval warfare generally. For the degree of divergence, contrast J. F. Verbruggen, "The Role of the Cavalry in Medieval Warfare," *Journal of Medieval Military History*, 3 (2005), with Bernard Bachrach, "Verbruggen's 'Cavalry' and the Lyon-Thesis," *Journal of Medieval Military History*, 4 (2006).

3 A Scotsman of the next century referred to Edward I as having "reduced [Wales] to such serfdom." John Barbour, *The Bruce*, ed. A. A. M. Duncan (Edinburgh, 1997), p. 50. Since this chapter is intended mainly for an audience of military historians and strategists rather than medievalists, it generally cites English translations of source material rather than the Latin originals, when such translations exist. It also cites older public domain editions (available on the internet) in preference to more accurate recent editions unless the older edition is problematic at the point referenced. An exception is the text of Gerald's *Description of Wales*, for which I have generally cited the Penguin edition rather than the public domain Forester translation.

4 Frederick C. Suppe, *Military Institutions on the Welsh Marches: Shropshire, A.D. 1066–1300* (Woodbridge, 1994), pp. 13–14; *Anglo-Saxon Chronicle*, s.a. 1097; *The Chronicle of Florence of Worcester*, tr. Thomas Forester (London, 1854), pp. 198, 201 ("he was scarcely able to take or kill one of them, while he lost some of his own troops and many horses"); *History of Gruffydd ap Cynan*, ed. and tr. Arthur Jones (Manchester, 1910), pp. 141–3 ("He did not take with him any kind of profit or gain except one cow. He lost a great part of the knights and esquires and servants and horses and many other possessions.")

5 *History of Gruffydd ap Cynan*, pp. 86–7, 151–3.

even nominal subordination to the English crown.[6] In his first major expedition into Wales, Edward I's great-grandfather Henry II (whom Jordan Fantosme flatteringly described as the greatest conqueror since Charlemagne) restored English over-lordship. However, his second campaign, reportedly aimed at actual conquest, failed completely.[7]

The Offensives of Henry II

In 1157, Prince Owain of Gwynedd, whose principality in north and north-west Wales was one of the three main native-ruled regions, along with Deheubarth in the south and Powys in between, drove out his brothers, occupied their territories, and attacked the English royal castle of Tegeingl in northeast Wales. In response, Henry II collected a substantial army and fleet and advanced into Powys by the Dee valley.[8] About a dozen miles into Welsh territory, Owain blocked Henry's advance with an entrenched position and offered battle, an unusual choice for a Welsh ruler. The English king ordered his main force towards the Welsh lines, but meanwhile led a detachment through the woods in an outflanking maneuver. His men, who seem to have been mostly light troops, stumbled into an ambush led by Owain's sons, resulting in an "extremely sharp fight," in which the Welsh killed the constable (the leading military officer) of the great Marcher earldom of Chester. Henry himself barely escaped.[9]

Quitting while he was ahead, Owain withdrew and returned to traditional Welsh guerilla tactics. Henry turned north, towards the coast, and proceeded to Rhuddlan, where he began the construction of a castle and sent his fleet to attack the large island of Anglesey. Here too the Welsh chose to stand and fight, and here too they won, inflicting a "great slaughter" on the disembarked English, killing

6 The *Acts of King Stephen* notes that Stephen's first effort to restore his lordship in the area after a widespread rebellion involved great expense for horsemen and archers, but nonetheless his paid troops, "after many of their number were slain fighting gloriously, the rest, shrinking to encounter the ferocious enemy, retreated in disgrace after fruitless toil and expense." He then sent a second force which was checked by the Welsh and ultimately "withdrew in poverty and disgrace." Stephen then concluded "that he was struggling in vain, and throwing away his money in attempting to reduce them." *Acts of King Stephen*, in Thomas Forester, trans., and ed., *The Chronicle of Henry of Huntington . . . also, the Acts of Stephen* (London, 1853), pp. 329–33.

7 Jordan Fantosme, *Chronicle of the War between the English and the Scots in 1173 and 1174*, ed. and tr. Francisque Michel, vol. 2 (London, 1840), pp. 112–114. For the following paragraph, see John D. Hosler, "Henry II's Military Campaigns in Wales, 1157 and 1165," in the *Journal of Medieval Military History* 2 (1994), pp. 53–71.

8 This was described by Robert of Torigni as "an extremely large invasion force," including one third of all the knights in England. *The Chronicle of Robert of Torigni*, in Richard Howlett, ed., *Chronicles of the Reigns of Stephen, Henry II, and Richard I*, vol. 4 (London, 1889), p. 193.

9 Hosler, "Military Campaigns," p. 64; quotation from John Williams ab Ithel, ed., *Annales Cambriae* (London, 1860), p. 46 (*acerrimo certamine*).

among others Henry FitzHenry, Henry II's own half-Welsh uncle.[10] If he ever had ambitions of significant conquests, the king abandoned them after this second setback. Instead he accepted Owain's offer of fealty, backed up by the provision of hostages. Owain also restored the lands he had confiscated from his brother, whose call for assistance had prompted the campaign. Henry could thus count the campaign a success, but not an easy, inexpensive, or one-sided victory. Indeed, less than a decade later, apparently in response to Henry's efforts to convert his acknowledged over-lordship into a more strictly defined feudal homage, Owain's son Dafydd led the Welsh in a bid to "throw off the rule of the French." In return, Henry launched another major invasion in 1165. The Welsh, at least, believed that this time the English, frustrated by the pattern of nominal Welsh submission, followed by "rebellion," intended to destroy their nation, or at least drive them from their homeland.[11]

Once again, a Welsh army, relying on a strong defensive position in a wooded valley where Henry could not use his cavalry effectively, blocked the English advance. After a period of stalemate, the invaders tried chopping down the trees to clear their way against the Welsh, but the latter made a brash attack on Henry's men. Welsh sources indicate the fighting was a tactical draw, with heavy casualties on both sides. From a strategic perspective the combat was a clear defeat for the English, who had failed to advance and who took a serious blow to their martial prestige when they failed to beat the Welshmen in an open fight. Subsequent events compounded the failure: Henry turned onto a different route, which took him into even more difficult terrain, where his troops suffered heavily from bad weather and lack of supplies. "Seeing that he could not at all arrange things according to his will," wrote the author of the *Annales Cambriae*, Henry killed or mutilated the Welsh hostages he held, then dismissed his army and "returned in shame into England."[12]

Giraldus Cambrensis (Gerald of Wales) and the Strategic Problem

It was after and with full cognizance of these events, but also with a detailed knowledge of the successful conquest of much of Ireland by his FitzGerald cousins, that the distinguished cleric Gerald of Wales – archdeacon of Brecknock, the

10 Ibid., p. 47 (where FitzHenry is given the name FitzGerald after his step-father, Gerald FitzWalter). FitzHenry's mother was Nest, daughter of the last king of Deheubarth, Rhys ap Tewdwr, and thus he was an uncle of both Gerald of Wales, who was 12 years old at the time of FitzHenry's death, and of Henry II (Henry I's grandson).

11 Hosler, "Military Campaigns," 68n; *Annales Cambriae*, 50 ("planning the overthrow or destruction [*excidium*] of all the Welsh"); R. R. Davies, *Domination and Conquest: The Experience of Ireland, Scotland and Wales, 1100–1300* (Cambridge, 1990), pp. 76–7.

12 *Annales Cambriae*, p. 50. According to Hoveden, Henry blinded the boys and cut off the noses and ears of the girls. Roger of Hoveden, *The Annals*, trans. Henry T. Riley, 2 vols. (London, 1853), p. 278.

youngest son of a powerful Marcher lord and the grandson of a Welsh princess – wrote his famous *Description of Wales* (1194). This text culminates with a chapter entitled "How This Nation May Be Conquered": the earliest prospective strategic document (that is, a written plan for how to win a war, prepared in advance) still extant in any medieval source.[13] Remarkably, considering its originality, Gerald's text is highly sophisticated in its approach.

Gerald was one of the most significant thinkers and writers of a period of remarkable intellectual efflorescence, often referred to as the "Renaissance of the twelfth century." Like most of the other important men of letters of his day, he studied at the undisputed center of advanced education in Latin Christendom, the University of Paris. This was the home of "scholastic" education, which focused on honing students' logical thinking and rhetorical skills through competitive, public debates. In these battles of the mind, it was crucial to see both sides of a question, for it was difficult to dispute an argument one had not anticipated and reflected on. This training was excellent preparation for the study of the law, which Gerald also learned and taught in Paris. It was equally good preparation for the formulation of strategy. Gerald also mastered the discipline of rhetoric, the core of which was the ability to form a coherent, logical argument and express it clearly, so well that he lectured on the subject to packed halls.[14]

Having come from a warlike knightly family, Gerald was both interested in and well informed about military affairs. As the "kinsman of all the princes and great men in Wales," and one of the chief ecclesiastical officers of the region, he had free access to the best sources of information.[15] All of these elements of his background contributed to the high quality of his *Description of Wales*, which historian Robert Bartlett describes not only as "the high point of twelfth-century ethnography," but also as extremely innovative in conception.[16] Bartlett emphasizes the exceptional quality of Gerald's work as a cohesive anthropological study, but he neglects the equally extraordinary and well-conceived nature of the strategic plan contained within the same work.

In devising his plan for the conquest of Wales, Gerald expressly takes into account the lessons of history. He wisely notes that anyone "who is really prudent and provident must find out what pitfalls are to be avoided by taking note of the disasters which have befallen others in the same position. It costs nothing to learn

---

13 Gerald's slightly earlier *The History of the Conquest of Ireland* does include a short prospective chapter entitled "In What Manner Ireland is to Be Completely Conquered," but its contents are almost entirely about what sort of troops should be used, rather than what they should do. In *The Historical Works of Giraldus Cambrensis*, tr. Thomas Forester and Sir Richard Cold Hoare, ed. Thomas Wright (London 1894), pp. 320–22.

14 Gerald of Wales, *The Autobiography of Giraldus Cambrensis*, ed. and tr H. E. Butler (London, 1937), p. 37. He also taught the other two elements of the *trivium*: Latin grammar and, significantly, logic.

15 Ibid., p. 60.

16 Robert Bartlett, *Gerald of Wales, 1146–1223* (Oxford, 1982), pp. 175–82.

from other people's experience."[17] He also draws on his own deep knowledge of Wales and contemporary events elsewhere. He considers English strengths and weaknesses, Welsh advantages and disadvantages, and situational factors. Moreover, Gerald takes into account military, political, cultural, topographical, and economic considerations. Rather than taking force structure as a given, he reflects on the troops best suited to accomplish the mission, and implicitly recommends changes in recruitment, tactics, and equipment. Perhaps most impressively, after describing how to conquer Wales, he adds a second chapter on how to rule it after its conquest. He then concludes with yet another strategic chapter, this one on how the Welsh might best resist English conquest. This he pitches as being for the benefit of the Welsh, a sort of testimony to his own intellectual impartiality.[18] Whether intended so or not, it rounds out his strategic analysis for an English audience, since any competent planner must reflect on and appreciate the enemy's most effective parry and riposte.

Gerald was something of a celebrity; he premiered his *Topography of Ireland* with a three-day public reading at the University of Oxford in 1188, and he was an important figure in the courts of Henry II, Richard I, and John. It is entirely likely that Edward I, whose military career began with fighting in Wales a quarter-century after Gerald's death, was familiar with the *Description of Wales* and indeed studied it with great care. He was far from averse to learning about warfare from books: before his coronation, while on crusade in the Holy Land, he had commissioned the first known translation of Vegetius' *De re militari* into the vernacular.[19] If he did not himself study Gerald's work, it seems his advisors had. In any case, when Edward decided to undertake the complete conquest of Wales, he followed Gerald's strategic prescriptions quite exactly. The king clearly understood the strategic problems involved in a conquest of Wales, as Gerald had described them.

An invader "can never hope," Gerald wrote, "to conquer in one single battle a people [who] will never draw up its forces to engage an enemy army in the field, and will never allow itself to be besieged inside fortified strong-points."[20] The Welsh had no major towns, no fixed economic hubs the English could threaten to

---

17 *Description*, p. 272. Gerald's remark is reminiscent of the *bon mot* attributed to Bismarck, that "fools say that to learn, one must pay with mistakes . . . I, however, have always striven to learn from the mistakes of others." Comte Emile Kérartry, *Le Dernier des Napoléon* (Paris, 1872), p. 240.

18 It would be natural to think this was part of his program of portraying himself as someone favorable to Welsh interests, as part of his campaign to become bishop of St. David's, the chief primate of Wales. However, the timing does not fit that interpretation, because the *Description of Wales* was written at a time when he seems to have been interested only in a rich English bishopric and not a Welsh see. Bartlett, *Gerald*, p. 48.

19 The miniature embellishing the manuscript "shows Vegetius, the philosopher, inviting a group of young knights to come to him with the words . . . 'come to me, lord knights who wish to have the honor of chivalry.'" Christopher Allmand, "The *De re militari* of Vegetius in the Middle Ages and the Renaissance," in Corinne Saunders, Françoise le Saux and Neil Thomas, eds., *Writing War. Medieval Literary Responses to Warfare* (Cambridge, 2004).

20 *Description*, p. 267.

force them to fight. Even if an invader did somehow come to grips with and crush a Welsh army, the Welsh "do not lose heart when things go wrong, and after one defeat they are always ready to fight again."[21] That meant it was practically impossible to beat them quickly. But in a long war too "they are difficult to conquer ... for they are not troubled by hunger or cold, [and] fighting does not seem to tire them." It is not easy to employ a strategy of exhaustion against foes who are fierce, agile, courageous, highly mobile, "passionately devoted to their freedom and to the defense of their country," and who, far from dreading conflict, "in peace ... dream of war, and prepare themselves for battle" by exercising constantly with their weapons and by hunting and mountain-climbing.[22]

Three principal synergistic advantages enjoyed by the Welsh redoubled the inherent difficulty of conquering them. First, their countryside was covered with woods, marshes, and mountains, terrain "where foot-soldiers have the advantage over cavalry" and where the standard Norman battle tactics were "no good at all."[23] Second, they had large numbers of soldiers relative to their small population, because (unlike in England) "the entire nation, both leaders and the common people, are trained in the use of arms."[24] Thanks to constant internecine fighting – like contemporary Irish warfare, focused on cattle-raiding – the Welsh were skilled in ambushes, night attacks, and hit-and-run raids. Thus, on their home ground, they were ideal troops for executing a defensive strategy aimed at defeating an invading army by harassing it, depriving it of forage and profit (both relatively scarce in Wales anyway), and ultimately exhausting it. Third, while the Welsh could hardly have more pressing concerns than dealing with an invasion of their homeland (a cause for which they would "willingly sacrifice, suffer, or die"), England was a power with significant strategic interests in Scotland, Normandy, and Aquitaine, each more important than Wales. Each frequently required English attention and military resources.[25]

The roster of difficulties facing an English strategist planning to conqueror Wales would not have required too much modification to apply to a British commander in the American Revolution, while some of the principal English advantages would fit with analogies to innumerable wars: larger numbers of well-equipped, paid soldiers; general superiority in open battle; and far greater economic and fiscal resources. However, that match-up of advantages against disadvantages did not suffice to guarantee success for the English against Wales, as Gerald recognized.[26]

Gerald did think England's superior wealth and manpower could bring victory, but only if used properly for a sustained effort. A king who wanted to conquer

---

21 Ibid., p. 260.
22 Ibid., pp. 233–4, 236.
23 Ibid., p. 269.
24 *Description*, tr. Forester, p. 490.
25 *Description*, pp. 233, 267.
26 R. R. Davies refers to the war of 1276–77 as a struggle between David and Goliath (*Age of Conquest*, p. 335); of course, the point of that Biblical story is that Goliath does not always win.

Wales, he insisted, "must be determined to apply a diligent and constant attention to this purpose for at least one year" without distractions by other business in England, France, or elsewhere. The king would also have to face heavy losses of men and large expenditures, equal to many times the annual revenues expected from the province.[27] And he would have to follow a sound strategic plan to make the best use of his superior resources.

Gerald's Plan for the Conquest of Wales

Well before the start of an invasion, Gerald suggested, the English should employ economic sanctions to prepare the way: they should prevent the Welsh from bringing in the grain, cloth, and salt (for preserving meat and fish) they usually imported from England. Garrison forces patrolling the land border and ships patrolling the coast should enforce a blockade. Next, an army should invade the coastal lowlands, where the English could take advantage of their naval superiority and receive supplies by sea, forcing the Welsh to take refuge in the highlands of Snowdonia. Third, the invaders should ring this area with infantry, drawn as much as possible from the Anglo-Welsh border zone known as the Marches and equipped with light armor, so that they would be better protected than the normally unarmored Welsh, but agile enough for fighting in difficult terrain. These men should prevent the Welsh from collecting additional supplies. Once winter set in and the trees had lost their leaves, English troops should push into the forests to probe and harass the Welsh. Reinforcements should replace casualties and allow exhausted troops to pull back and recuperate. Thus the English, drawing from their larger population, would keep their forces up to strength, while the Welsh would have difficulty replacing losses.

All this military action, in Gerald's scheme, was to be accompanied by a culturally sensitive political offensive. Three long-standing customs of the Welsh aristocracy, Gerald argued, severely weakened their nation. First, their inheritance customs involved dividing patrimonial lands among brothers, including illegitimate ones, instead of conferring them on the eldest son. Second, Welsh noble youths were generally raised by foster families, with parents sending each of their sons to a different lord to establish bonds between families – which, however, weakened the bonds of affection between brothers who later had to agree on the division of the family lands, or to fight over them. Third, the Welsh princes refused to subordinate themselves to a single king (who, if he had existed, might have resolved many inheritance disputes by law rather than force).

These three factors contributed to constant family feuds over inheritances, in which there were winners and losers; the losers often went looking for someone to help recover what they saw as their rights. This meant there were always opportunities for the English to divide and conquer. Gerald's exact advice was to "stir

---

27 *Description*, p. 270.

the Welsh up to internecine feuds by bribery and by granting away each man's land to someone else." "They will quarrel bitterly with each other," he predicted, "and assassinations will become an everyday occurrence." Although Gerald did not quite say so, he suggested that these bloody internal struggles would not only deprive the Welsh of the unity and strength necessary for effective resistance, they would also leave them disheartened and, in some cases, eager for the return of law and order, even if imposed by a foreign power. "In a short time," he concluded, "they will be forced to surrender."[28]

Between Gerald and Edward

In the eighty-two years between the writing of Gerald's plan and its execution, there was no lack of warfare between Wales and England. A review of the history of the conflicts of this period both reveals the soundness of his judgment and supports the conclusion that the ultimate defeat of the Welsh was indeed the result of implementation of a sound strategic conception, and not simply the inexorable consequence of England's superior resources.

In Henry II's day, the principality of Gwynedd in North Wales had been the rock on which his two invasions had broken, and it long remained the seat of Welsh ambitions for some degree of autonomy. When Edward I launched his conquest of "Wales," it was principally a conquest of Gwynedd, the ruler of which had recently acquired the more sweeping title of "Prince of Wales." In South Wales, the Norman Marcher lords and creation of English and Flemish towns on the coast had begun the process of undermining the power of the native rulers, but it was the death of Lord Rhys of Deheubarth in 1197 that left South Wales incapable of mounting independent resistance to English domination. The basic reason was one Gerald had emphasized: the Welsh practice of partible inheritance. With 15 of Rhys's sons competing for their shares of his estates, fragmentation and weakness were inevitable. In Powys, the main principality of central Wales, the same problem emerged, though not so dramatically. In 1160 it was split between two heirs; the chances of heredity thereafter allowed the southern half to remain united under the dynasty that later became English earls as de la Poles, but in 1236, northern Powys was divided among five brothers, with each receiving no more than the equivalent of a minor barony.[29]

If the English had left the Welsh to their own devices, the centripetal tendencies of the inheritance system might have been counteracted by the power of warfare to unite; it was this possibility, after all, that led to much of the internecine warfare that Gerald noted. But the English did not want that to happen, and Welsh law, along with the English kings' status as suzerains of the Welsh princes, enabled the English to wrap themselves in the mantle of justice as they stepped in to support

28 Ibid., p. 272.
29 David Walker, *Medieval Wales* (Cambridge, 1990), pp. 90–92.

the weak against the strong, thereby perpetuating disunity. Thus, after Llywelyn ap Iorwerth had led his cousins against his uncles to gain power in Gwynedd, the natural tendency of English policy would have been to undermine him. But in 1205 he made an astute marriage to Joan, the illegitimate eldest daughter of King John. For once, the Welsh tradition of treating natural-born children as equal to those born in wedlock proved advantageous, since Llywelyn could give John the benefit of a princely marriage for his favorite daughter without shaming himself.

Nonetheless, when Llywelyn, having consolidated his power in Gwynedd, occupied the former kingdom of Ceredigion in western Wales, and took control of southern Powys, John recognized the threat. If Llewelyn could solidify his control of those regions, he would be so superior to the other Welsh princes that, even with English support, they could not serve as effective foils to the power of Gwynedd. Llywelyn would be in a position to become *de facto* prince of Wales, rather than merely of North Wales (a title he actually used). In 1209 John aided the attempts by several Welsh lords to recover lands and lordships that had fallen to Llywelyn, most importantly southern Powys. The prince of Gwynedd, unwilling to accept this, struck back. John, according to the *Brut y Tywysogion*, "became enraged, and formed a design of entirely divesting Llywelyn of his dominion." For the purpose he led a "vast army," including six Welsh *tywysogion* ("princes," or "leaders"), against Gwynedd.[30]

The result, however, was almost a replay of 1165. By the time the English army reached Deganwy, "the army was in so great a want of provisions, that an egg was sold for [three quarters of an English foot soldier's daily wage]; and it was a delicious feast to them to get horse flesh." John returned to England in shame.[31] This was an intolerable result, and he quickly raised an even larger army. With better logistical preparations, he was more successful, capturing Bangor and building fourteen or more castles (doubtless of timber, not stone) to secure his gains. Llywelyn sued for peace, agreeing to give up significant portions of his lands, secure the capitulation of the other Welsh lords, and pay a large indemnity.[32] This was accounted a great success for John at the time, and modern historians have also generally seen it that way. Nevertheless, the effort to actually conquer Gwynedd, rather than simply securing the subordination of its leader, had once again failed.[33]

---

30 *Brut y Tywysogion*, ed. and tr. John Williams ab Ithel (London, 1860) [hereafter *ByT*], p. 267.

31 Ibid., pp. 267–9; *Annales Cambriae*, p. 67. The whole expedition seems to have taken place during May. John Edward Lloyd, *A History of Wales from the Earliest Times to the Edwardian Conquest*, 2d ed. (London, 1912), vol. 2, pp. 634–5.

32 *Annales Cambriae*, pp. 67–8; 67n. This was from July to August. Lloyd, *History of Wales*, vol. 2, p. 635.

33 R. R. Davies overstates the situation when he characterizes the campaign as bringing Llywelyn to "the most abject surrender," and showing that conquest "was well within the king of England's reach." *Domination and Conquest*, p. 79. The fact that John planned another campaign in 1212 and failed to launch it for causes that at least partly resulted from the strain the 1211 campaign had placed on relations with his magnates suggests that a complete conquest of Wales was more difficult than Davies implies.

The extent of the difference was made clear in the following year, when John, evidently intending to secure his gains, ordered the construction of numerous castles inside Wales. From the new castle at Aberconway, the king's men pushed their authority to the point that Llywelyn "could not brook the many insults done to him." Once again he rallied the other Welsh lords against the English, and within two years captured all the castles just ceded to John.[34] Llywelyn then cooperated with the English barons who had risen in rebellion against John's tyrannical rule, gaining even more territory in the ensuing conflict, occupying southern Powys again and greatly strengthening his position in western and southern Wales with the capture of Cardigan, Carmarthen, and even Swansea. He even carried the war into England, seizing Shrewsbury and joining the rebels at the Battle of Lincoln. However, the royalist victory in that engagement, along with the death of John and the accession of the young Henry III, led many English magnates to make peace with the crown, and Llywelyn followed suit. The Treaty of Worcester in 1218 left him far stronger than he had been before John attacked him in 1211, and Wales more united than it had ever been. The Welsh princes still did homage to Henry III but did so only with Llywelyn's permission.[35]

This settlement endured until 1223, when the earl of Pembroke led an offensive to regain his and the crown's lost lands in southwest Wales. He regained Cardigan and Carmarthen, as well as Montgomery in the middle March. Henry's government tried to follow up this success with a more ambitious royal expedition in 1228. Yet again the crown aimed "to subjugate Llywelyn, son of Iorwerth, and all the Welsh princes." But the English offensive, which began only in late September, ground to a halt in Ceri, near the border, where Llywelyn seems to have won a small battlefield victory. The Welsh captured the great Marcher baron William de Braose, who was later exchanged for the important castle of Builth.[36] The normal result followed: the English king returned to England without any success more concrete than receiving the fealty of the Welsh *tywysogion*. Neither Llywelyn nor Wales was "subjugated" except in the loosest sense, which is much less than what Henry seems to have intended. Another spate of conflict broke out from 1231–1234, with Llywelyn again coming out on top.[37]

"If their princes could come to an agreement and unite to defend their country," Gerald had warned, "or better still, if they had only one prince and he a good one ... no one could ever beat them."[38] The career of Llywelyn ap Iorwerth, later known as Llywelyn Fawr (Llywelyn the Great) bears out his prediction. Llywelyn

34 *ByT*, pp. 271–9.

35 Davies, *Age of Conquest*, pp. 242–3.

36 *ByT*, p. 317; *Calendar of Close Rolls, 1227–3* (hereafter *CCR*), p. 115 (military summons issued on 3 September).

37 In 1231 an English army advanced to Painscastle and covered the reconstruction of the fortification in mortared stone. In 1233 he sided with various Marcher barons against the king and destroyed the town of Brecon. In 1234 he and his allies secured a peace that left him in secure control of all his territories. Lloyd, *History of Wales*, vol. 2, pp. 675–81.

38 *Description*, p. 273.

himself appreciated the importance of Welsh unity under a single strong leader and went to great effort to ensure that his principality would pass intact to his legitimate son Dafydd, rather than being split with Dafydd's elder brother Gruffydd, as Welsh tradition demanded. Henry's government, however, issued letters forbidding Dafydd to receive the homages of the nobles of Gwynedd and Powys, thereby sabotaging Llywelyn's effort.[39] The Welsh prince apparently could not bring himself to take the only step that would have prevented Henry III from playing the two brothers off against each other to weaken Wales – executing Gruffydd for treason, which he had opportunities to do. The result was that after the prince's death in 1240, Gruffydd became a pawn of the English, and Dafydd's position was consequently weak.[40] This, combined with Dafydd's lack of Welsh allies and sheer bad luck – an extraordinarily dry summer made the marshes and rivers that usually played such a large role in defending the region easily passable – made Henry's invasion of Gwynedd in 1241 unusually successful. To buy peace, Dafydd had to give up all his father's gains since 1216, pay an indemnity to cover the costs of Henry's offensive (a debt subsequently discharged by the surrender of Deganwy), and, most gallingly, accept the judgment of the English court as to what share of Llywelyn's inheritance his brother Gruffydd should receive.[41] Nonetheless, this was still part of the usual pattern of Anglo-Welsh warfare and a long way from the "conquest" of Gwynedd.

The death of Gruffydd in 1244 freed Dafydd to launch a counterstroke against the English, which he immediately did. Most of the leading men of Wales joined, and by year's end he had forced the three hold-outs to do the same. Within Wales, the countryside was his; the power of the English hardly extended beyond the walls of the castles they garrisoned.[42] Welsh raiders came "swarming from their lurking places, like bees," carrying fire and sword into England and ambushing English forces sent to check their incursions.[43] Henry III's counterattack of 1245 followed the path of 1165 and the first campaign of 1211. As the Welsh chronicle summarizes it, "Henry assembled the power of England and Ireland, with the intention of subjecting all Wales to him, and came to D[e]ganwy. And after fortifying the castle, and leaving knights in it, he returned to England, having left an immense number of his army dead and unburied, some having been slain and others drowned."[44] This was hardly a favorable result for the English monarch, but

---

39 Thomas Rymer, ed., *Foedera, conventiones, literae,* etc. (Hague edition), I:1:132.

40 Matthew Paris, *English History, From the Year 1235–1273,* tr. J. A. Giles, 3 vols. (London, 1852–1854), vol. 1, p. 372.

41 Lloyd, *History of Wales,* 2:697–8; *Foedera,* I:1:138–9.

42 *ByT,* p. 331.

43 Paris, *English History,* vol. 2, pp. 27–8, 45–7.

44 *ByT,* p. 331. This is not simply Welsh propaganda. A newsletter from a noble in the English army notes four "knights" (including a Gascon crossbowman) and about a hundred of their retainers killed, aside from an uncertain number who drowned, during just one skirmish over a supply-ship that grounded on the wrong side of the river separating English and Welsh forces. Paris, *English History,* vol. 2, pp. 109–10. Henry entered Wales on 21 August, reached Deganwy on the 26th, and stayed there, fortifying the position, for two months. Ibid., and Lloyd, *History of Wales,* vol. 2, pp. 703–5.

the campaign had also been hard on the Gwynedd. The Welsh had destroyed crops within reach of the English, as part of their logistical strategy, and troops from Ireland had ravaged the island of Anglesey, the bread-basket of Wales.[45] This step was particularly effective because the agrarian sector of the Welsh economy was increasing markedly in this period.[46] In addition, the English cut off the livestock-for-grain trade that was important for feeding the Welsh even when their own crops were left intact.[47]

The next spring, moreover, Dafydd died of natural causes. Two of his four nephews, Llywelyn and Owain ap Gruffydd, divided the portions of his lands still under native control. Without a single dominant figure, Welsh unity fractured further, with two minor *tywysogion* siding with the English to dispossess two others. The English drove Llywelyn and Owain and their allies temporarily into the mountains and the wilderness. But Henry faced serious troubles in France and England, so the English chose to accept Dafydd's nephews' proffers of peace. By the Treaty of Woodstock in 1247, they ceded the portion of Gwynedd east of the Conwy – nearly half the principality – to Henry, marking a new low-point for the fortunes of native Wales.[48] It has rarely been appreciated, however, that this was a compromise peace. By the terms of the agreement of 1241, Henry had a double right to take all of Gwynedd into his own hands, since Dafydd had not left an heir of his body, and because of his rebellion.[49] Instead, the majority of the principality, including its most valuable parts (Anglesey and Snowdonia), remained under native Welsh rule.

Prince Edward and Prince Llywelyn ap Gruffydd

A decade later events drove home the great difference between a humbled principality of Gwynedd and a completely conquered one. Once again the Welsh were able to make an effective counter-attack when they united behind a single heir of Llywelyn the Great – in this case Llywelyn ap Gruffydd, Dafydd's nephew, who in 1255 defeated two of his brothers in battle, imprisoned them, and thereby gained control of all the principality except the portion that had been ceded to England in 1247.

King Henry III's son and heir, the future Edward I, had recently received the border earldom of Chester and the royal lands in Wales as part of his appanage. In 1256, perhaps anticipating trouble as a result of Llywelyn's victory over his brothers, 15-year-old Edward paid an unusual visit to his new lordship of eastern Gwynedd. Gerald's strategic plan, as noted earlier, had included a section on how

45 Gerald notes an old proverb that Anglesey "is the mother of Wales" and could provide enough grain to feed all the Welsh. *Description*, 230; *Journey through Wales* (tr. Thorpe), p. 187.

46 Suppe, *Military Institutions*, p. 15.

47 *Foedera*, I:1:155; Paris, *English History*, vol. 2, pp. 114–5, 244.

48 Davies, *Age of Conquest*, pp. 302–4; *Foedera*, I:1:156.

49 *Littere Wallie, Preserved in Liber A in the Public Record Office*, ed. J. Goronwy Edwards (Aberystwyth, UK, 1940), no. 8.

to occupy and govern Wales once conquered. He advised that a firm and just official who would obey the laws and exercise moderation towards those who accepted royal rule, and in particular flatter the Welsh lords with honors, which they craved, should govern the land.[50] Henry III had not heeded this sage counsel; on the contrary, he sold the right to extract revenue from his Welsh lands to the highest bidder, first to John de Grey and then Alan de la Zouche. Since the latter paid 1,100 marks for the fee-farm, more than double the amount offered by the former, it should have been obvious to the king that his tax gatherers would have to squeeze the Welsh population immoderately and bend or break their laws to recoup his costs.[51] Moreover, de la Zouche publicly boasted how he had brought the Welsh to submit to *English* laws, in place of their traditional codes.[52] When Edward received the area from his father, rather than ameliorate the situation, he turned the administration over to Sir Geoffrey Langley, a *parvenu* royal official already infamous for his rapacity and oppression of the king's English subjects, who was working to extend the English structure of government into occupied Gwynedd.[53]

Moreover, far from being honored, the Welsh under Edward's control were treated in what they considered high-handed and demeaning ways. Such indignities would have been bad enough coming from Henry III, whom they had (however unwillingly) accepted as their lord, but the Welsh perceived them as much worse coming from young Edward. For men who viewed their status as deriving principally from their distinguished lineages rather than land or wealth,[54] slights coming from a mere *officer* of relatively low birth (such as Geoffrey Langley) were doubly humiliating.[55] The English did little better with regard to another of Gerald's suggestions, to reward those Welsh who had assisted in the king's conquest.[56] The princes of Deheubarth had hoped to use England's might to throw off the loose subordination to Gwynedd they had experienced under Llywelyn the Great. However, they were disappointed to find themselves under the stricter tutelage of a lord who made no pretense of respecting their independence.

It was a sign of the dissatisfaction with the crown's management of Welsh affairs that by 1251, Owain and Llywelyn of Gwynedd had brought three of the

50 *Description*, 2, p. 21.

51 Paris, *English History*, vol. 2, pp. 435, 486.

52 Paris, *English History*, vol. 2, p. 486.

53 Paris, *English History*, vol. 2, pp. 358–9, 531–2, and vol. 3, pp. 28, 200–1; *Annales monastici*, vol. 3, pp. 200–1.

54 *Description*, tr. Forester, p. 505.

55 Langley, for example, though he had become wealthy and powerful in the service of Henry III, was the son of a "modestly endowed Gloucestershire knight," who had inherited only one manor, and gained one more by marriage. P. R. Coss, "Sir Geoffrey de Langley and the Crisis of the Knightly Class in Thirteenth-Century England," in T. H. Aston, ed., *Landlords, Peasants and Politics in Medieval England* (Cambridge, 1987), pp. 167, 168n10.

56 For Henry III's mistreatment of his Welsh supporters see *ByT*, p. 333; Davies, *Age of Conquest*, pp. 225, 228; and Lloyd, *History of Wales*, vol. 2, pp. 710–11.

other main Welsh princes into a secret sworn brotherhood aimed at resisting further erosion of native Welsh power.[57] The greater part of Wales then united behind Llywelyn ap Gruffydd when, in late 1256, his former subjects in eastern Gwynedd came before him to announce "that they would rather be killed in war for their liberty, than suffer themselves to be trodden down by strangers in bondage."[58] Within a short time, eastern Gwynedd, Deheubarth, Ceredigion, and Builth were all under the control of Llywelyn and his allies.

Despite his extensive possessions, Edward had neither the cash nor the men to put up effective resistance. He went to his father for aid. Henry dismissed him, saying his coffers were empty and he had more pressing business. Edward's uncle loaned him 4,000 marks, but that was a drop against the tide. The Marches suffered heavily from raiders, and the Welsh routed a substantial English force led by John Lestrange, Rhys Fychan, and Gruffydd ap Gwenwynwyn, marching under Prince Edward's banner, near Montgomery. They crushed an even larger army at Cymerau, reportedly with the loss of 3,000 men, and captured and destroyed numerous castles held by the English or their allies.[59] Anticipating that this would provoke a large-scale response, the Welsh took precautions: they "prudently sent away their wives, children, and flocks into the interior of the country, about Snowdon and other mountainous places inaccessible to the English, ploughed up their fields, destroyed the mills in the road which the English would take, carried away all kinds of provisions, broke down the bridges, and rendered the fords impassable by digging holes, in order that, if the enemy attempted to cross, they might be drowned."[60]

Because a large portion of the Welsh economy rested on animal husbandry rather than farming, and the slopes of Snowdon offered excellent pasturage, these steps were not as painful as the equivalent would have been for the English. Moreover, the war allowed the Welsh to recoup some of their losses by plundering their neighbors over the border.[61] The rainy season made any English reprisals difficult, since it rendered the roads through the wetlands impassable for the invaders, though the locals could make still make their way through.[62] According to Mat-

---

57 *Littere Wallie*, nos. 261 (between Llywelyn and Gruffydd of North Powys) and 284 (between Llywelyn and Owain on the one hand and the two Rhyses on the other: *in conseruando ius suum et acquirendo quod iniuste ablatum est pro loco et tempore iuuabit alium et manutenebit contra omnes viuentes ac si essemus fratres conterini*). Technically Gruffydd was not bound to the Rhyses, but by transitivity he was, in effect.

58 Maredudd ap Rhys Gryg, Rhys Fychan, and the prince of northern Powys. *ByT*, 341, 343. By this point, however, Maredudd ap Owain had replaced Rhys Fychan in the alliance; they could not both be accommodated because their territorial claims were incompatible.

59 In addition, two smaller defeats elsewhere cost the English 194 and 130 soldiers, respectively. *Annales Cambriae*, pp. 92–5; Lloyd, *History of Wales*, vol. 2, p. 720. Two more minor Welsh victories followed in 1258. *Annales Cambriae*, pp. 95–6.

60 Paris, *English History*, vol. 2, p. 238.

61 *Description*, 233; Paris, *English History*, vol. 2, pp. 243, 267, 269.

62 Paris, *English History*, vol. 2, pp. 204, 217.

thew Paris, though Edward threatened to "crush them like a clay pot," the Welsh "only laughed at . . . and ridiculed" his efforts, until the English prince, seeing how little he could accomplish, was tempted to give up Wales and the Welsh as untameable, while Henry was "overcome with grief . . . at the slaughter of so many of his liege subjects."[63]

Eventually rallying, the king summoned troops from all his lands, and meanwhile took the drastic step of destroying the harvests in the fields of the border-lands to prevent the Welsh from gaining access to the crops by force or by commerce. This caused serious shortages in his army, as well as for the population. Henry's army, once assembled, advanced as far as Diserth and Deganwy, the only two fortresses within Gwynedd still under his control, to break Llywelyn's sieges and resupply the castles. But without accomplishing anything else, Henry dismissed his forces and returned to London, with Welsh warriors nipping at his heels and killing any stragglers.[64] By year's end, the three main Welsh *tywysogion* who had not immediately joined the rebellion had either fallen in line or (in the case of Gruffyd ap Gwenwynwyn) gone into exile in England. Between his actions, the Welsh raids, and the end of normal cross-border trade, the economy of the Marches collapsed to the point of causing a serious famine.

When Henry began to prepare to renew the war, his knights protested at the prospect of another costly and useless campaign, and parliament took the extremely unusual step of refusing a grant of taxation.[65] Many Englishmen were in some respects sympathetic towards the Welsh cause, since they too felt abused and oppressed by royal ill-government and the greed of the king's officials and foreign relatives. Henry's inability to wage effective military operations combined with many other grievances to provoke what soon turned into a full-scale rebellion led by the earl of Leicester, Simon de Montfort. The disunity that had so often plagued Wales now struck England in full measure, and Llywelyn quickly seized the opportunity. He allied with the earl and barons, just as his grandfather had done with the opposition to John, and ultimately with greater success. The Treaty of Montgomery in 1267 conceded to Llywelyn both the title of "Prince of Wales" and the feudal status that it implied: henceforth the other leading men of Wales, and not just Gwynedd, would do homage to him, rather than directly to the king of England.[66] Llywelyn, in turn, would do homage to Henry.

63 Ibid., pp. 238 (trans. modified), 241.
64 Ibid., pp. 245–7, 269, 291; Lloyd, *History of Wales*, 721–2. The army left Chester on 19 August, and seems to have begun the march back to England as soon as 4 September.
65 Ibid., pp. 267–9, 273.
66 Davies, *Age of Conquest*, pp. 314–5; *Littere Wallie*, no. 1. Llywelyn's possession of various lands he had conquered was also confirmed. One lone Welsh prince, Maredudd ap Rhys, remained a tenant-in-chief of the crown, but these were minor considerations compared to the creation of a real principality of Wales. This was a reversal of the agreement of 1241, by which Dafydd conceded to Henry III the homage of all the nobles of Wales. *Littere Wallie*, no. 4.

The Importance of Strategy

Perhaps influenced too much by the eventual outcome, historians have tended to view the eventual English conquest of Wales as inevitable, or nearly so.[67] The argument is essentially that the general path of development of European concepts of monarchy and sovereignty ensured that the English crown would seek a more thorough-going control of Wales; that the disparity of resources between the two countries provided England the strength needed; and that the loss of most of England's continental possessions led the English to focus those resources on military efforts within Britain. In this view, John's campaigns of 1211–12 "had laid the ground-plans of a military conquest and settlement of the country which it only remained for Edward I to copy and put fully into operation." John had failed because of political miscalculations and lack of patience, but nonetheless had demonstrated "that the native Welsh kingdoms had little chance of withstanding the military might of the English state."[68] The author of those words was the greatest historian of medieval Wales of his generation, but he was not a military historian, and had a somewhat exaggerated view of what the simple application of resources and conventional military power can accomplish.

In reality, the humiliations of Gwynedd in 1211–12, 1241, and 1247 no more show that Wales was ultimately doomed than the English disasters of 1218, 1256–57, and 1257 prove the opposite. When the Welsh were divided, the English had the advantage; when the English were distracted by civil strife or foreign wars, the Welsh made gains. Even when a king of England applied his realm's military strength fully towards the conquest of Wales, he could be stymied entirely, or limited to minor gains such as the construction of a single new castle (all that was accomplished in 1157, 1165, 1228, and 1245). It would take a large number of such campaigns to subdue Wales, and each expedition involved "heavy and quite alarming expenses."[69]

Indeed, since the Welsh could capture or destroy several castles in a single campaigning season with less effort and expense than it took the English to build one new one, this sort of offensive could lead to regression rather than advance in English territorial control. In 1231, for example, the Dunstaple annals record that Llywelyn Fawr destroyed ten castles of the March, while Henry III's army remained immobilized in order to rebuild just one.[70] Moreover, even when feudal service (and the shield-tax levied on knights who chose not to muster) paid for much of the manpower, extracting the soldiers needed for an unpleasant and unprofitable campaign drained the king's political capital. Henry III's campaign

67 E.g. Davies, *Age of Conquest*, p. 330.
68 Davies, *Age of Conquest*, pp. 292–3, 297, 295.
69 *Description*, p. 270 ("as much as is levied in taxes from the Welsh over a whole series of years").
70 *Annales monastici*, vol. 4, p. 127.

of 1245 best illustrates this point. "His majesty the king is staying with his army at Gannock [Deganwy]," wrote a nobleman:

> for the purpose of fortifying a castle which is now built in a most strong position there; and we are dwelling round it in tents, employed in watchings, fastings, and prayers, and amidst cold and nakedness. In watchings, through fear of the Welsh suddenly attacking us by night; in fastings, on account of a deficiency of provisions, for a farthing loaf now costs five pence; in prayers, that we may soon return home safe and uninjured; and we are oppressed by cold . . ., because our houses are of canvas, and we are without winter clothing . . . . Whilst we have continued here with the army, being in need of many things, we have often sallied forth armed, and exposed ourselves to many and great dangers, in order to procure necessaries, encountering many and various ambuscades and attacks from the Welsh, suffering much . . . . There was such a scarcity of all provisions, and such want of all necessaries, that we incurred an irremediable loss both of men and horses. There was a time, indeed, when there was no wine . . . amongst the whole army, except one cask only; a measure of corn cost twenty shillings [four months of wages for a foot soldier], a pasture ox three or four marks [160 or 240 days' wages], and a hen was sold for eight pence [four days' wages]. Men and horses consequently pined away, and numbers perished from want.[71]

When the army returned to England after ten weeks, the king's coffers were empty – indeed, he had to impose on his brother for a loan of 3,000 marks, which the latter secured by pawning his jewels. The army had suffered substantial losses, the soldiers had gained no plunder or glory, and, as one chronicler notes, Henry was "*unable*, as well as unwilling, to make any longer stay" [emphasis added].[72] The chronicler nonetheless claimed that the king returned "crowned with good fortune," convinced that the devastation wrought by his forces had brought the Welsh to the edge of ruin, and planning to return in the spring to finish the job. This, however, seems to be a distortion of hindsight, since the king's subsequent actions bespeak more desperation than confidence:

> in order that the Welsh might not obtain provisions from the neighbouring [English] provinces . . . he caused the inhabitants of that country, and those in subjection to him, to be impoverished, and especially deprived of food, so much so, that, in Cheshire and other neighbouring provinces, famine prevailed to such a degree, that the inhabitants had scarcely sufficient means left to prolong a wretched existence.[73]

71 Matthew Paris, *English History*, vol. 2, p. 113
72 Ibid., p. 114.
73 Ibid.

This action fits better with the perspective of the Welsh chronicle, which depicts Henry's campaign as a costly failure, and by implication attributes the subsequent collapse of Gwynedd's fortunes instead to the death of Dafydd and the consequent defection of Maredudd ap Rhys Gryg and other Welsh princes.[74]

The resources of the English crown were large, but the king's territories were much smaller at the start of Edward I's reign than in Henry II's day, due to the loss of most of the wealthy Angevin lands in France. Llywelyn ap Gruffyd, on the other hand, controlled a larger portion of Wales than Owain had in 1157 or 1165, and the Welsh had in the meantime adopted elements of the Anglo-Norman style of war. They had added barded cavalry to their forces, developed sophisticated siege methods for attacking castles, and built their own castles in substantial numbers to strengthen their defensive capabilities – without losing their native martial tradition.[75] For Edward to achieve greater success than his predecessors against a stronger foe, he would have to conduct his war not just with greater determination, but also with a better plan.

As Clausewitz points out, an offensive that culminates and halts before it renders the enemy incapable of counter-attack leaves the initial attacker in a dangerous position. Until he fully consolidated his gains, he would continue to suffer the disadvantages of being on the offensive (logistical difficulties, long lines of communication, the disadvantage in information that comes from operating in the midst of a hostile populace) without benefitting from its advantages (concentration and initiative).[76] Moreover, an invader seeking to hold on to any substantial territory suffers from the principal disadvantage of the operational-level defense: if he wants to control the countryside and exploit its resources, he must disperse his forces, but scattered troops are vulnerable to defeat in detail. Therefore, an effective offensive strategic plan aimed at the defeat of the enemy has two

---

74 *ByT*, pp. 331, 333.

75 Gerald says one of the things that would contribute to making the Welsh nation unconquerable would be if they "were more commonly accustomed to the Gallic mode of arming, and depended more on steady fighting than on their agility." By his time the nobles had become good horsemen, but were still not numerous and remained light-armored, typically with small coats of mail (*loricis minoribus*). *Description*, tr. Forester, pp. 521, 491; *Opera*, ed. James F. Dimock, vol. 6 (London, 1868), p. 190. By 1245, according to Matthew Paris's probably exaggerated report, the Welsh fielded 500 well-armed knights on iron-clad horses (along with 30,000 footmen). *English History*, vol. 3, p. 217. Although the sources give little tactical detail, Welsh victories in six consecutive clashes of 1257–58 indicate a high level of tactical proficiency had been achieved. *Annales Cambriae*, pp. 92–6. The Welsh had also made progress in siege warfare. The *Brut y Tywysogion* has an isolated (and possibly anachronistic) mention of the use of "engines" for breaching castles by the Welsh as early as 1113, but common reference to the use of "engines" in siege warfare begins in 1196. *ByT*, pp. 113, 243, 253, 277, 321 (in 1231 an army led by Llywelyn ap Iorwerth, among others, "broke the castle" of Cardigan [Aberteivi] with engines). Miners are noted as used against the Welsh in 1196, and by a mixed Anglo-Welsh army in 1214. Ibid., pp. 243, 277.

76 This is a major and under-appreciated theme in Carl von Clausewitz *On War*, trans. and ed. by Michael Howard and Peter Paret (Princeton, NJ, 1976). See particularly Books VII.5 and VIII.4, and cf. VIII.7.

requirements. First, it must identify physical targets that, if captured or destroyed, will either break the enemy's *will* to continue, or eliminate his *ability* to do so. Second, it should ensure the attacker has sufficient strength and staying-power to reach those centers of gravity before the campaign drains his ability to continue the offensive.

In most cases, a defender should consider the attacker's situation as well as his own: if the defender can see that the attacker has reached the point of exhaustion, he has a favorable negotiating position and is likely to limit the concessions he is willing to make accordingly. Until the reign of Edward I, some English invasions of Wales were so ineffective that they failed to impel the defenders to offer any concessions at all for peace; others gained the English a forward position or induced Welsh leaders to renew the homage that they in any case generally acknowledged they owed to the English king; only in a few cases had the princes of Gwynedd surrendered any territory. Indeed, since the capitulation of 1247 followed the death of Dafydd rather than being a direct result of the prior campaign, only the invasions of 1211 and 1241 clearly had that last result, and only briefly. In any case, not one of the English attacks left the Welsh so badly defeated in military terms that they would not have been *able* to continue the war. In no case had the invaders succeeded in engaging and destroying the army of Gwynedd or occupying Snowdonia.

Until 1276, the strategies the English had employed to attempt the conquest of Wales reflected too much concern with how best to use what they *had*, rather than how to get what they *needed*. To force the complete surrender of Gwynedd, as Gerald had made clear, required pushing the Welsh into their mountain refuges and keeping the pressure on them for a full year to wear them away by attrition and ultimately hunger. Anything less allowed the defenders to wait out the offensive, and then counterattack to recover territory and destroy castles held by the English, or at least launch raids on enemy territory to resupply themselves and encourage the enemy to accept a compromise peace. Yet English king after king had begun a campaign in the summer or even the fall and returned home before winter. John, unusually, had begun his campaign early and returned for a second attack in the same year, and that was a major reason for his greater success. In most cases, however, not only had the English failed to sustain the offensive in winter, but they had also declined or failed to mount a new attack the following spring.

The basic reason was simple: because it was *difficult and expensive* to do what Gerald's plan required, namely to sustain an offensive effort to the point where the Welsh had to choose among starvation, complete surrender, or an attempt to restore their situation with a near-hopeless effort to crush the invaders in a direct battle.[77] In the thirteenth century the English had geared their military system,

77 Modern historians have tended to downplay these issues, emphasizing instead that the English kings were generally satisfied to leave their Welsh vassals in power, provided they offered due fealty. This was doubtless true in some instances, but especially given the frequent chroniclers' assertions of more ambitious aims, and indeed common sense, it seems likely that in some other

like that of all contemporary Western powers, towards mounting major campaigns over the summer, spearheaded by armored cavalry who owed service at their own expense for relatively short periods – 40 days, in England. It was possible to bring feudal contingents into operation in rotation, but even England's resources were not so disproportionately great that one eleventh of her strength could be counted upon to defeat a counterattack by the concentrated forces of the princes of Gwynedd and their allies. The latter could draw on the martial ardor and skill of much of their population, not just its upper stratum. In any case, English knights had little enthusiasm for the rigors of camping in the cold and rain of a Welsh winter under constant harassment. Knights and footmen willing to do what was needed could nonetheless be found in sufficient numbers, provided that the king had the money and was willing to pay them, but that required straining the fiscal system to its limits, as only a strong king could do, and then only if he could avoid dividing the proceeds between the conquest of Wales and other conflicts.

Even a king who raised great sums and was willing to devote them to Welsh matters would likely not have had much success, had he spent the money to keep twice as many armored horsemen in his employ from July through October and ensure they had adequate supplies. In fact, although there are no specific figures from the earlier campaigns to compare them to, by later standards Edward I's highly successful campaign of 1276–1277 was not exorbitantly expensive; the expenditures he made would have been within the reach of Henry I, John, or Henry II, for example.[78] The success of Edward's conquest of Wales between 1276 and 1282 did owe something to favorable circumstances as well as his military competence, but it was also the result of his application of a better strategy, which, in all key respects, was the one that Gerald of Wales had outlined almost a century earlier, but that none of Edward's predecessors had implemented.

The War of 1276–1277

Gerald's principal recommendations, as already noted, were: (1) divide the Welsh, especially taking advantage of the family feuds that arose from the practice of partible inheritance, promising some lords the lands of others; (2) use economic warfare, preparing the ground for conquest by cutting off all imports; (3) invade by the coastal lowlands, ensuring supply by sea; (4) employ light-armed troops drawn from the Marches and within Wales, led by locals who knew the terrain and

---

years the English monarchs would happily have dismembered the principality of Gwynedd entirely, or indeed depopulated it entirely, if they thought it could be done easily or cheaply. It may be true – though the chronicles say otherwise – that, as R. R. Davies puts it, "for much of the twelfth and thirteenth centuries the kings of England seemed neither capable nor anxious to deliver the *coup de grâce*," but then, there is little point in being anxious to do what one is not capable of doing. *Domination and Conquest*, p. 24.

78 The total cost of the war was somewhere around £23,000 – well less than the amount raised by the war-tax granted by Parliament to pay for it. Morris, *Welsh Wars*, pp. 140–2. The war of 1282, by comparison, cost £120,000 or more. Michael Prestwich, *Edward I* (Berkeley, CA, 1988), p. 200.

Welsh tactics, and were hardier and more accustomed to campaigning in wilderness areas than most Anglo-Norman soldiers. Use them to ring Snowdonia, apply "patient and unremitting pressure," and wear down the Welsh with probes, raids, and skirmishes, and by denying them food. Bring up fresh troops as necessary; and (5) keep the king's attention focused on Wales for a full year, without becoming distracted by strife elsewhere.

It was part of England's standard repertoire for Welsh wars to execute steps 1 and 3, and in Henry III's reign the value of step 2 had been well appreciated, though in 1245 the English had taken the most stringent measures along these lines only after the end of the main army's campaign. English kings had also made substantial use at times of Marcher troops and Welsh allies, though not in the ways Gerald called for. However, most military offensives against Wales were begun in late summer and finished by early fall. In John's two-stage invasion of 1211 the second phase was merely an ad-hoc response to the embarrassment of the first. Even so, the two campaigns lasted only from May to August. By contrast, Edward I opened his first war as king in November of 1276, and finished it in November of 1277, requiring just three days less than Gerald's estimate of one year.[79]

Most historians see the war of 1276–1277 as arising from arrogance and intransigence on both sides. Edward had sheltered Llywelyn's younger brother Dafydd and Gruffydd ap Gwenwynwyn of Southern Powys after the two had plotted to assassinate Llywelyn, had made difficulties about Llywelyn's rights to certain disputed lands, and had supported the bishop of St. Asaph against Llywelyn over the latter's encroachments on the bishopric's revenues. Llywelyn, feeling that Edward was acting in bad faith, tried to bring diplomatic pressure to bear by declining to give the new king the homage he was due, as well as withholding installments of the large sum he owed to the crown by the terms of the Treaty of Montgomery.[80] This was poor policy, however. Although Llywelyn had reason to be dissatisfied, nothing Edward did should have been an unbearable affront, or as painful as what war could bring, whereas Llywelyn's refusal to do homage was a challenge Edward simply could not ignore.

In 1275, both parties could see war coming. A group of Irish "kings" asked Edward to aid them in suppressing a rebellion against his authority in Ulster, while the king of Castile requested help in the war against the Saracens, but Edward refused to allow himself to be distracted. He relied on peaceful diplomacy to pursue his step-mother's claim to the county of Agen and to resolve other tensions between England and France.[81] At the end of the year, Edward discovered that Llywelyn had, by proxy, married Eleanor de Montfort, daughter of the earl of

---

79 Thus even if we consider only the operations of King Edward's own army, from July through November, the campaign was not "brief" compared to previous efforts. Certainly the war as a whole should not be interpreted as a mere "military promenade, followed by successful negotiations." Cf. Prestwich, *Edward I*, p. 182.

80 See Davies, *Age of Conquest*, pp. 320–30 for a clear discussion.

81 *Foedera*, I:4:76–7; I:2:147–153.

Leicester who had so severely challenged royal authority in the reign of Henry III, and sister of Simon and Guy de Montfort, who had recently murdered Edward's own cousin, Henry of Almain, whom Edward had sent to the de Montforts as a peace envoy. Under the circumstances, Edward interpreted the betrothal as a provocation, even a threat. After capturing Eleanor as she sailed to Wales, he imprisoned her, which Llywelyn naturally considered a hostile act.[82] In spring and summer of 1276, both sides complained of raids and border skirmishes.[83] In October, Edward ordered defensive preparations in Montgomery and Oswestry on the Welsh border.[84]

The next month, Edward decided that Llywelyn had already received sufficient opportunities to deliver his homage, declared him a rebel, and began hostilities.[85] Llywelyn would gladly have continued to put off the conflict or avoid it altogether, so it was Edward's choice to begin the conflict in late autumn.[85] The English king initially set the date for the feudal muster as midsummer 1277.[87] He could have done the same even if he had allowed the diplomatic process to continue through mid-spring. The declaration of war in November, however, gave him the opportunity to begin the war over the winter, something none of his predecessors had ever shown the slightest inclination to do, but an action that made good sense in terms of Gerald's strategic plan. By opening hostilities at an unexpected time, Edward forestalled the Welsh from laying in supplies in anticipation of the war. Already in the declaration of war itself, the king forbade all communication between his subjects and Llywelyn's men, and ordered "that no one shall take into [Llywelyn's] land, or permit to be taken thither through their land or power, by land or by sea, victuals, horses, or other things that may be useful to men in any way."[88]

Within a week of the declaration of war, the king appointed three magnates as his lieutenants for the conduct of the war, each based in a royal fortress: the earl of Warwick in Chester in the north, closest to Gwynedd; Roger Mortimer at Montgomery in the middle March, opposite Southern Powys, and Pain de Chaworth in the south, at Carmarthen. Each was a Lord Marcher, who well understood

---

82 Edward seems to have been convinced, as he wrote at the time, that Eleanor was inspired by the desire to use her new husband's power to revive the Montfortian program of resistance to royal tyranny. Smith, *Llywelyn*, pp. 391–99, 402.

83 Ibid., pp. 402–6.

84 *CCR, 1272–9*, p. 315.

85 Ibid., pp. 359–61.

86 In early 1277 Llywelyn was pleading to be given the king's grace and proclaiming his willingness to do homage and to pay a substantial cash indemnity in exchange for a peace of goodwill (including the release of Eleanor). Smith, *Llywelyn*, pp. 410–11.

87 Ibid., p. 360. The writs of summons were issued 12 December. *CCR*, p. 410. On 24 January, several dozen men received letters of protection for service through midsummer as they were "going" on service in Wales; from 3 March additional letters were mostly for those who had "gone." *Calendar of Patent Rolls, 1272–81* (hereafter *CPR*), pp. 189–92.

88 *CCR*, p. 361. Cf. the orders of 12 Dec. 1276 and 3 January 1277 to prevent "corn [grain], wine, honey, salt, iron, arms," or other goods from moving into Wales. Ibid., pp 410, 366.

the land and the enemy, just as Gerald had advised.[89] The king provided them with moderate-sized strike-forces of cavalry at royal wages and gave them direction over the feudal contingents and arrayed infantry of regional landholders as needed. Each captain, in pursuance of the divide-and-conquer strategy advocated by Gerald, received authority to receive into the king's peace any Welshmen willing to submit.[90] All sorts of other preparations were initiated: collecting a tax granted by parliament, taking loans from Italian bankers, purchasing and stockpiling grain, importing warhorses, summoning Gascon crossbowmen, repairing border fortresses, purchasing large numbers of crossbow quarrels, and so on.[91] On 10 February, the archbishop of Canterbury excommunicated Llywelyn on the basis of his failure to fulfill his oaths to his lord. The archbishop subjected his followers to the same sanction, if they did not leave his service within a month. This made it possible for the war against the Welsh prince to be presented as a holy war, and gave any Welshman who considered defecting to the English a convenient excuse for doing so.[92]

Edward did not allow winter weather to delay the initiation of active campaigning against the outer ring of Llywelyn's dominions. Each of Edward's three field forces was sufficiently strong to be secure against anything but a full-scale attack by Llywelyn, and that was not much of a threat, since the requirement of mobilizing for a major counterattack would have ruined any real chance of surprise. In any case, the English could afford whatever losses fighting might entail, whereas the Welsh could not, so Llywelyn ordered his men to avoid major clashes. The force operating from Chester, which included Llywelyn's brother Dafydd, seems to have occupied the neighboring district of Maelor or Bromfield in northern Powys before the end of the year.[93] The lord of that region turned against his brother, Madog ap Gruffudd, who then also surrendered to avoid losing all his

89 John E. Morris untangles the narrative in *Welsh Wars*, an outstanding book, especially considering its age. However, especially since his approach is mainly from the English perspective, his work should be read in conjunction with Smith, *Llywelyn*. Chaworth had under royal pay both infantry and 100 cavalry under John de Beauchamp, custodian of Carmarthen and Cardigan castles. Smith, *Llywelyn*, p. 419n. He also had substantial contingents of cavalry provided by the Marcher lords, who did not receive royal pay.

90 *CPR*, 186 (Mortimer); *Foedera*, Hague ed., I:2:158 (Chaworth); it is scarcely conceivable that Warwick was not given the same authority. Dafydd ap Gruffydd and the justiciar of Chester were given similar but more limited powers. *CPR*; see also pp. 201, 219. A petition of Trahaern ap Madog from this period refers to his coming into the peace of Bohun (the earl of Hereford) and the king. *Calendar of Ancient Petitions Relating to Wales*, ed. W. Rees (Cardiff, 1975), p. 468; Smith, *Llywelyn*, p. 417n.

91 *Welsh Wars*, p. 115; *CPR*, pp. 193, 195–6.

92 Smith, *Llywelyn*, pp. 407–9, 412–13, 428 [*expediccione votiva*]. It is interesting to note that Gerald did not mention this step in his strategic plan, perhaps because he was unwilling to advocate the use of the Church as a tool of the crown.

93 The power given on 26 December to Dafydd and to the justiciar of Chester to receive the submission of Llywelyn ap Gruffydd of Maelor and all his men seems clearly to have been the result of a negotiation with the *tywysog*, not merely an anticipatory tool. *CPR*, p. 186.

lands. Dinas Brân, one of the strongest castles in Wales, had fallen to the English by early May.[94] In December Llywelyn in person opposed Mortimer's detachment in the Middle March, but before 5 February 1277 the English had driven the prince out of the region with the help of the earl of Lincoln and the men of Shropshire and Hereford. Moreover, Southern Powys was largely under control of the English and the most loyal of the Welsh *tywysogion*, Gruffyd ap Gwenwynwyn.[95] Dolforwyn castle, recently constructed by Llywelyn to block the route up the Severn, fell to Mortimer's men on 8 April after a short siege. The same force quickly asserted control over Cydewain, Ceri, and Gwertheyrnion. Before the month's end they occupied the ruins of Builth castle and dominated that district as well.[96]

Pain de Chaworth, meanwhile, employed a combination of military pressure and negotiations to secure submission of Llywelyn's ally-by-compulsion, Rhys ap Maredudd. Rhys had a claim to the castle of Dinefwr, traditional capital of Deheubarth, and the surrounding district, then held by his cousin Rhys Wyndod, Prince Llywelyn's nephew. In accordance with Gerald's advice to divide the Welsh by promising one man another's lands, Chaworth promised, or at least strongly implied, that once Edward conquered Dinefwr and the surrounding area, he would ensure Rhys ap Maredudd received his due rights to those lands, or if Edward chose to keep them as royal lands, he would compensate Rhys ap Maredudd for his losses.

In exchange, Rhys put his castles and men at Chaworth's disposal.[97] Gruffydd ap Maredudd, the principal Welsh lord in Ceredigion, also entered the king's peace.[98] The three *tywysogion* combined might have been a fair match for royal forces in the Tywi valley, but with two in English service and Llywelyn in no position to lend effective assistance, Rhys Wyndod had little choice but to surrender. By the first week of June, his castles of Dinefwr, Llandovery, and Caercynan were in Chaworth's hands as royal property, and the English immediately began strengthening their fortifications.[99] The remaining Welsh nobles of Deheubarth soon came into the king's peace, generally losing some of their lands and keeping others, just as Llywelyn ultimately would.[100] Meanwhile, the earl of Hereford used

94 Smith, *Llywelyn*, 423–4; *Calendar of Ancient Correspondence Concerning Wales* (Cardiff, 1935) (hereafter *CAC*), 83; *CCR*, pp. 398–9.

95 Morris, *Welsh Wars*, pp. 120–21; *CPR*, p. 192.

96 *ByT*, p. 365. Refortification of the castle of Builth had begun by 3 May. Smith, *Llywelyn*, p. 418.

97 Smith, *Llywelyn*, p. 419–21; *Foedera* (Hague), 1:2:158. The language is a bit ambiguous: it could be interpreted to leave the king room to decide that Rhys did not have the right to Dinefwr in the first place, in which case Edward's retention of the castle would not involve his "demanding" Rhys's rights ["ius suum deposcet"], and therefore would not require "restitution." Rhys agreed to do homage to Edward, and Edward agreed that he would never alienate Rhys's homage to another, except by the latter's free will – i.e. Rhys would not be subjected again to Llywelyn.

98 Smith, *Llywelyn*, p. 421; *CAC*, pp. 55–6, 71–2.

99 *CPR*, p. 212. Pain de Chaworth was in charge of all three castles. £285 10s. were disbursed for the strengthening of the "king's castles" of Dinefwr and Caercynan. The National Archives, Kew (hereafter TNA), E101/3/20.

100 Smith, *Llywelyn*, pp. 421–2.

a combination of his resources and royal assistance to make good his claim on the three cantrefs of Brycheiniog (Brecknock), thus securing the rear of Mortimer's and Chaworth's advances.[101]

The bill for wages for the winter and spring operations of the three advance forces amounted to nearly £3,000 sterling – a far from trivial sum, but nonetheless an excellent bargain, considering the extent to which they had weakened Llywelyn and prepared the ground for the summer's main push.[102] Not only had Llywelyn lost lands and access to their men, money, and supplies, but he had also been compelled to commit his forces to the field in winter, which, despite the legendary disdain of the Welsh for difficult conditions, began the process of wearing his men down. The same was true, no doubt, for those of the king's men involved in these preliminary actions, but unlike the Welsh, large contingents of fresh troops were about to join them.

Edward had summoned the feudal host for 1 July. Most joined at Chester, from where the English army would follow the traditional invasion route along the northern coast, while a few hundred more troopers, now under the king's brother Edmund, advanced on Aberystwyth and began a large new castle at Llanbadarn, from which they could control the country right up to the southern border of Gwynedd, at the river Dyfi. The main army at Chester, under Edward himself, had approximately 1,000 heavy cavalry – more than enough to ensure superiority in open battle, but only one third of what the kingdom could have supplied.[103] However, Edward had provided for a much larger force of foot soldiers, as well as carters, craftsmen, and laborers by the thousands.[104] Gerald had called for the English to hold Wales after its conquest by building castles and ensuring access to them by clearing wide paths through the woods to hinder ambushes. Edward decided to begin the process during the campaign itself, but (unlike most English kings before him) neither to settle for construction a short distance into Wales, nor to drive in deep without a secure line of communications to the rear.[105]

Instead, during his advance, he would stop each 15 miles or so, begin construction of a fortress, and start by digging large ditches to protect a camp. As soon as the defensive works were sufficient to prevent the camp from being overrun by a sudden assault, he would garrison it with infantry and cavalry. Meanwhile, the main strength of the army would keep Llywelyn at bay and provide cover for a large force of axe-men – over 1,500 men during August – as they burned and cut

---

101 *ByT*, 365; Morris, *Welsh Wars*, p. 123; Smith, *Llywelyn*, pp. 416–8.

102 Morris, *Welsh Wars*, pp. 118, 141.

103 Ibid., p. 127. Edward III mustered some 3,000 men-at-arms for the 1346 campaign, and a similar number for 1359–60, even though these were overseas campaigns and, in the latter instance, after the Black Death. Clifford J. Rogers, *War Cruel and Sharp: English Strategy under Edward III, 1327–1360* (Woodbridge, 2000), Appendix 1.

104 Morris, *Welsh Wars*, pp. 127–32, 139. The abbot of St. Werburg's in Chester sent 100 laborers for the construction at Flint at his own expense; if others of the king's men did similarly, the total of workmen may have been substantially larger even than the pay records indicate. *CPR*, p. 226.

105 Though on the scanty evidence available John may have done the same in 1211.

down the forests to create an "extremely broad" open way to the next destination, advancing at a rate of something like half a mile per day.[106]

Once the way was clear, the army advanced to Flint. In addition to the quantities of wood made available by the forest-clearing, Edward had a great deal of prepared timber brought in by sea, allowing for construction of more lasting works and, probably more importantly, speeding up the process. With over 500 skilled carpenters and masons on hand, the labor went quickly.[107] Still, it took until 26 July for the army to reach Flint. Moreover, the 40 days' unpaid service owed by the feudal cavalry had ended before the army was ready to advance from Flint to Rhuddlan on 20 August. The earls, most of whom still brought their retinues at their own expense, remained with Edward nonetheless. With their men, approximately 125 volunteers from the feudal contingents retained at pay, and the king's own household, the strength of the cavalry was approximately 500 men, probably at least double what Llywelyn (deprived of his allies) could muster, and better equipped as well.

The mass of Edward's army at Rhuddlan, however, consisted of infantry (bowmen, spearmen, and a few crossbowmen), whom he had recruited in unprecedented numbers. Of these, approximately 3,000 came from Lancashire, Derbyshire, and Rutland, but the great majority, as Gerald had recommended, were from Wales (9,000) and the English Marches (3,500).[108] A good portion of the English infantry would have had light or medium armor (a quilted gambeson or short mail shirt and metal helmet), as Gerald had also called for.[109] Another 2,000 men or so, perhaps 10 percent cavalry and the rest infantry, patrolled the roads from bases at Chester and Flint. In addition, over 700 sailors manning 26 ships were on station to support the campaign.[110] In later wars, Edward used the royal right of purveyance (forced

106 Morris, *Welsh Wars*, pp. 131–135. Wykes's chronicle, in *Annales monastici*, vol. 4, p. 273 (quotation); Osney annals, in *Annales monastici*, vol. 4, p. 272 (burning).

107 Morris, *Welsh Wars*, pp. 130, 138–9. This required substantial advance planning. At the start of May a knight of the royal household was sent to supervise the taking of oaks from the forest of Chester for the construction at Flint (which the king would not reach until July). TNA, E101/3/15. Royal officials had been sent to diverse parts of England as early as mid-June, specifically to collect masons and carpenters, "as many as [they] can get, and in whosesoever's works or service they may be." When the clerks brought in one contingent, they were sent back out to gather more workers. *CPR*, p. 213; TNA, E101/3/16. Thomas Wykes's chronicle says both Flint and Rhuddlan were strengthened so well they became impregnable. *Annales monastici*, vol. 4, p. 273.

108 Morris, *Welsh Wars*, pp. 131–33.

109 The Assize of Arms promulgated by Henry III in 1252 required lesser gentry with freeholds worth 5 to 10 pounds of land revenues to own a purpoint and iron helmet; those with 10 to 15 pounds were to have a mail haubergeon and a helmet; but only those with more than 15 pounds had to maintain a horse. William Stubbs, *Select Charters*, 8th ed., (Oxford, 1905), pp. 371–2. Even earlier, in 1231, a levy of infantry from Gloucestershire for a Welsh campaign was limited to those with metal armor. Michael Prestwich, *Armies and Warfare in the Middle Ages. The English Experience* (New Haven, CT, 1996), p. 122.

110 Morris, *Welsh Wars*, pp. 128, 132–3.

purchase of goods at fixed, moderate prices) to acquire supplies for his armies.[111] But in 1277 he seems to have used it only for his own household troops and relied on profit-seeking merchants to keep the rest fed.[112] This served the purpose well enough: there are no reports of shortages, starving soldiers, or the butchering of prized horses for food.[113]

While the work of fortification and road-clearing proceeded, Edward's soldiers made frequent raids into Welsh territory. This was doubtless done partly to collect supplies and partly to keep the troops occupied. Nevertheless, Thomas Wykes's chronicle notes that Edward's purpose was to leave the Welsh destitute and use hunger to push them into their final refuge on Snowdonia, until they gave up all thought of resistance.[114]

Edward did not ignore the political aspects of the war during military operations. Llywelyn's brother Dafydd was in Edward's army, with 20 horsemen and 200 infantry at English pay.[115] In a document sealed at Flint on 23 August, Edward promised to provide him and his brother Owain (imprisoned by Llywelyn) with the share of Gwynedd they were due under Welsh law. This was a clear example of Gerald's recommendation to divide the Welsh by promising to one man what another held. However, the document also laid the groundwork for the post-combat occupation policy; it made clear that all Wales was forfeited to the crown, so that Dafydd and Owain would hold their lands as fiefs graciously provided by the king and carrying the same obligations connected to English fiefs, including attendance at Edward's parliaments, rather than as inherited properties. Moreover, the agreement included the proviso that Edward could retain for himself Anglesey and parts of Snowdonia, thus making future rebellions impossible, or at least impractical.[116] Gruffydd ap Gwenwynwyn of Southern Powys and Madog ap Llywelyn ap Maredudd, exiled claimant to the lordship of Meirionydd (held by Prince Llywelyn since 1256), were also in Edward's service.[117]

111 Ibid., pp. 84, 197.

112 This is indicated by the relatively small expenditures recorded on the relevant account roll – just £416 17s. 10d. ob., including the clerks' expenses – even though the heading of the document says the sums were spent on supplies to sustain the king's *army* ("*exercitus*") in the Welsh War. TNA, E101/3/16. Numerous letters of protection were issued to merchants bringing food to the army. *CPR*, pp. 224, 226, 227, 230.

113 In September the king was prepared to order that any goods taken from the priory of Daventry should be restored to the monks, and a whole group of similar orders was made as late as 16 October, which shows that the supply situation was still far from desperate. (Ibid., pp. 228–9, 232; note also the various protections "notwithstanding the need of the king and the others in the army of Wales," ibid.) Cf. Prestwich, *Edward I*, p. 181. That a messenger was sent in September to hasten the collection of supplies by royal purveyance does indicate some concern, but not necessarily actual "difficulties," of which there are no mentions in the chronicles or letters (in sharp contrast to most previous expeditions).

114 *Annales monastici*, vol. 4, p. 274; cf. Gerald, *Description*, p. 268.

115 Morris, *Welsh Wars*, p. 128.

116 *Littere Wallie*, pp. 103–4 ("pro securitate nostra et pacis populi seruande").

117 Smith, *Llywelyn*, pp. 425–6. So also, it would appear from the letters of protection they received in October on departing the army, were Rhys Fychan and Cynan ap Maredudd. *CPR*, p. 229.

At the end of August the army advanced to Deganwy on the Conwy River, the border between western and eastern Gwynedd. The English had tightened the ring around Snowdonia from both the east and south. The Conwy was a difficult barrier, but Llywelyn could not both defend it and protect the Isle of Anglesey in his rear. After reducing his logistical burdens by dismissing the troops in surplus of his needs (mainly Welsh infantry), Edward sent roughly half his remaining force by sea to Anglesey. Along with the soldiers went 360 harvesters with scythes and sickles to collect the island's grain. The wheat Llywelyn was surely counting on to sustain his people through the winter would instead feed the English garrisons occupying eastern Gwynedd and Ceredigion.[118]

Normally, if they succeeded in continuing resistance until the hungry invaders retreated for the winter, the Welsh would "burst out like rats from their holes" to raid the marches or otherwise obtain provisions "as they would commonly do, even in time of war, either by purchase, or by robbery, through friendship, relationship, or kindred."[119] With all Wales except Gwynedd under English control, and Edward's men holding Deganwy, Rhuddlan, Flint, Ruthin, and Llanbadarn, and also Anglesey, it would be practically impossible for the Welsh to bring in significant quantities of food. Rhuddlan and Flint, moreover, were being fortified on a scale that "outshadow[ed] anything that had gone before," including with waterside works to ensure that they could be supplied and reinforced by sea.[120] The principal route towards England would, thus, be blocked securely.

So far, Edward had followed Gerald's strategy practically to the letter. The next phase in the archdeacon's plan was to wait for the leaves to fall and then send detachments of Marchers and friendly Welsh to harry the refugees in the hills until they surrendered. For the first time in the Anglo-Norman wars against Wales, the English were in a position to do just that. By 20 September, the king had decided against further major field operations in winter,[121] but even after Edward had disbanded much of his army, he kept sufficient troops in the castles encircling Llywelyn's remaining dominions to "besiege Snowdon," as the contemporary chronicler Bartholomew Cotton put it, and harry the Welsh with continued raids.[122] Edward was even bringing in relays of fresh troops to replace worn-out soldiers, as Gerald had advised: 1,930 infantry joined the army as late as 23 September.[123] A few days

---

118 Morris, *Welsh Wars*, pp. 134–5.

119 Roger of Wendover, *Roger of Wendover's Flowers of History: Comprising the History of England from the Descent of the Saxons to A.D. 1235*, tr. J. A. Giles (London, 1849), vol. 2, p. 539; Paris, *English History*, vol. 2, p. 114, translation modified.

120 Norman J. G. Pounds, *The Medieval Castle in England and Wales: A Political and Social History* (Cambridge, 1993), p. 169. At Rhuddlan a canal was dug to facilitate supply by water.

121 He sent his tents back to the Tower of London on 20 September. *CPR*, p 229.

122 Bartholomew Cotton, *Historia Anglicana*, ed. Henry Richards Luard (London, 1859), p. 155. Another contemporary similarly described "Wales encircled and besieged." *Opus chronicorum*, in *Chronica monasterii S. Albani. Johannis de Trokelowe*, etc., ed. Henry Thomas Riley (London, 1866), p. 38.

123 Morris, *Welsh Wars*, p. 136. These troops only stayed one week, probably because they were deemed unnecessary and a logistical burden.

before that, two merchants were given letters of protection to bring food in to the army, lasting until Christmas.[124] An extension of the campaign into winter proved unnecessary, however, since by early November Llywelyn was ready to capitulate.

The Settlement of 1277

The Treaty of Aberconwy, sealed on 9 November 1277, represented a disastrous defeat for Llywelyn and a clear victory for Edward. Nevertheless, the prevailing view is that it represented a "negotiated settlement" rather than a "total submission" by the Welsh prince, and that Edward made concessions to Llywelyn due to his difficulties and to avoid the costs and dangers of continuing the war until the English had completely overrun Gwynedd.[125] The evidence is not sufficient to determine if this is correct, but it appears to this author that the treaty did indeed represent a "total submission": perhaps a negotiated *surrender*, but not a mere negotiated *settlement*.[126] Indeed, Edward seems to have designed the treaty's terms specifically to make clear that it was *not* the result of a compromise imposed on him by Llywelyn's continued resistance. Rather, Edward intended the Treaty of Aberconwy to reflect the status of Wales as a conquered land, his to dispose of however he wished.

By its terms, Llywelyn retained possession of the lands he still controlled, and even regained Anglesey, but he held the island only as a life grant, for which he would pay an annual fee to the royal treasury.[127] On the prince's death, if he had legitimate children, they were to receive part of Anglesey (the rest falling to Edward) and part of western Gwynedd (the rest going to Dafydd). If, as seemed fairly likely, Llywelyn died without heirs of his body, all his lands would revert to the crown, rather than his next of kin. Llywelyn's elder brother Owain Goch was to be freed from prison and settled on suitable lands, which were to come out of

---

124 *CPR*, p. 230; cf. also p. 222 for protections from September and October to last until Easter.

125 Smith, *Llywelyn*, p. 434; Prestwich, *Edward I*, pp. 181–2.

126 I thus disagree with the view that Wales was "not conquered" in 1277. Prestwich, *Edward I*, p. 182. In evaluating the Treaty of Aberconwy it should be remembered that during the course of the war of 1276–1277 Edward gave similar terms to lesser Welsh *tywysigion* who sought to enter his peace, even when he clearly had the ability to dispossess them completely if he chose. A lord who had learned the price of rebellion through defeat and partial confiscation, but who on the other hand still retained enough land to be a useful servant, and one with something to lose if he rebelled again, could be an ideal vassal. Llywelyn had argued, in his last-ditch effort to avoid the war of 1276–1277, that "he would be of greater service to the king than those who, even though they waged the king's war, sought their own advantage rather than the king's honour." (Smith's paraphrase, *Llywelyn*, p. 410). This did not carry much weight when the prince was unwilling to do homage, but once he had fully admitted his subordination to the English crown, it might well be persuasive.

127 This was to make a point that they were held by the king's grace; once the point was made by inclusion in the treaty, Edward immediately granted that the sum need not be paid. *Calendar of the Welsh Rolls* [hereafter *CWR*], in *Calendar of Various Chancery Rolls: Supplementary Close Rolls, Welsh Rolls, Scutage Rolls* (London, 1912), p. 158.

Llywelyn's remaining possessions. Eastern Gwynedd and Ceredigion were to be retained by the king, as were Builth, Cydewain, Ceri, and Dinefwr and the other castles of the Tywi valley, and whatever else he or his men had occupied. Although the Welsh were to retain their own law, Edward and his judges were to be supreme in the interpretation of that law, with even cases arising inside Gwynedd subject to appeal to the king. The homages of all the major Welsh princes, which under the Treaty of Montgomery had been due to Llywelyn as Prince of Wales, were now to be rendered directly to Edward.[128]

On the other hand, Edward allowed Llywelyn to retain the title of "Prince of Wales" and the homages of five minor *tywysogion* of distinguished lineages but little territory, "since," as a contemporary summary of the text explains, "he could not call himself 'prince' if he had no barons under him."[129] These concessions, however, were personal and would lapse with Llywelyn's death. The king also allowed the prince to complete his already-contracted marriage to Edward's first cousin, Eleanor de Montfort, grand-daughter of Henry II. In evaluating the significance of these sops to the prince's dignity, one should remember Gerald's admonitions that their conqueror should treat the Welsh magnanimously after their surrender, since they valued honors above all else, and especially treasured the opportunity to marry into illustrious bloodlines.[130]

Nevertheless, apparently to ensure that neither Llywelyn nor anyone else would conclude these concessions reflected the strength of the Welsh negotiating position, or any weakness on Edward's part, Llywelyn also had to agree to pay the immense sum of £50,000 as a fine for his "disobedience." Since he could not possibly pay such an amount, he could only ask for the king's "grace and pity" regarding it. Edward promptly forgave the debt, but he had made the point: he had the right to take for himself everything Llywelyn possessed. If he did not do so, it was only by choice, his "grace and mercy" to a prostrate foe.[131]

Even within the portion of Gwynedd retained by Llywelyn, moreover, his lordship over his own men would be expressly subordinated to the loyalty they owed the English crown. Ten of the most noble of Llywelyn's followers, delivered to Edward as hostages, arrived at Chester to swear fealty to Edward on the fragment of the True Cross – a relic for which the Welsh were known to have particular veneration. They had to promise they would never bear arms against the king or in any way oppose him, and that if Llywelyn or other Welsh leaders should again rebel,

128 *Littere Wallie*, pp. 118–22 for the Latin; *Annales monastici*, vol. 4, pp. 272–4, for an interesting summary in French. Related documents are summarized in *CWR*, pp. 158–9.

129 From the French summary of the agreement, *Annales monastici*, vol. 4, p. 273. For the small practical value of this concession, see Smith, *Llywelyn*, p. 443.

130 *Description*, pp. 271; 251 ("The Welsh value distinguished birth and noble descent more than anything else in the world. They would rather marry into a noble family than a rich one."), p. 263.

131 "En la grace e la pite del rey" in the French summary of the treaty (*Annales monastici*, vol. 4, pp. 272–3); "graciam et misericordiam" in the Latin (*Littere Wallie*, p. 119).

they would serve the king with all their strength to crush the rebellion.[132] Even more gallingly, 20 leading men from each cantref remaining to Llywelyn had to swear a similar oath *every year*, for an indefinite period.[133] These provisions only reinforced the point already implicit in the treaty's territorial terms: everything Llywelyn retained, he owed to Edward I's munificence, not his hereditary rights or power of continued resistance. He was not to be allowed to doubt that the English had indeed conquered Wales, and that his dream of a semi-autonomous native state under a Welsh prince had been broken "like a clay pot," as young Edward had once threatened. Except for privileged areas held by English Marcher Lords, all Wales was subject to the crown's authority just as thoroughly and directly as the counties of England were, and moreover a much larger portion of Wales than of England was under the immediate lordship of the king, rather than his vassals.[134]

Epilogue

Edward had subdued Wales by following the strategic plan laid out long before by Gerald of Wales, but he was somewhat less attentive to the archdeacon's advice for post-combat occupation. Admittedly, he did, by his own lights if not by those of the conquered, take into account Gerald's observation that the Welsh "want more than anything else to be honoured," and his advice that "once they have paid the penalty for their wrongdoing and are at peace again, their revolt should be forgotten as long as they behave properly, and they should be restored to their former position of security and respect, for 'The quarrel over, it is wrong to bear a grudge.'"[135] In addition to allowing Llywelyn to marry into the royal family, Edward himself gave away the bride and, what is more, paid for the wedding. He also almost immediately allowed Llywelyn's hostages to return home once they had sworn fealty to him, explicitly as an expression of Edward's faith that Llywelyn would remain faithful to his obligations – probably not at all what they had expected when delivered to English custody![136]

Even before the end of the war, Edward had begun to implement Gerald's recommendation to "build castles, [and] widen the trackways through the woodlands," and indeed he did more in these respects than Gerald could probably have imagined. He vigorously continued to build up Flint, Rhuddlan, Builth, and Llanbadarn outside Aberystwyth, creating royal boroughs attached to each, and retaining and strengthening Dinefwr, Carreg Cennen, and Llanymddyfri (Landovery) in

---

132 *CWR*, p. 169; *Littere Wallie*, p. 121. In 1279, the bailiff of Ruthin was ordered by Edward to allow four Welsh lords to remain in Llywelyn's service "for so long as it shall please the king, saving the king's faith." *CCR*, p. 564.

133 *Littere Wallie*, p. 121.

134 Especially after Edward in November 1279 acquired Carmarthenshire and Cardiganshire from his brother Edmund, who was given English estates in compensation.

135 *Description*, p. 271.

136 *CWR*, 169. Remember the Welsh hostages executed or mutilated by John and Henry II when their lords or family-members rebelled.

Deheubarth. Edward spent more on these massive works, over five years, than on the conduct of the war itself.[137] Although he did make use of some Welsh officers, the castles, as Gerald had recommended, remained under the control of English captains and garrisons.[138] In addition to clearing the road from Chester to the Conwy during the war, after the peace treaty Edward ordered Marcher barons and native rulers alike "that the passes through the woods in diverse places in Wales should be enlarged and widened, so that access might be more open to those travelling through." Commissioners were sent to ensure that this was done, and all royal subjects ordered to assist them.[139]

However, though Edward himself might have disagreed, a modern historian cannot say that the conquered lands kept under royal control were ruled "with great moderation." Gerald had emphasized the importance of dealing justly with the subject population. "The governor appointed must be a man of firm and uncompromising character," he wrote. "In times of peace he will observe the laws, and never refuse to obey them; he will respect his terms of appointment and do all in his power to keep his government firm and stable." Gerald noted how easy it was for officers in such situations to "turn a blind eye to lawlessness, allow themselves to be influenced by flattery . . . rob the civilian population in time of peace, [and] despoil those who can offer no resistance."[140] The problem started at the top. Although Edward presented himself as the fount of justice, and is remembered in history as "the English Justinian," in the case of his first major dispute with Llywelyn over lordship of the district of Arwystli, he "made a mockery of justice by turning the law into an instrument of his own power."[141] As historian R. R. Davies notes, "high-handed and tyrannical officials" in the newly occupied royal territories received "too free a hand to bully native society into submission." All too often, even where the Treaty of Aberconwy had guaranteed the natives their ancient liberties and customs, the English failed to uphold these promises.[142]

The result was a great rebellion, launched in March 1282 by Dafydd ap Gruffydd and several other Welsh princes, but soon enough coming under Llywelyn's leadership.[143] This required a second conquest of Wales, employing the same methods as the first, but on an even larger scale, and pressed to the bitter end. Once again Edward created three military commands to begin operations in the

---

137 Rhuddlan alone ultimately cost some £11,000, almost half the cost of the 1276–1277 war. Morris, *Welsh Wars*, p. 145. Edward's expenditures on his Welsh castles over the course of his reign amounted to around £80,000. Very full details can be found in *The History of the King's Works*; the relevant sections have been published separately as Arnold Taylor, *The Welsh Castles of Edward I* (London, 1986).

138 *Description*, p. 271.

139 *CWR*, pp. 164, 168, 171, 173, 188.

140 *Description*, p. 271.

141 Davies, *Age of Conquest*, pp. 345–7, quotation at p. 346.

142 Ibid., p. 348

143 As with the war of 1276–77, this war can best be understood by reading Morris's *Welsh Wars* in conjunction with Simth's *Llywelyn*.

north, the Middle March, and the south-west. These forces restored royal control in Ceredigion and the Tywi Valley, and kept the Welsh in check in Powys. An even larger force of infantry than in 1277, supported by an even greater fleet, deployed for the main operations in the north. Dafydd's castles of Ruthin and Denbigh were captured and eastern Gwynedd secured. The English again subdued Anglesey with a large amphibious force. This detachment suffered a minor disaster when the Welsh attacked and defeated the head of the column crossing the Menai straits by a pontoon bridge, but the English retained control of the island. Warm clothing was provided for the troops, and the war continued into the winter without remission.

It was December when Llywelyn himself was killed in a skirmish near Builth. Dafydd continued the struggle in his place, but in January Edward's main force crossed the Conwy and moved into Snowdonia to capture Dolwyddelan. The troops on Anglesey crossed over and moved into the mountains from the west. By April all of Snowdonia had fallen, and the fortress of Castell-y-Bere in Merionydd, Dafydd's last stronghold, surrendered to the king's mercy. In June, Dafydd and his entire family were captured. Edward had Dafydd hanged, drawn, and quartered and refused his niece (Llywelyn's daughter) and children permission to marry, which practically extinguished the line of Llewelyn Fawr. Anglesey, Meriontheshire, and western Gwynedd (renamed Caernarfonshire) became direct royal lordships, secured by massive and architecturally state-of-the-art castles at Beaumaris, Cricieth, Harlech, and Castell-y-Bere. Edward largely dispossessed the Welsh rulers of Powys, the Middle March, and Deheubarth, except for two *tywysogion* who had remained loyal to the Crown. "For the native dynasties of Wales, the disinheritance of 1282–83 . . . was as traumatic as were the events of 1066–70 for the Anglo-Saxon aristocracy."[144]

There would be other revolts, under Llywelyn ap Gruffydd's distant cousin Madog ap Llywelyn in 1294–95, and Owain Glyndwr in 1400–1412, but they had no chance from the start. The experiences of 1276–77 and 1282–83 had demonstrated that the crown's resources, when mobilized by a determined ruler and *applied in accordance with a sound strategy*, were sufficient to overcome even the difficult problems posed by a warlike people skilled in guerilla tactics, fighting to defend a region fortified by nature with formidable swamps, great forests, and rugged hills, and inspired by "the sheer joy of being free."[145]

144 Davies, *Age of Conquest*, p. 361.
145 *Description*, p. 270.

8

THE ANGLO-BURGUNDIAN ALLIANCE AND GRAND STRATEGY IN THE HUNDRED YEARS WAR

The Formation and Significance of the Anglo-Burgundian Alliance

From 1419 to 1435, the alliance between England and the nascent state of Burgundy, an agglomeration of territories spanning the border between France and the Holy Roman Empire, deeply influenced the course of the great war between France and England known as the Hundred Years War (1337–1453).[1] The alliance with Burgundy fundamentally changed the English approach to the war. Before 1419 King Henry V of England broadly adhered to the diplomatic position staked out by his great-grandfather Edward III during the first phase of the war: although the crown of France belonged to him by hereditary right, in the interest of peace he would be willing to compromise and accept sovereign rule over just a half or a third of the kingdom, with the rest (and the title king of France) passing to his opponent, the Valois king Charles VI.[2]

Even though England had at best a third of the population and wealth of France, the two kingdoms were fairly well matched in military strength, as demonstrated by the inability of either side to end the conflict in victory from 1337 to 1419. The French always had the advantage in number of soldiers, but the disparity was not as great as the difference between the national populations. The English royal government was more efficient than the French bureaucracy, and parliament, as a mechanism for enlisting national consensus behind the war effort, provided the kings of England with a substantial edge in the mobilization of resources. Moreover, English qualitative military superiority usually offset, sometimes

---

1 My thanks to Professor Craig Taylor for assistance with this study. Considering its intended audience, I have where possible cited translations rather than texts in Latin or French. The version printed here is practically unchanged from the previously published version.
2 For Edward III, see Clifford J. Rogers, "The Anglo-French Peace Negotiations of 1354–1360 Reconsidered," in *The Age of Edward III*, ed. James Bothwell (York, 2001), pp. 193–213. For Henry, see Christopher Allmand, *Henry V* (Berkeley, CA, 1992), pp. 68–74; and Anne Curry, *Agincourt: A New History* (Stroud, 2005), pp. 40–49.

DOI: 10.4324/9781003399971-11

dramatically, French advantages in quantity. That qualitative advantage had enabled them to win every full-scale battle of the war up to the formation of the Anglo-Burgundian alliance – including the famous victories of Crécy (1346), Poitiers (1356), and Agincourt (1415) – and most of the lesser fights as well.

Three qualitative factors are most important for explaining these tactical successes. First, the English archers, whose extraordinarily powerful longbows could drive four-ounce arrows with enough accuracy to hit an individual target at 220 yards and enough kinetic energy to penetrate most armor, were by far the best infantry of their day.[3] Second, the whole English army, from foot-archers to army commanders, was deeply imbued with what Clausewitz calls "the military spirit."[4] Third, in most periods of the war the top English commanders demonstrated far greater tactical and strategic competence than their French adversaries. When the French found good generals such as Constable Bertran Duguesclin and "Le bon duc" Louis of Bourbon, resources tipped the balance marginally in their favor, but the advantage of the defensive in the siege-based campaigns that constituted the period's predominant form of warfare prevented them from driving the English off the continent and ending the conflict.[5] The strength of the defensive also, however, meant that even when they enjoyed a large advantage in political and military leadership, as they often did, the English simply did not have the manpower to conquer all of France, or even to defend, in the long run, any substantial gains they might temporarily make.

So long as English qualitative advantages balanced French quantitative ones, there was little prospect of either side bringing an end to the conflict by achieving complete military victory. There were times when political leaders on both sides appreciated this and worked to resolve the conflict through diplomacy and negotiations. Time and again, these efforts foundered on the fact that the minimum acceptable terms of the two sides were fundamentally incompatible. The essential issue of the war was not how much land within France the English should rule – on that there could have been compromise – but rather the question of whether the English territories in France should be sovereign possessions of the kings of England, or should remain legally and politically subordinated to the crown of

---

3 The ability of longbow arrows to penetrate armor remains controversial. The previous statement is not meant to imply that armor was of no use against arrows; arrows often glanced off plate surfaces, and even when an arrow did penetrate, armor could still mean the difference between a grave wound and a relative pinprick. See "The Battle of Agincourt," *The Hundred Years War (Part II): Different Vistas*, ed. L. J. Andrew Villalon and Donald J. Kagay (Leiden, 2008), pp. 109–113; for the best statement of a more conservative view of the longbow's efficacy, see Matthew Strickland and Robert Hardy, *The Great Warbow: A History of the Military Archer* (New York, 2005), ch. 15.

4 Carl von Clausewitz, *On War*, ed. and trans. Michael Howard and Peter Paret (Princeton, NJ, 1984), pp. 187–89.

5 See Clifford J. Rogers, "The Medieval Legacy," *Early Modern Military History*, ed. Geoff Mortimer (London, 2004), pp. 6–24, and idem, "The Artillery and Artillery Fortress Revolutions Revisited," in Nicolas Prouteau, Emmanuel de Crouy-Chanel and Nicolas Faucherre, eds., *Artillerie et fortification, 1200–1600* (Rennes, 2011), pp. 75–80 (chapter 4 in this Variorum volume).

France. The political theories of the earlier Middle Ages had been able to accommodate divided jurisdictions and shared suzerainties comfortably, but the conception of sovereignty that had developed over the course of the thirteenth and early fourteenth centuries, under the influence of the revival of Roman law, insisted on the absolute and ultimately inalienable power of the state. There was therefore no middle ground on which the English or French could construct a durable peace.[6]

The impossibility of finding a diplomatic compromise derived in part from the dynastic aspect of the struggle – the fact that the kings of England, as heirs of Edward III, claimed to be the rightful kings of France. If an English king accepted anything less than sovereignty over a share of France, he would in effect be dishonoring himself and his ancestors, and admitting that they had for generations waged an unjust war, at the cost of countless lives, immense treasure, and terrible devastation. If a French king surrendered sovereignty over any portion of France, he would be doing the same.

So long as compromise was impossible and the military balance between England and France was too even to allow either side to overcome the strategic inertia imposed by the superiority of the defense to achieve complete victory, there was no way to bring the war to an end. But in the early fifteenth century, the intermittent madness of King Charles VI of France led first to intense factionalism in the court and then to outright civil war as two parties struggled to gain control over the royal government and revenues. One of these factions was led successively by the king's brother, the duke of Orléans, then that duke's son and successor, then by Bernard d'Armagnac, constable of France (after whom this party is often referred to as "the Armagnacs"), and finally by the dauphin Charles, the future Charles VII. The other party was headed by King Charles VI's cousin, John the Fearless, duke of Burgundy, then by John's widow, Margaret of Bavaria, and his son and successor, Philip the Good of Burgundy.

The dukes of Burgundy were direct rulers of two large, wealthy, and compact territorial blocs. One, southeast of Paris, comprised the duchy of Burgundy, the county of Charolais, and the county of Burgundy (the Franche-Comté). The other bloc, northeast of the capital, included Flanders and Artois, among the most urbanized and wealthiest areas of France outside Paris. In addition, the comparably rich and populous Imperial duchies of Brabant and Namur and counties of Hainault, Holland, and Zeeland were bound firmly to Burgundy throughout this period, and eventually acquired by Philip himself. The duke's chief adviser reckoned in 1431 that the Burgundian territories, including the lands listed previously as well as the counties of Auxerre and Mâcon and the castellany of Bar-sur-Seine (acquired in 1424), collectively contained from a third to a half as many parishes as the entire kingdom of France.[7] The whole area between these two blocks was throughout

---

6 For the negotiations after 1360, see J. J. N. Palmer, "The War Aims of the Protagonists and the Negotiations for Peace," in Kenneth Fowler, ed., *The Hundred Years War* (London, 1971).

7 The memorandum actually favors the figure of one half, but takes one third to be conservative. *Oeuvres de Ghillebert de Lannoy, voyageur, diplomate et moraliste*, éd. Ch. Potvin et J.-C. Houzeau

the period an area in which the duke of Burgundy had great influence. He also had many supporters in Paris and the surrounding towns.

Until 1417, the Orléanist or Armagnac faction generally controlled all of France west and southwest of Paris, except for the English duchy of Guienne or Aquitaine (centered on Bordeaux), and Brittany, which followed an independent policy. Starting with the capture of Harfleur in 1415 and continuing until 1422, but mainly between 1417 and 1419, Henry V of England reduced the dominions of the Orléanist or Armagnac faction by the military conquest of Lower Normandy and various adjacent areas.

In the fourteenth century, the English had taken advantage of local civil wars within Flanders and Brittany to gain temporary dominance in those provinces, and later the conflicts among Charles of Navarre (a member of the French royal house with extensive possessions in the north of the realm as well as king of his own small Pyrenean kingdom), the bourgeoisie of Paris, and the house of Valois had contributed substantially to the collapse of French resistance that led to the Treaty of Brétigny in 1360.[8] But the civil war between Orléans and Burgundy was different: the two contenders were sufficiently balanced with each other and with England that what had been a game of two main players (the kings of England and the kings of France) and a variety of pawns became a three-way contest for dominance. Partly because of this change in dynamics, the stakes changed too.

As late as the spring of 1419, the two main questions were: first, what portion of France, if any, would be detached from the kingdom and transferred *de facto* or *de jure* to English sovereignty; and second, which French faction would dominate the remainder of the kingdom? That changed in September 1419, when, despite the recent reconciliation between the two French factions, Duke John the Fearless was murdered while meeting with the dauphin.[9] This naturally drove his son and successor, Duke Philip the Good, into the arms of the English. Since Philip's faction controlled Paris and mad King Charles VI, he was able to offer a great deal in exchange for English help in securing himself against his enemies and in gaining vengeance for his father's murder. The result was the Treaty of Troyes (1420), by which Charles VI and his queen disinherited their only surviving son (the dauphin), provided for the marriage of their daughter Katherine to Henry V, and made their new son-in-law the regent of France and heir to the French throne. Henry was to rule France along with a council of French nobles of the Burgundian party.[10]

(Louvain, 1878), p. 488; Richard Vaughan, *Philip the Good. The Apogee of Burgundy* (Woodbridge, 2002), p. 260.

8 Jonathan Sumption, *The Hundred Years War*, vols. 1–2 (London, 1990–1999), or more concisely, Clifford J. Rogers, *War Cruel and Sharp: English Strategy under Edward III, 1327–1360* (Woodbridge, 2000).

9 Allmand, *Henry V*, 132–35, offers a good summary of these events; for a fuller treatment, see Richard Vaughan, *John the Fearless: The Growth of Burgundian Power*, 2d ed. (Woodbridge, 2002), ch. 10.

10 The Treaty of Troyes is in E. Cosneau, *Les grands traités de la Guerre de Cent Ans* (Paris, 1889), pp. 102–115; for the council, see clauses 7, 27. Armstrong's conclusion that clause 27 "reserved to

As the terms of the treaty indicate, Henry no longer had any interest in pursuing a partition of France along the lines of the Treaty of Brétigny. The combination of recent English military successes and the alliance of England and Burgundy created prospects that Henry V might make good the long-standing English claim to the French throne and end the war in complete victory rather than merely a favorable compromise. From that point onward, for the first time since 1337, none of the main protagonists was pursuing the strategic goal of dividing France. The prize at which all aimed was not the largest share of French territory, but the largest share of power within a united France. This fundamental change in the nature of the war resulted directly from the creation and cementing of the Anglo-Burgundian alliance.

The Maintenance of the Alliance

To achieve their new strategic goal, the English had to do two things. First, they had to retain the support of Burgundy. Second, they had to continue to capture the fortified towns and castles that brought with them dominion over the surrounding agricultural countryside, one after another, until the balance of resources tipped so far that a complete Lancastrian victory seemed inevitable. At that point the Armagnac leaders would have to submit to Henry's authority and the war would end. From 1415 to 1419 English armies had made fairly rapid progress towards that latter goal, despite having to keep a wary eye on Burgundian territory. The crushing English victory at Agincourt had left the dukes of Orléans and Bourbon prisoners in England, Brittany intimidated into English-leaning neutrality, and a whole generation of French men-at-arms worse than decimated.[11] John the Fearless, meanwhile, had also gained substantial ground (including control of Paris, and of Charles VI himself) despite needing to defend the line of the Seine against the English.[12] Under the new circumstances, the complete military defeat of the dauphin's party was not an unrealistic goal, even if the English and Burgundians

---

the duke of Burgundy the second place [after Henry V] in the government of the realm of France" rests on a misquotation of the text; to be precise, clause 27 only gave Burgundy that role in governing the arrangements for Charles VI personally. Nonetheless, there is no doubt that the overall spirit of the treaty did imply that Philip, as the head of the Burgundian faction, would have a leading role in guiding royal policy. C. A. J. Armstrong, "La double monarchie France-Angleterre et la maison de Bourgogne (1420–1435). Le déclin d'un alliance," in *England, France and Burgundy in the Fifteenth Century* (London, 1983), no. 11, p. 343.

11 The number of noble families in France at this time can be estimated at around 30,000; the number of noble adult males of fighting age (which was a wide range at that time) can be roughly approximated at around 35–40,000: Philippe Contamine, "The French Nobility and the War," in Fowler, ed., *The Hundred Years War*, pp. 136–139. The number of nobles killed at Agincourt was probably at least 5,000–6,000: Rogers, "Agincourt," pp. 104, 104–5 n. 234. In World War I, French military dead amounted to roughly 10 percent of the adult male population of 1914. Among the nobility, therefore, the single day of Agincourt was a greater demographic catastrophe than the entire First World War was for France as a whole.

12 Vaughan, *John the Fearless*, pp. 216–27, 263.

cooperated only in the loosest fashion: each sparing the other the need to guard their mutual frontier, and both pressing the war wherever it suited their interests.[13]

On the other side of the equation, it was practically impossible for the Armagnacs to win the war outright so long as the alliance between Burgundy and England held.[14] Their priorities, therefore, had to be to stave off Anglo-Burgundian conquests and buy time, hoping for the rescue of their cause by extraneous factors – probably a break between England and Burgundy, or perhaps a civil war within England. The latter seemed like a possibility, considering that Henry V was the son of a usurper and not, from a genealogical perspective, the rightful heir of Richard II, who had been deposed in 1399 and murdered in 1400.[15] The French also clearly had to do whatever they could to lure Philip the Good to reconcile with the dauphin, or at least to minimize his support to the English war effort.[16] Despite the murder of John the Fearless, this was not out of the question. More than once before over the course of the Hundred Years War, the bitterest enemies within France had patched up their differences to check English advances.

Once the three-way game was rolling and the English had, in effect, committed to pursuing the rule of all France rather than the acquisition of a part of it, all three players needed a partner to succeed, but only two pairings were possible, since the ambitions of the dauphin and Henry V were polar opposites and irreconcilable. It would either be Charles and Philip against Henry, or Philip and Henry against Charles. In other words, both Henry and Charles had only one choice of partner, but Philip had two, and the ultimate possessor of the crown of France was therefore likely to depend on which side he supported.[17] Both English and French

13 I am thus in agreement with Mark Warner, "The Anglo-French Dual Monarchy and the House of Burgundy, 1420–1435: The Survival of an Alliance," *French History* 11 (1997), pp. 103–7, rather than Allmand, *Henry V*, pp. 439–42, 146, who sees the Treaty of Troyes and Henry's commitment to insist on the throne of France as a "serious error of judgment," and seems to think the "rapidly growing sense of national spirit" in France made the prospect of a Lancastrian dynasty ruling France chimerical. See also Philippe Contamine, "La 'France Anglaise' au XVe siècle. Mythe ou réalité?" in *La 'France Anglaise' au moyen âge. Actes du 111e congrès national des sociétés savantes (Poitiers 1986)* (Paris, 1988), pp. 24–29, and A. Leguai, "La 'France bourguignonne' dans le conflit entre la 'France française' et la 'France anglaise' (1420–35)," ibid., pp. 41–52.

14 John de Waurin *A Collection of the Chronicles and Ancient Histories of Great Britain, Now Called England (1422–31)*, tr. Edward L. C. P. Hardy (London, 1891), p. 202.

15 The seriousness of this possibility was underlined just before Henry V sailed to reopen the war in 1415; a plot headed by the earl of Cambridge aimed at murdering Henry and replacing him on the throne with Edmund Mortimer, earl of March, who was descended from Edward III's second son rather than (as Henry was) from his third son. Allmand, *Henry V*, pp. 74–77.

16 There is a significant distortion involved in referring to the dauphinist party as "the French," since many Frenchmen, including the duke of Burgundy and most of the citizens of Paris, were on the side of Henry V and Henry VI for most of the period under discussion. However, historians often do so in order to keep their prose readable, and I have sometimes done the same in this essay.

17 Waurin summarizes the whole period from 1420–1431 by saying Henry V "reigned mightily in his own name in the kingdom of France, and that principally through the favour and alliance of the noble duke Philip of Burgundy." Waurin, *Chronicles*, p. 244.

strategy for the war, therefore, needed to aim at retaining or gaining his support. That required some understanding of his objectives.

What did Philip the Good want? One can summarize his goals under five headings.[18] First, and most important, came security. His father had effectively built Burgundy into an independent polity; Philip wanted above all other priorities to retain his *de facto* sovereignty over his own lands. That meant defending their territorial integrity and also protecting his own person, his family, and his supporters against his enemies in the civil war: the partisans of Orléans and Armagnac, and (though less inexorably) the dauphin.[19] Second, he sought to expand his territorial holdings, meaning not so much increasing the area he controlled *de facto*, but rather adding new hereditary lands to his own and his family members' patrimonies. Third – partly as a means to the first end, but partly as another goal in itself – he wanted power and influence within the kingdom of France.[20] Fourth, he demanded vengeance for his murdered father. And fifth, he wanted to be respected as a knight and a military leader.[21] A *sine qua non* constraining his pursuit of these

18 In this paragraph I largely follow the analysis of Vaughan in *Philip the Good*.

19 On the primacy of this consideration, note that the point that declining to ally with England would mean allowing Philip's mortal enemy to come to the throne of France was used as the clinching final argument by the counselors arguing in favor of an alliance with Henry V. See the summary of the debate in Allmand, *Henry V*, p. 140.

20 In 1418 John the Fearless had controlled Paris and Charles VI, and with the support of Queen Isabeau had therefore controlled the royal government of France itself, though due to the civil war his power did not in practice make itself felt in all parts of the realm. Philip clearly had less interest in being the power behind the French throne than his father had demonstrated, and ultimately he put much less effort into this goal than into the expansion and consolidation of his own territories, but that seems to have been a contingent rather than a predetermined development. Chance developments gave him a series of favorable opportunities to gain hereditary territories; had he not put so much energy into pursuing those opportunities, he might have made more effort to dominate the Parisian government instead. It should be noted that before the Treaty of Troyes Philip sought appointment as lieutenant-general of the realm from Charles VI and Queen Isabeau, and in the Treaty of Troyes itself Philip ensured a by-name place for himself in overseeing the "government" of Charles VI's person and household, and that the government of the realm would be undertaken "with the counsel of nobles and wise men" who were "obedient" to Charles V – that is, Frenchmen of the Burgundian party. Leguai, "'France bourguignonne,'" p. 44; Cosneau, *Grands traités*, pp. 105, 112–13.

21 Two key pieces of evidence reveal how important this was to him; it was not merely a matter of conventional rhetoric. First, when he accepted and went into intense training in preparation for a duel with Humphrey of Gloucester, his actions backed up his professed willingness to defend his knightly honor at the risk of his own life. Second, a memorandum of Philip's counselor Hughes de Lannoy, who knew him well, shows that Lannoy felt there was a risk that the duke would act contrary to his own interests if he felt that "his honor ha[d] been tarnished" by his failed siege of Calais. Vaughan, *Philip*, 38–39, 105–6. (It is indicative of the one major weakness of Vaughan's otherwise excellent work that he found it "bizarre" that "Philip really was naïve or impetuous enough to entertain serious dueling intentions.") Also, Waurin, a Burgundian soldier who was present at the time, described Philip in 1423, when he arrayed with the English to resist an expected French relief army outside Mâcon, as "considered the most chivalrous prince in the world" and affirms that "the thing that duke Philip most desired was to find himself in arms." Waurin, *Chronicles*, 48.

five objectives was to "guard his honor, against which he would not take any action, regardless of any consequences."[22]

The creation, maintenance, and disruption of the Anglo-Burgundian alliance was, therefore, the business of rulers, their councils, and their diplomats. Modern historians – unlike contemporary observers – have usually treated it from the perspective of diplomacy and high politics, sometimes adding in biographical and interpersonal considerations, but often largely divorcing the subject from military events and actions. To be more precise, scholars have attended to how changes in the alliance affected the war, but said relatively little about how the war affected the alliance. This perspective is well summarized in the remarks of the leading Anglophone historian of Valois Burgundy, Richard Vaughan, about the period when the Anglo-Burgundian alliance was being forged:

> For the Burgundian chroniclers, and perhaps for the participants, Philip the Good's French campaigns in the years after 1420 seemed of paramount interest. But for us, viewing the whole long reign in the perspective of history, these military activities assume a secondary importance.

22 Those familiar with the literature on Philip, which tends to depict him as a rather shifty or duplicitous figure, may be surprised to see this assertion, but the words are from private Burgundian diplomatic documents. They are from a point in 1423 when Philip first opened negotiations with the court of the dauphin over a possible reconciliation between him and Charles. Vaughan, typically, saw these negotiations as a sign that Philip, "far from appearing as an outright English partisan . . . now emerged as a sort of *tertius gaudiens*, prepared to negotiate seriously with either side, and intent on extracting material advantages for himself by playing one off against the other." (*Philip*, 8). But in fact in these documents Philip does show himself a continued partisan of Henry VI (referring to Charles as the dauphin rather than the king, and calling the dauphinists "la partie adverse"), and insistent on sustaining the alliance he had committed to with the Treaty of Troyes. At his instructions, Philip's ambassadors emphasized to the duke of Savoy, who was acting as a mediator, that Philip expected him to "guard his honor, against which he would not take any action, regardless of any consequences" ["contre lequel il ne vouldroit rien faire pour chose qu'il lui pust avenir"], and provided Savoy with copies of his treaties, promises and oaths towards the king of England so that Savoy could know what would be necessary to keep his honor safe. Then Burgundy's chancellor, Nicolas Rolin, told the dauphin's representatives directly that Philip would not "break his oath and promises" for any reason, and that he was only willing to proceed with negotiations insofar as they did not violate his treaty with the English. Rolin did suggest terms on which reconciliation between Charles and Philip might be possible – but only, as he explicitly said, as part of a three-way peace with England as well. Arsène Perier, *Nicolas Rolin, 1380–1461* (Paris, 1904), pp. 98–107. Even when in 1435 Philip had become inclined to break the treaty with England, the strongest argument that his pro-English counselor Hughes de Lannoy mustered to dissuade him from doing so was that breaking the Treaty of Troyes would stain the duke's honor, "which honor is something that all princes and knights should have their principal regard for, above all wordly things." The pro-French party had to counter this argument by agreeing that the duke "should desire to seek good renown, founded on virtue" above all else, then turning the argument on its head with the threat that if he did *not* break the Treaty of Troyes (which, they argued, was illegal anyway), the cardinals overseeing the negotiations would report him as being false to his promise to seek peace. Joycelyne Gledhill Dickinson, *The Congress of Arras, 1435. A Study in Medieval Diplomacy* (New York, 1972), pp. 70, 74. See also ibid., p. 77, and note 43.

It was the diplomatic system [developed by Philip] which ensured the peace and security of Philip's lands in these years, not the battles and sieges.[23]

And yet, of the five goals adumbrated earlier, the *results* of military operations had large and direct influences on the first (Burgundian security), the second (Burgundian expansion), and the fourth (vengeance), while Burgundian *participation* in active warfare was absolutely essential to the fifth (martial honor). As to the third, Burgundy's power in the France depended in the first instance on the power of his Lancastrian ally in the game of thrones, which in turn depended on military success. Moreover, the strength of the Anglo-Burgundian alliance was to a substantial degree shaped not just by the outcomes of various military operations, but also by what happened in the *process* of conducting coalition warfare.

Another major factor that has not received quite the attention it deserves (though it has been given more notice than the military dimensions of the alliance) is how the chances of births, deaths, and marriages and their implications for family politics affected the rise and fall of the alliance, and thereby the fate of the Lancastrian kingdom of France. Historians have long recognized that the 1423 marriage of Philip the Good's sister Anne to John, duke of Bedford, Henry VI's uncle and his regent in France, was an important factor in binding England and Burgundy together, until her death in 1432.[24] (The infant Henry VI succeeded to the dual monarchy of England and Lancastrian France after Henry V's and Charles VI's deaths in 1422.) What has not received adequate emphasis is that a key strategic weakness of England in the 1420s was Henry VI's lack of marriageable relatives. Philip the Good's first wife was the daughter of Charles VI and the sister of both the French dauphin and Henry V's queen, Katherine of Valois; in 1421 this meant Philip was uncle by marriage, but not by blood, to the future Henry VI, and brother-in-law to the Lancastrian queen of France, Katherine.[25] Thus, it was through his wife that Philip had two fairly strong family ties to the Lancastrian dynasty, but she died in 1422, the same year as Henry V, leaving Philip both without an heir and without close family ties to the infant Henry VI. Duke Philip remarried in 1425, and then again in 1430. Had a suitable English bride been available to tie Burgundian and Lancastrian family interests together, and had Henry VI's counselors made wise arrangements, the collapse of the alliance in 1435 might perhaps have been avoided, with incalculable consequences for the war and the subsequent history of Europe. On the other hand, the marriage of Jacqueline of Hainault with Henry V's brother Humphrey of Gloucester, and the

---

23 Vaughan, *Philip*, p. 10.
24 For example E. Carleton Williams, *My Lord of Bedford* (London, 1963), pp. 100–105, 125, 174, 222; B.-A. Pocquet du Haut-Jusse, "Anne de Bourgogne et le testament de Bedford (1429)," *Bibliothèque de l'école des chartes*, 95 (1934), pp. 284–326.
25 He was also brother-in-law to the dauphin, but the latter's role in the murder of John the Fearless made that connection moot.

events that followed from it – including Philip's challenging Humphrey to a duel, and Humphrey's acceptance of the challenge – probably did more than any other single factor to harm Anglo-Burgundian relations. Moreover, Bedford's hasty and probably ill-advised choice of a second wife, after the death of Philip's sister, was another significant factor in undermining the alliance.[26] This should remind us of the power of personality, chance, and contingency in the realm of grand strategy and alliance politics as well as in other areas. It should also remind us that culturally determined values, in this case the high priority placed by fifteenth-century rulers on family ties, dynastic inheritances, and martial honor, vary across time and place. Analysts who ignore them in favor of supposedly invariable geostrategic principles and *raison d'état* are likely to misunderstand the motivations that can sustain or break alliances.

The First Crisis: Transition of Leadership

If it were possible to quantify the strength of the Anglo-Burgundian alliance and to graph it over time, one peak of cohesion would be in late 1421 or early 1422. In December 1421 the future Henry VI was born; this seemed practically to ensure a long-term set of close dynastic ties between the Lancastrian monarchy and Burgundy, since this made Philip, already the brother-in-law of the current king and queen of England, also uncle-by-marriage of the heir apparent. Thus, the expected future "king of England and France" would be first-cousin to any children born to Philip and his wife, Queen Katherine's sister.[27] Moreover, Henry V and Duke Philip, working together, had recently advanced a whole series of steps towards the ultimate defeat of the dauphin and the solidification of Henry V's status as regent and future king of France: the Treaty of Troyes, the occupation of Paris and the creation of an effective Anglo-Burgundian government based there, and the elimination of most dauphinist strongholds in northern France. Henry seemed invincible in battle or in siege, was respected and feared by all and loved by many. It seemed clear that by supporting him Philip would be backing the winning horse.[28] Henry had an extremely strong reputation as a strict upholder of justice and law, and he clearly appreciated the importance of the Burgundian alliance and

26 These are all discussed in the following.

27 He would also, admittedly, be first cousin to any son of the dauphin. But the dauphin as yet had no children, and the degree of hostility between the houses of Valois and Burgundy was still so high that the familial bonds could not weigh against it. The degree of dynastic linkage between Burgundy and Lancaster, in other words, was a question of absolutes rather than of comparatives.

28 The chronicler of St. Denis, for example, described him as throughout his reign "worthy in arms, prudent, [and] wise," adding, "No prince of his time appeared to surpass him in capability to subdue and conquer a country," by reason of his prudence and other good qualities. [Michel Pintoin,] *Chronique du religieux de Saint-Denys: contenant le règne de Charles VI, de 1380 à 1422*, ed. and tr. L. Bellaguet, vol. 6 (Paris, 1840), p. 480.

in almost all ways did everything in his power to strengthen it.[29] So Philip could feel confident of the security of his long-term relationship with the comrade-in-arms beside whom he had conducted the sieges of Sens, Montereau, and Melun in summer 1420 (to the benefit of his own martial reputation).

But in the middle of 1422 the alliance hit its first period of crisis, triggered by the early death of Henry V and compounded by the passing of both Charles VI and his daughter Michelle (Philip the Good's wife and Queen Katherine's sister), at about the same time. These developments posed three basic problems for the Anglo-Burgundian coalition. First, the trio of deaths almost entirely dissolved the structure of family ties binding together the dynasties of Lancaster and Burgundy. In a matter of months, Philip went from being the son-in-law of the (Lancastrian-controlled) King Charles VI, the brother-in-law of the king's regent and heir to the crown (Henry V), and the uncle-by-marriage of the expected future king (Henry VI), to being nothing but the uncle-by-a-previous-marriage of the infant king, with no particular connection to the regent and heir apparent (Bedford).[30] Second, especially in combination with the death of Charles VI, Henry's death called into question the Burgundians' calculation that they had joined the winning side in the war between Lancaster and Valois.[31] Third, the loss of Henry's firm hand released his younger brother Humphrey of Gloucester to take actions that seriously threatened the alliance. Humphrey soon married Jacqueline of Bavaria and moved to occupy her inheritance of Hainault, steps that put Gloucester (and so England) directly at odds with the policies of Duke Philip, who expected to inherit that valuable province himself.

Under these circumstances there was perhaps some small chance that, despite his personal hatred for the dauphin, whom he blamed for his father's death, Philip

29 The one major exception being his reception of Jacqueline of Bavaria, heiress of Hainault, who was Burgundy's vassal, first cousin, and first cousin's wife, who had fled from her husband to England, contrary to Burgundy's wishes and interests. But, despite this, Henry *did* appreciate the importance of the Burgundian alliance; on his death-bed (according to Monstrelet, and this has been accepted by most historians) he enjoined the guardians of his legacy to offer the regency of France to Burgundy and emphasized as strongly as he could the need to protect the Anglo-Burgundian alliance, "for should any there be any ill-will between you, which God forbid, the affairs of this realm, which are now in a very promising state for our side, could go very much worse." Enguerrand de Monstrelet, *Chroniques*, ed. L. Douët d'Arcq, 6 vols. (Paris, 1857–62), vol. 4, p. 110; Enguerrand de Monstrelet *Chronicles*, tr. Thomas Johnes, vol. 1 (London, 1840), p. 483; Jenny Stratford, *The Bedford Inventories* (London, 1993), p. 6.

30 Though I should acknowledge here that the status of "uncle" by way of marriage to a maternal aunt who died before the nephew was old enough to have known her may have meant more to the parties involved than it likely would under similar circumstances to a modern person. In diplomatic correspondence Henry VI and Philip always referred to each other as "most dear uncle" and "most dear nephew," and they may have felt that way. We know Philip was close to his blood nephew, John of Cleves: see his letters in Vaughan, *Philip*, p. 131. He was also close to his sister Anne (Pocquet du Haut-Jusse, "Anne," p. 318 et passim), who was Henry VI's aunt by marriage to Bedford, which may have enhanced his avuncular feelings towards Henry.

31 Leguai, "'France bourguignonne,'" p. 45.

might switch sides. A much more likely risk was that he might simply pull back from the war effort to formal or *de facto* neutrality and let the contenders for the rule of France fight it out, weakening each other, while he benefitted from peace to build up his resources and focus on his policy of territorial acquisition in the Low Countries. The first sign of the seriousness of the danger to the Anglo-Burgundian alliance came when Philip declined the efforts of the new leader of the house of Lancaster, the duke of Bedford, to bring him back to a leading position in the Parisian court, which probably included an offer to make Philip the regent of France.[32] The second was when Philip declined membership in the prestigious Order of the Garter, on the grounds that he could not commit himself (as the statutes required) to refrain from fighting other members of the order, suggesting he was getting ready to fight Humphrey of Gloucester, who after Henry V's death headed the regency council in England.[33] By the end of the year, he had even opened negotiations with the court of the dauphin, including seeking a partial truce on his borders.[34]

Yet within a year of Charles VI's death, the health of the Anglo-Burgundian alliance had been fully restored and England seemed back on the path to winning the war. This change was due almost entirely to the efforts of the man who succeeded Henry V as regent of France: his brother John, duke of Bedford. Fully aware of the importance of the Burgundian alliance, Bedford immediately set about doing all he could to strengthen it. In pursuit of this objective he had two major cards to play, one military and one dynastic. Although the chronology of the steps Bedford took to resolve the crisis is complicated, and each strand of policy interacted with the other two, one can best understand his actions by looking at them in terms of the three basic problems he had to address.

The first was the restoration of family ties between the two dynasties. The obvious way to address that problem was with a new marriage, but here Bedford faced a difficulty that would continue to plague English diplomacy for the rest of the war: a distinct scarcity of marriageable men and women close enough to the royal family to serve as significant bonds between the two dynasties. In 1423, King Henry VI had no siblings, no offspring, no nieces or nephews, and not

---

32 Armstrong, "Double monarchie," p. 344; but cf. Leguai, "'France bourguignonne,'" pp. 46–7. Henry V himself seems to have worried that his premature death would doom the Treaty of Troyes, since on his deathbed he urged his followers to continue to fight for it, but in effect authorized them to accept, if necessary, a settlement by which the English would give up the crown in exchange for retaining Normandy (and presumably Guienne) as sovereign possessions. Monstrelet, *Chroniques*, vol. 4, p. 110. Since the single most important reason for Burgundy's backing of Henry had been to ensure that the dauphin, the murderer of John the Fearless, did not become king of France (Vaughan, *Philip*, 4), this provision may have been worrisome to the Burgundians, and may be one reason that by the end of the year Philip was looking towards a three-way peace. (See note 21.)

33 He had temporized on this subject since before Henry V's death, but finally gave this answer between 22 April 1423 and 6 May 1424. George Beltz, *Memorials of the Order of the Garter* (London, 1841), pp. lxii-lxiii.

34 Perier, *Rolin*, p. 99.

even a single first cousin of Lancastrian blood. The young king, thus, had only two close, dynastically relevant, unmarried relatives: his widowed mother and his uncle, Bedford.[35] To appreciate the significance and unusualness of this fact it is necessary to contrast it with the situation of earlier English kings at key diplomatic moments of the Hundred Years War. In 1337, when the war began, Edward III's unmarried close relatives included his mother, his eldest son and heir, another son, two daughters, two blood nephews, twelve other nephews and nieces, a sister-in-law, and nine first cousins of Plantagenet descent (of whom five were English), a total of 29 individuals. In 1420, before the Treaty of Troyes, Henry V's unmarried close relatives included three brothers, a half-uncle, and nineteen Lancastrian first cousins (fourteen of them English) – a total of 23 individuals.[36]

Since he could not realistically have offered the marriage of Henry VI or the remarriage of the Queen Mother (neither of which would have been allowed by Gloucester and the council in England) Bedford had only one dynastic card to play: himself. But it was a powerful card. As a man in his prime, regent of Lancastrian France, one of the greater territorial magnates of Europe, and heir presumptive to the dual Lancastrian crowns, Bedford may have been the single most dynastically desirable bachelor in Christendom. Since Henry VI was an infant and many children died before producing heirs,[37] there was more than a small chance that Bedford, and ultimately his son, would be king of England and France. Even if that did not transpire, Bedford in his own right not only held extensive lands in England,[38] but also could be expected to carve a substantial apanage within

35 His one maternal uncle was Charles VII; his one maternal aunt and three maternal first cousins were Valois rather than Lancastrian, so of little use to bind Burgundy to Lancaster over Valois.

36 These numbers are based on various genealogical references and may not be exactly accurate since, for example, we do not know the marriage year of some of the cousins. I have included as "unmarried" those few who were betrothed or in holy orders, since it was not unknown to cancel a betrothal or to pull someone out of a convent in order to make a dynastic marriage alliance, and because it seemed best to make no exceptions rather than to decide what factors should remove an individual from the count. I have also counted many individuals too young to consummate a marriage, as such were often used for dynastic marital alliances. Note that sharing a common grandfather makes two individuals first cousins even if they have a different grandmother, hence Henry's Beaufort cousins (descended from John of Gaunt, duke of Lancaster) are not distinguished from his cousins via Philippa and Elizabeth of Lancaster.

37 For example, Duke Philip at this point had had only one child, who had died in infancy; later he would have three more legitimate children, of whom only the third (Charles the Rash) survived to adulthood; of Henry IV's sons, his first died in infancy, his second (Henry V) barely lived long enough to produce one child, his third (Clarence) died young before producing legitimate offspring, and his fourth and fifth (Bedford and Gloucester) also died without legitimate progeny.

38 As duke of Bedford, earl of Richmond, and earl of Kendal, his income in 1418 had been estimated as on the order of £5,000 (making him one of the two or three richest magnates in England), and since then he had acquired other valuable offices and estates. Jenny Stratford, *The Bedford Inventories*, 4; *Proceedings and Ordinances of the Privy Council of England*, ed. N. H. Nicolas, 7 vols. (London, 1834–7), vol. 2, p. 243.

France out of lands confiscated from Henry VI's enemies, as he indeed did do in mid-1424, when he took the titles of duke of Anjou and count of Maine.[39]

So, in order to restore the family linkages between England and Burgundy, Bedford took the strongest possible step by marrying Philip's sister Anne. This boosted the alliance in a fourfold way. First, it turned Bedford and Philip into brothers-in-law and made Anne, who was dear to Philip and became so to Bedford, a permanent force for harmony between them. Second, it amounted to giving a great gift to Philip: the hand of one of most eligible bachelors of Christendom, which could be expected to inspire some gratitude (especially since Bedford not only married Anne, but made her a happy bride).[40] Third, the gift was valuable not just for its own sake, but also because the decision to give it to Philip showed clearly how extremely highly the regent valued Burgundy's friendship, and thus gave Philip something else he wanted, which was public and irrevocable affirmation of his importance as a partner in the alliance. And fourth, it created the prospect of tight bonds between Lancaster and Burgundy into the next generation. The two men would be uncles to each other's children (assuming they had children in the future); those children would be each other's first cousins. Moreover, the marriage of Bedford and Anne created a possibility of an actual unification of the dynasties of Lancaster and Burgundy. Philip's thirteen-year marriage to Michelle of Valois had not produced any surviving children, he had no legitimate brothers, and he had not yet remarried. His heirs were therefore his sisters and their progeny; if Anne and Bedford had a son, that son would stand to inherit the Burgundian county of Artois, and possibly other shares of the principality of Burgundy after the death of his aunts (Philip's other sisters, of whom only one as yet had children), as well as standing to inherit England and Lancastrian France, if Henry VI had no children.[41]

The same round of marital diplomacy that led to the rebuilding of familial ties between Lancaster and Burgundy also helped address Bedford's second problem: to re-establish the credibility of his side as a potential winner in the civil war within France. The succession of a helpless infant in place of a seemingly unstoppable warrior-king, combined with the loss of the legitimizing force of control over the mad King Charles VI (which had made it easy for those ready to accept defeat or contemplating switching sides to do so), seemed to leave the English in a much weaker position in 1423 than they had been in at the start of 1422. But Bedford undertook a series of mutually reinforcing military and diplomatic

---

39 Stratford, *Bedford Inventories*, p. 9.

40 Williams, *Bedford*, pp. 103–4. Philip had been angling for a marriage between one of his sisters and one of Henry V's brothers since before the Treaty of Troyes, and indeed that prospect had been emphasized by his counselors in making the case for the creation of the Anglo-Burgundian alliance in the first place. Allmand, *Henry V*, p. 140.

41 The prospects of this dynastic extension into the next generation were enhanced by the fact that both Bedford and Burgundy already had illegitimate children and so were known to be fertile. On the planned division of Burgundian territories among Philip's sisters if he died without heir of his body, see Pocquet de Haut-Jusse, "Anne," p. 307.

efforts that both benefitted from, and in turn added strength to, the Burgundian alliance. One reason Duke Philip was willing to renew and strengthen his bonds with England through the marriage of his sister to Bedford was because he and the regent had significantly improved England's chances for ultimate victory by turning one more major player in the game from enemy to ally: the duke of Brittany had agreed to switch sides and join Burgundy and England in a triple alliance, fortified by the marriage of Philip's only other unmarried sister, Margaret, to the Breton duke's brother, Arthur de Richemont.[42]

The successes of Bedford's marital diplomacy were due in part to the successes of his simultaneous martial endeavors. In the aftermath of Henry V's death, the dauphin's captains had undertaken several minor operations in hopes of benefitting from English disarray, but in most cases the gains they made were quickly undone by Anglo-Burgundian responses. Most significantly, the French captured the strategic bridge-town of Meulan, on the Seine between Rouen and Paris. Bedford (with some Burgundian as well as English and Anglo-French troops) went in person to take it back, and conducted the siege so adroitly that he not only impressed the chronicler Jean de Waurin (a veteran soldier) and recaptured the place but also

---

42 At the same time, two of the principal barons of the Gascon frontier, the counts of Foix and Comminges, also switched to the English side. Foix had been the leading figure in dauphinist Languedoc. G. du Fresne de Beaucourt, *Historie de Charles VII*, vol. 2 (Paris, 1882), p. 12; Vaughan, *John the Fearless*, p. 265. Mark Warner sees the triple alliance among Brittany, Burgundy, and Lancaster as "the opening of an elaborate double game on the part of the duke of Burgundy" because "the day after the main treaty, the dukes of Burgundy met without their ally and brother of Bedford to conclude a 'secret' treaty whereby it was agreed that their alliance would remain in force in the event that either party should decide to commence negotiations with Charles VII." This would seem to be a prime piece of evidence for the argument that the Anglo-Burgundian alliance was never strong; that the traditional view of Philip as a duplicitous politician is correct; and that an eventual reconciliation with between Burgundy and the dauphin was nearly inevitable. But, first, we do not actually know that this was a "secret" treaty. Suggesting the contrary, it opens with the standard diplomatic language of an open treaty or letter patent: "Philip duke of Burgundy and John duke of Brittany, to all those who may see and hear the present letters: Greetings. We inform you that . . ." (Plancher, *Bourgogne*, 4, p.j. no. XXIII, p. xxvij.) Second, more importantly, Warner mistakes the terms of the document. In it the two dukes promise each other that if one makes peace or agrees to a truce with the dauphin, he will nonetheless continue to support the other *against the dauphin* (or anyone else who attacks him). It was a version of the standard diplomatic clause prohibiting the parties in an alliance from making a separate peace and hanging their partners out to dry. In the text Burgundy actually refers to the *triple* alliance ("of which alliance the tenor is as follows: 'John [duke of Bedford], regent of the realm of France, &c . . .'") and promises to adhere to that alliance, vis-à-vis Brittany, regardless of any treaty or accord made with the dauphin. There is no indication that the document was looking towards a peace that would violate Burgundy's treaty with Bedford; rather, it probably was envisioning the possibility of a three-way Anglo-French-Burgundian peace. (See note 23.) Burgundy's dealings with the dauphinists appear much less "two-faced" (cf. du Fresne de Beaucourt, *Histoire*, pp. 413–14; Warner, "Dual Monarchy," p. 122) if one refrains from presuming that Philip knew that no Anglo-French compromise peace was possible. In fact, as late as 1432 and 1433, despite military setbacks and the failure of the English to provide him with the support he expected (as discussed later), he refused to accept any separate peace treaty with Charles that did not include the English. Du Fresne de Beaucourt, *Histoire*, pp. 450–51, 456–57.

compelled the French lords inside to surrender other fortresses they held, as part of their capitulation. As a result, Montl'héry and Étampes, among other places, fell under English control. In the same spring and into the summer, other English and Burgundian detachments also captured Montaiguillon, Noyelles-sur-Mer, Rue, Le Crotoy, Ivry, La-Charité-sur-Loire, Compiègne, Sedan, and other places. Thus the regent demonstrated that the military tide was still flowing in favor of the English, despite Henry V's death.

The strongest proof of that fact came in the summer of 1423. The dauphin's party made its largest military effort of the year in an attempt to capture Cravant. The town was in an important strategic location, and was made more important because a group of prominent Burgundian captains had hastily entered the town in order to cast out a French force that had seized it by a ruse. Capturing these Burgundian nobles along with the town, argued one French captain, would mean the capture of Auxerre and control of that whole region, and would "give the Burgundians so much to do, that they will not know on which foot to dance."[43] Because Cravant was badly provisioned, it had to be rescued rapidly, but the French had sent a full-scale army against it at a time when the English and Burgundians had many troops occupied in other operations. In this emergency, Bedford managed to collect some 2,500 troops to join a Burgundian force of similar size in crushing a much larger Franco-Scottish army at the battle of Cravant. The defeated dauphinists lost some 5,000 men killed or captured, including both of the army's commanders. This was a major setback for French, who were already losing ground. It practically ended the threat to Philip's southern domains and provided Philip with a potent reminder of the military value of English military assistance. After the battle, English and Burgundian soldiers worked together to clear several more dauphinist strongholds from the borders of Philip's territories. At one point the duke of Burgundy himself formed up in battle array alongside the English captain Sir William Glasdale, ready to fight a force that Charles of Bourbon was expected to bring to raise the siege of the castle of La Roche outside Mâcon.[44] The Burgundian chronicler Jean de Waurin, who was in the army as a man-at-arms, remarks that there had been much brotherly affection between the English and Burgundians even before they fought and won the battle of Cravant together.[45] The strength of comradely feeling by the end of the successful joint campaign is easy to imagine.

It is a good example of the interrelationships of diplomatic and military developments that the Anglo-Burgundian victory at Cravant seems to have been a

---

43 Waurin, *Chronicles*, p. 40. The besieged, in pleading for help from the dowager duchess of Burgundy (who was in charge of the region while her son was in the north) wrote that if they were not supported, "the country of Burgundy would be destroyed." *Livre des trahisons de France envers la maison de Bourgogne*, in Kervyn de Lettenhove, ed., *Chroniques relatives à l'histoire de la Belgique sous la domination des ducs de Bourgogne*, vol. 2 (Brussels, 1873), p. 169.

44 Waurin, *Chronicles*, pp. 47–48.

45 Jean de Wavrin [Waurin], *Recueil des croniques et anchiennes istoires de la Grant Bretaigne, à present nommé Engleterre (1422–1431)*, ed. William Hardy (London, 1879), p. 66: "se misrent auz champz en grant fraternite."

crucial factor in another diplomatic coup that added significantly to the probability of the English outright winning the Hundred Years War. Since the death of Henry V, his son's regency council in England had been working to settle the century-old conflict with Scotland. King James I was more than amenable to peace. Although he had been a prisoner of the English since 1406, by 1419 James had effectively been co-opted by Henry V, to the point of serving as one of Henry's senior captains in the war against France.[46] Again matrimonial and personal considerations were important here: as James's famous poem *The Kingis Quair* reveals, he was deeply in love with Henry's first cousin, Joan Beaufort, daughter of the earl of Somerset.[47] But although diplomacy working towards a "final peace" between England and Scotland had been initiated as early as February 1423, it was not until 19 August – probably within days of receiving word of the outcome of Cravant (fought on 1 August) – that Murdoch Stewart, the regent of Scotland, authorized the Scottish chancellor to begin the formal negotiations that produced the Treaty of York, agreed on 10 September.[48] Although it was not immediately effective, this agreement represented the prospect of removing the dauphin's most significant foreign ally from the playing board.

Through these actions and his general conduct, Bedford had proven himself a man of stature and a worthy successor to Henry V as co-leader of the Anglo-Burgundian war effort. He "was a man of energy and worth," remarked a contemporary Frenchman, "humane and just, who greatly loved those French noblemen who adhered to him, virtuously striving to raise them to honor. So for his whole life he was greatly admired and cherished by the Normans and French of his party."[49]

Thus, by the end of the summer of 1423, Bedford had done an extraordinary job of restoring English prestige and solidifying the Anglo-Burgundian alliance. But he still faced a third danger to the strength of the coalition: the reckless actions of his younger brother, Duke Humphrey of Gloucester, Protector of England and head of Henry VI's regency council in England. In January 1423, Humphrey had married Jacqueline of Bavaria, heiress of Hainault (and, disputably, of Holland and Zeeland), who had earlier fled to England to escape ill-treatment by her first husband, John, duke of Brabant. This marriage, and Gloucester's consequent claim to lordship over those three provinces, was an intolerable affront to Philip

---

46 At the siege of Dreux in 1421. Michael Brown, *James I* (East Linton, 2000), p. 23.

47 Skeptics inclined to presume that this was actually a simple marriage of state romanticized by poets and chroniclers, and that *The Kingis Quair* is merely a conventional work of literature that reveals little about the poet's actual romantic feelings, should note that while James and Joan had eight children, James had no known bastards. By contrast, his royal father and grandfather had at least two and nine illegitimate children, respectively.

48 *Foedera, conventiones, litterae etc.*, ed. Thomas Rymer (The Hague, 1739–1745), 4:4:97–8; note also 4:4:85–6; 4:4:93 (March; Scots to Pontefract); 4:4:97 (English ambassadors and James to Pontefract; Douglas "for a final peace with Scotland"); 4:4:111 (truce for seven years starting 28 March 1424, neither to assist enemies of the other).

49 Thomas Basin, quoted Williams, *Bedford*, p. 108.

of Burgundy. Both Jacqueline and John of Brabant were Philip's first cousins, and the Burgundian duke had arranged to inherit the couple's lands if they had no children, which he had probably also arranged with Brabant to ensure would be the case. Thus, Gloucester was both sticking his nose into Burgundian family business and setting himself directly in opposition to Philip's most important policy agenda, his pursuit of territorial expansion into the Low Countries.[50] The duke of Burgundy clearly felt that Bedford's failure to prevent Gloucester's action was something of a stab in the back, one that called into question the regent's commitment to the alliance, and hence its value.

This challenge too was met by Bedford's diplomacy, at least for the time being. He and Philip had both recently been reminded by the scholars of the University of Paris that if war broke out between Burgundy and Gloucester, it would likely lead to the rupture of the Troyes settlement, the devastation of France, and the prompt victory of the dauphin, who had been responsible for the murder of Philip's father. To avert this disaster, Bedford and Burgundy leaned hard on Gloucester and Brabant, respectively, to accept their joint mediation in their dispute, and to refrain from resorting to force in the interim. This did not resolve the issue, but for the time being it was satisfactory to Burgundy, since his protégé was left in possession of the disputed lands.[51]

With this crisis averted, or at least deferred, Bedford had successfully responded to all three of the dangers to the Anglo-Burgundian alliance resulting from Henry V's death: he had re-established family ties between the two houses, demonstrated clearly that England and Lancastrian France were still allies with good prospects for ultimate victory, and had prevented his brother from causing a rupture between England and Burgundy. He had moreover made positive steps to strengthen the alliance even further, most significantly by providing substantial military forces to aid Burgundy against the dauphin.[52] The joint Anglo-Burgundian military operations not only contributed to the solidity of the alliance by their successful results, but they also built up ties of affection between leading men on both sides in the process of achieving those results, as Waurin testifies.[53]

50 There has been no full study of Humphrey since K. H. Vickers, *Humphrey Duke of Gloucester: A Biography* (London, 1907), which remains useful.

51 For Bedford and Burgundy's judgment essentially passing the matter on to the pope, dated 19 June 1424, see *Cartulaire des comtes de Hainault*, ed. Léopold Devillers, vol. 4 (Brussels, 1889), p. 391; note also Joseph Stevenson, ed., *Letters and Papers Illustrative of the Wars of the English in France during the Reign of Henry the Sixth, King of England*, 2 vols. (London, 1861–1864), vol. 2 pt. 2, pp. 388–9. For the eventual result (on 27 February 1426) see *Cartulaire*, pp. 539–41.

52 For another example of Bedford's conciliatory diplomacy (which left the valuable castellanies of Péronne, Roye, and Montdidier in Burgundy's hands) see Dom U. Plancher et al., *Histoire générale et particulière de Bourgogne, avec les preuves justificatives*, vol. 4 (Dijon, 1781), preuve justificative [hereafter p.j.] no. XXV, pp. xxviij-ix (8 September 1423).

53 Waurin, *Recueil*, p. 66.

The Second Crisis

In May and June of 1424, Bedford took further steps to strengthen the alliance. First, he secured from both Jacqueline and Humphrey an extension of the period for arbitration of their dispute with Brabant, until the end of June. Then, over three days, Duke John and Duke Philip agreed on a general settlement of various matters. First, on 19 June, they jointly issued their conclusion on the dispute over Jacqueline's lands. In essence, they agreed that the matter should be decided, if the two sides could not come to a mutual agreement, by a papal decision on the validity of Jacqueline's marriage to Brabant.[54] This was not an ideal outcome from Burgundy's perspective, but it was good enough, since it left his cousin and ally in possession, and since Philip probably expected the pope's decision to go in his favor, which is what eventually happened. Second, Bedford accepted from Henry VI's regency council the titles of duke of Anjou and count of Maine. Although these areas were not yet under English control, they were next on the agenda for conquest. This grant, provided it could be made a reality, stood to make Bedford (and thus eventually, it could be hoped, his heir, Philip's blood nephew) roughly on a par with the duke of Brittany and the duke of Burgundy as one of the three great magnates of northern France.[55] Third, Bedford essentially handed over the principal spoils of the recent Anglo-Burgundian military operations by signing over to Philip the counties of Auxerre and Mâcon and the castellany of Bar-sur-Seine, together a considerable addition to Burgundy's domains.[56]

Events demonstrated the excellent health of the alliance shortly thereafter. The dauphin, despite the defeat at Cravant, managed to assemble the largest French army collected since Agincourt, perhaps 14,000 men, to attack Lancastrian Normandy. Returning in full measure the aid he had received at Cravant, the duke of Burgundy rushed his best commander, the sire de l'Isle Adam, to Bedford's assistance with a force estimated at 3,000 men, including 1,000 men-at-arms[57] — a very large contingent for a Burgundian army, considering that the duke himself had mustered only about 1,500 soldiers for the battle of Mons-en-Vimeu two years earlier, where he had fought in person, and that Henry V had only had 1,000 men-at-arms at Agincourt.[58] Bedford himself took the field at the head of about 9,800

---

54 *Cartulaire*, pp. 391–92.

55 According to William of Worcester, well informed on such matters, Maine alone, after its conquest, produced an annual revenue of £10,000 in support of the war effort. William of Worcester, *The Boke of Noblesse. Addressed to King Edward the Fourth on His Invasion of France in 1475*, ed. John Gough Nichols (Edinburgh, 1860), p. 19.

56 These were officially granted for a limited term only, to be restored to the royal domain if Philip failed to demonstrate that they were compensation due him for debts owed to his house by the Lancastrian government. Plancher, *Histoire*, p.j. nos. XXXIV–XXXV.

57 F. W. Brie (ed.), *The Brut or the Chronicle of England*, vol. 2 (Oxford, 1908), p. 564.

58 Mons-en-Vimeu: Vaughan, *Philip*, 14 (figuring one *gros valet* per man-at-arms); Juliet Barker, *Conquest. The English Kingdom of France in the Hundred Years War* (London, 2010), p. 78. Burgundy would later again have about 4,000 at Brouwershaven, where he fought in person. Vaughan, *Philip*, p. 43.

soldiers,[59] so that with the Burgundians, the opposing armies were nearly equal. The French seemed ready to fight a general engagement, even without a large numerical superiority, something unusual for them, and perhaps a sign of their strategic desperation. Shortly before a battle was expected, Bedford paid honor to this Burgundian contribution by delivering the royal banner of France to be carried by its leader, the sire de l'Isle Adam.[60]

Yet within seven weeks of the day the Burgundians joined Bedford's army, the situation had changed radically. From an active cooperation of two powers against one, the alliance between England and Burgundy had become little more than a non-aggression pact; Philip had signed a truce with Charles and was effectively leaving the English to fight the French by themselves.[61]

The reasons for this shift are complex. One major factor was the intervention of Humphrey of Gloucester in Hainault, against Philip's wishes and interests, which is further discussed later. But historians have largely ignored another important cause. When the vanguard of the Anglo-Burgundian army was just twelve miles distant from the French and battle seemed imminent, Bedford took an astounding step: he detached the entire Burgundian element of his army, nearly a quarter of his strength, and sent the troops marching for Picardy, reportedly saying that he had enough men to defeat the French without them. Two days later, now facing odds of almost 3:2 instead of roughly 1:1, Bedford commanded during a hard-fought battle at Verneuil. The combat was much tougher than Agincourt, according to a soldier who served in both engagements, and Bedford came near disaster when well-armored Lombard cavalry broke one of his wings of archers, but the English eventually won the battle quite decisively.[62]

The whole episode of Bedford's dismissing l'Isle-Adam's contingent before the battle is not even mentioned by Philip the Bold's biographer, or in Juliet Barker's recent history of this phase of the war.[63] No surviving evidence offers much insight into the cause of Bedford's surprising action, and the regent's biographer, E. Carleton Williams, devotes just two unenlightening sentences to it: "In view of the disparity existing between the two armies his action is difficult to understand. Perhaps the duke was emulating Henry V's defiant exhortation before the battle of Agincourt, or it might be for the graver reason that he felt the Burgundians were untrustworthy."[64] Given Bedford's reputation for prudence, and the fact

59 Waurin, *Chronicles*, p. 67.
60 Ibid., p. 69.
61 Plancher, *Bourgogne*, p.j. no. XXXVII.
62 Waurin, *Chronicles*, pp. 72–79; for the departure of the Burgundian troops (not identified as such), note also *Brut*, 2:564.
63 Vaughan, *Philip*, p. 11; Barker, *Conquest*, p. 78. (Barker notes l'Isle Adam's arrival but not his departure.)
64 Williams, *Bedford*, p. 133. The chronicler Waurin, who was in the army, puts the best face on the matter by saying Bedford commanded the Burgundians to return to their siege of Nêle, since "he was well able to spare them" and "they had great need and very legitimate cause for returning to their said siege of Nelle." Yet, he notes that the Burgundian leaders "would have liked better to

that since Agincourt he had witnessed the example of his brother Clarence, who underestimated the French and in consequence was killed at Baugé in 1421, the former hypothesis seems highly improbable. Even if Bedford were so confident in divine providence, English prowess, or both, that he sincerely believed he had no *need* of the Burgundians, he also would not have given up much of the glory of a potential victory by allowing them to participate, since (without Burgundy himself being present) he, Bedford, would unambiguously have been the commander of the army. It is also unlikely that he dismissed the Burgundians because he thought they were "untrustworthy" in the sense of being less reliable troops than his Englishmen. They had already shown at Cravant that they could be good battlefield comrades, and even if he did doubt their steadiness or effectiveness, and feared that they might make a weak spot in his battle line, he could have avoided that by using them as a reserve or enveloping force. Sending them away was a clear statement that he actively did not want them to participate. Williams's second guess thus seems the only credible explanation, but if it is the correct one, it means Bedford was gravely mistaken, since there is no indication in Burgundian archives that the Burgundians were plotting a betrayal, despite the high level of tension over Gloucester's intervention in Hainault.[65]

If we, with access to English archives, can only guess at Bedford's reasoning, the same must have been true of the scorned duke of Burgundy and his council.[66] It seems that the Lancastrian regent's action must have sent three basic messages to Burgundy, all of which ran directly counter to the diplomatic message Bedford had sent earlier by marrying Anne. First, and most explicit: we don't need you. Second, we don't trust you: either we don't trust you to fight courageously and well, or, more likely, we don't trust you not to stab us in the back.[67] Third, we don't want to be in your debt.

An alliance of convenience with the English still had clear advantages for the Burgundians in 1424 (they had profited greatly from English aid in 1423), but by definition could only be expected to remain in force so long as it remained advantageous for them. Only a stronger alliance, fortified by sentiment and mutual investment, could be expected to weather hard times for the English, when it would

---

remain with him to accompany him to the battle" and "left with great regret," and in fact there is no credible way to argue that the siege of Nêle was anywhere near as important as increasing the odds of victory at Verneuil. Moreover, Waurin states earlier in his narrative that l'Isle Adam had left the force maintaining the siege "well provided for." Waurin, *Chronicles*, pp. 67–72. Realistically, the only thing that would have prevented its fall would have been an English defeat at Verneuil.

65 Vaughan, *Philip*. However, in Bedford's defense, it is worth noting that the night before the battle a number of Frenchmen in his army deserted and went over to the dauphin's side; it was the morning after that event that Bedford sent the Burgundians away. Waurin, *Chronicles*, pp. 72, 80–81.

66 Even if Bedford gave some now-lost explanation or excuse to them, they would have had no way to be certain of its sincerity.

67 "More likely" because, as noted already, if Bedford had only been worried about Burgundian competence rather than Burgundian intentions, he could have employed the Burgundian force as a reserve or an enveloping detachment.

be needed most, and which (given the history of the war thus far and of human affairs generally) could be expected before the conflict was finally resolved. But the value of a long-term, solid alliance rested, for Philip, on his ability to be confident that his views and interests would receive due weight in the counsels of the Lancastrian government of France, and England. That confidence in turn rested on the knowledge that the English needed and valued his assistance, and that Henry VI's government trusted him. Trust is not always rewarded, but distrust almost always undermines relationships. National alliances, like marriages, may survive on the basis of mutual need even when mutual respect and affection are absent, but in such cases fidelity is too much to expect. Alliances of this sort are rarely sound bases for the inherently *long-term* considerations of grand strategy.

The shared victory of Cravant had built up a reservoir of good-will in Burgundy towards the English; the slap in the face Bedford delivered to the Burgundians on the eve of Verneuil must have significantly drained that reserve. It probably also led Philip to reassess his situation vis-à-vis the other bone of contention between him and Bedford: the ambitions of the duke of Gloucester in the Low Countries. As already noted, as part of his effort to resolve the first crisis of the alliance, Bedford had gotten Gloucester and his duchess to accept himself and Burgundy as mediators in the dispute over Hainault, Holland, and Zeeland. But the truce included in that agreement expired in June, and the mediators had not been able to agree on anything beyond referring the matter to the pope. Immediately on receiving word of this non-resolution of the dispute, Humphrey set in motion plans to "pass over the sea with God's might in our own person for to receive and take into our hands our lands and lordships there with due obeisance of our subjects," borrowing whatever funds he could to that end.[68] His preparations, which also included collecting ships and soldiers, can hardly have remained unknown to the duke of Burgundy, who had already promised to back John of Brabant with military force if necessary. Philip thus confronted the possibility of a two-front war, against Humphrey and against the dauphin. Since Bedford seemed unable to leash the former, while Charles was eager for a truce with Burgundy (as a first step towards rupturing the Anglo-Burgundian alliance so that he could concentrate on fighting the English), Philip proceeded to negotiate a truce on his southern borders, the "truce of Chambéry," with the dauphinists. Significantly, and doubtless as part of the price of getting the truce he wanted, in the document the Burgundian duke referred to Charles as "the king," a step away from the previously invariable

---

68 *Historical Manuscripts Commission, Twelfth Report, Appendix, Part IX. The Manuscripts of the Duke of Beaufort, the Earl of Donoughmore, and Others* (London, 1891), p. 395: a letter dated in London on 28 June, to the prior of Ely, requesting a loan of £200. Remarkably, in a sign of either gross dishonesty or complete self-deception, Gloucester asked for the help "considering that ye know of . . . this our voyage as likely to turn to right great ease and welfare of this realm." (Quotation modernized.) It is hard to imagine how he could have persuaded even himself of that likelihood.

Anglo-Burgundian diplomatic practice of completely denying Charles's right to that title.[69]

The two brothers-in-law were each disappointed and angry with the other: Bedford was furious about the Franco-Burgundian truce; Philip was outraged by Bedford's failure to keep his younger brother in check, as Gloucester prepared to use force against Burgundian interests rather than submitting to papal judgment of the matter, the only option that could prevent fighting between English and Burgundian troops without shaming one duke or the other. There was a shouting match between them in Paris in October. But although bruised, the alliance did not break. John and Philip continued to like and respect each other,[70] and to be bound together by their mutual love for Anne of Burgundy, ever the peace-maker between them, and by their mutual enmity towards Charles VII. Their argument may have been heated, but came in the midst of many consultations as they tried to find a way to peaceably settle the Hainault affair.[71]

The main causes of their mutual discontent, the truce of Chambéry and Gloucester's intervention in the Low Countries, looked like they might blow over. The truce was only valid until the start of the next campaigning season (1 May 1425), and Gloucester lacked the strength to sustain a position on the Continent against Burgundy's power unless he got backing from Bedford or from the royal government in England, neither of which had any intention of offering him support. If a Burgundian army deterred him from actually moving into Hainault from Calais (where he had just landed on 17 October), or persuaded him to accept the arrangement that Bedford and Burgundy had agreed on (letting the pope resolve the matter), the problem would be much ameliorated. If neither the regent nor Philip had gotten quite the full level of backing they would have liked from the other, they could each appreciate the other's situation: Philip could hardly expect Bedford to go so far as to send troops against his own brother, and Bedford could not blame Philip too much for seeking a truce in the south when he, Bedford, had failed to protect his ally from a threat in the north. And while each may have wanted more from the other, each also preferred the other's neutrality to the other's active

69 Printed Plancher, *Bourgogne*, p.j. no. XXXVII. It is not quite right, however, to say (as is said in various books) that Burgundy in this document called Charles "King of France." It is doubtless an intentional diplomatic nuance that Charles is simply called "le Roy" rather than "le Roy de France." In a later truce document (from 1431) Philip refers to Charles as "calling himself king of France." Ibid., p.j. no. CX, p. ciij.

70 This can be seen clearly in Burgundy's willingness to accept Bedford – Gloucester's own brother – as the judge of the duel he wanted to fight with Gloucester in 1425 (for reasons explained later), "for he [Bedford] is a prince of such character that I know and truly recognize that to you, to me, and to all others he would wish to be a righteous judge." Waurin, *Chronicles*, p. 99. I think Armstrong, "Double monarchie," p. 373, and Leguai, "'France bourguignonne," p. 49, underestimate the "amical" element of the Anglo-Burgundian alliance – or rather, the Bedford-Burgundy alliance – for most of its duration.

71 They continued to dine and socialize together, and even both participated in the jousting that accompanied the wedding of Burgundy's *maître d'hôtel*.

opposition. Philip needed Bedford to continue to refrain from assisting his brother militarily, and Bedford needed Philip to stay bound to England and not switch to supporting France. Moreover, if Bedford's conduct during the Verneuil campaign had been an insult to Burgundy, it was an insult that could be diplomatically papered over, and the crushing nature of the regent's victory, the second destruction of a French army in two years and the third in a decade, left the dauphin (as the English and Burgundians generally continued to call Charles VII) militarily weak. The Lancastrian dynasty still looked like the winning horse.

So the English still seemed to have a good chance of restoring their alliance with Burgundy to a full, active partnership working towards the defeat of the dauphin and his party. But neither marital nor martial circumstances offered strong opportunities for strengthening the alliance. Philip was at this time a widower, but already seeking papal dispensation for a marriage to Bonne of Artois, dowager countess of Nevers, a woman with dauphinist ties. The Lancastrians would doubtless have been happy to offer a "higher bid" and to reinforce their ties with Burgundy by supplying Philip with an English bride, but there simply was no suitable Englishwoman available.[72] Nor was there a suitable male relation of the king who could marry Burgundy's one remaining unwed sister, Agnes. She instead married the heir of the duke of Bourbon, a leading supporter of the dauphin.[73] In the short term, joint Anglo-Burgundian military operations that would benefit Burgundy were also impracticable, first because Philip was readying for the possibility of war with Bedford's brother Humphrey of Gloucester in Hainault, and secondly because of the truces between Burgundy and the French territories on his borders.

Although Duke Humphrey occupied Hainault with an army in late 1424, Bedford had a brief window to tamp down the resulting Anglo-Burgundian rift before it did much damage. Duke John did prevent the disaster of major fighting between Gloucester and Burgundy, by the simple expedient of refusing to back his brother against his ally. Without Bedford's support, Gloucester lacked the strength for a war with Burgundy over Hainault. However, Humphrey's diplomatic efforts to avoid that outcome took an unexpected turn when, in an exchange of letters, Gloucester and Burgundy accused each other of lying; Burgundy challenged Gloucester to a duel, and the latter accepted. This whole affair cannot be explained simply in

72 A letter of 1425, written shortly after Bonne's death, noted five possible brides for Philip; four were specific individuals, one was generically "une grant dame en Angleterre." Louis Stouff, *Catherine de Bourgogne et la féodalité de l'Alsace Autrichienne* (Paris, 1913), part 2, p. 181; Vaughan, *Philip*, p. 55. The usual interpretation is that Burgundy prudently chose not to strengthen his alliance with England through marriage (Armstrong, "Double monarchie," p. 369), or brushed aside advice to do so (Warner, "Dual Monarchy," p. 113 [misdated]). But it is more likely that the reason the letter does not identify the great English lady is because there was no one suitable to name, since the marriage of Joan Beaufort (Henry V's first cousin) to the king of Scots in February 1424.

73 John Beaufort, duke of Somerset, first cousin of Henry V, would have been suitable but was not available because he was a French prisoner from 1421–38.

terms of the political: it was also personal, and must be understood in terms of Philip's lifelong concern with avoiding any blemish on his knightly honor.[74]

Bedford succeeded in preventing the duel from taking place and also helped ensure that Gloucester withdrew, with his forces, from Hainault.[75] But he had not done so fast enough. At the end of January 1425, just after Philip had issued his challenge to Humphrey and while there was still an English army in Hainault, Philip agreed to prolong his truce with the dauphin until 25 December.[76] Then in February he arranged a new, open-ended truce with the duchess of Bourbon, tied to the forthcoming marriage of Agnes of Burgundy with Charles of Bourbon. Because of the geography, this practically ensured there would be no fighting between Burgundians and dauphinists for the foreseeable future.[77] In effect the duke of Burgundy withdrew, for five years, from active participation in the Hundred Years War, and turned his attention almost entirely to expanding his territorial control in the Low Countries.[78] The game of two against one was back to one against one.

During those years English arms continued to be successful, though only moderately so, against the concentrated efforts of Charles VII's France. In 1428, English forces began the siege of Orléans, the main dauphinist city north of the Loire. Even without active Burgundian assistance to the English[79] it seemed to many that a Lancastrian victory in the war had become imminent. As Marguerite la Touroulde recalled some years later,

> At that time there was such misery and such shortage of money in this kingdom . . . that things were in a pitiable state. And even those who

74 See Vickers, *Humphrey*, chapter 4, and Vaughan, *Philip*, chapter 2.

75 Bedford was deeply involved in preventing the duel; how much he had to do with causing Gloucester's withdrawal from Hainault is unclear, but at a minimum it included the important fact that he refused to support his brother's ambitions.

76 Then on 2 December the truce was further extended into the following year; further extensions followed later. Plancher, *Bourgogne*, p.j. no. XLVII, p. liij.

77 The area north of the Loire had been practically cleared of dauphinist territories; the Loire separated Burgundian Nevers from dauphinist Berry; all territories of Burgundy's southern border were fully shielded by the truce, which included Forez and Beaujolais. Very unusually, the proclamation of this truce did not include an end-date. Plancher, *Bourgogne*, p.j. no. XL, pp. xlviii–xlix.

78 I.e. from fall 1424 to late summer 1429. For practically this whole period Philip was occupied in a "long, hard, costly war" to gain control of Hainault, Holland, and Zeeland; Gloucester effectively abandoned Jacqueline, but she and her partisans nonetheless did not surrender her independence or her heritage without a fight. Vaughan, *Philip*, pp. 40–50.

79 Williams, *Bedford*, pp. 165–67, does say that Burgundy initially contributed a contingent to the siege of Orléans, then withdrew it in fury when Bedford refused a French proposal to turn the city over to Philip to hold on behalf of its duke, who had been a prisoner of the English since Agincourt. The offer and Bedford's refusal are in Monstrelet, *Chronicles*, pp. 551–52, and Waurin, *Chronicles*, p. 169, but neither Monstrelet nor Waurin makes any mention of Burgundians either participating in or withdrawing from the siege, which they almost certainly would have if there had been any. In fact, Waurin reports that Bedford's council objected to the idea of Burgundy gaining the city "without striking blow," implying his men had not participated in the siege. Williams's source is not clear, and I find the episode doubtful.

were loyal to the King [Charles VII] were all nearly in despair. I know this because my husband was at the time the Receiver General, and he had not four crowns all told, of the King's money or his own. . . . At that moment there was no hope except in God.[80]

Subsequent events, including the chance death of the best English military commander and the arrival on the scene of Joan of Arc, turned the tide against the English, though not decisively so until Burgundy went from passive support to active opposition in 1436. Still, it seems fair to say that, considering how close to complete defeat the French came, it is likely that the dauphin's position would have collapsed before 1428 if in 1424–7 he had been forced to meet active pressure from Burgundy at the same time as he faced the still-flowing tide of English conquests.[81] If one accepts this logic, it makes the Franco-Burgundian truce of September 1424 (the Treaty of Chambéry) as much of a turning point in the history of the Anglo-Burgundian alliance, and in the course of the war, as the much more famous Treaty of Arras in 1435, by which Philip formally renounced his bond with the English and re-entered the war on the side of Charles VII.

The Third Crisis

Before that final collapse of the Anglo-Burgundian alliance, there was a third brief period when the coalition seemed as tight as ever. The advent of Joan of Arc and the failure of the high-stakes English siege of Orléans in 1429, followed by the English battlefield defeat in a meeting engagement at Patay, then the dauphin's successful drive on Reims and his coronation there as Charles VII, posed an existential threat to Lancastrian France. The dauphinists' capture of a string of towns on the way to Reims also gave the French bases for operations against Burgundy's own lands, in an area not covered by Philip's southern truces.

80 Marguerite la Touroulde, in Régine Pernoud, *The Retrial of Joan of Arc: The Evidence at the Trial for Her Rehabilitation, 1450–1456*, tr. J. M. Cohen (London, 1955), p. 94, translation slightly modified; for the Latin (which reports the testimony in third person form), see Pierre Duparc, ed., *Procès en nullité de la condamnation de Jeanne d'Arc*, vol. 1 (Paris, 1977), pp. 376–77. Similarly Monstrelet, *Chronicles*, pp. 548, 545–46: "king Charles was in very great distress; for the major part of his princes and nobles, perceiving that his affairs were miserably bad, and everything going wrong, had quite abandoned him"; he believed "should [Orléans] be conquered, it would be the finishing stroke to himself and his kingdom." Jacques Gélu, archbishop of Embrun, wrote in 1429 of how "it was hard to find anyone who was obedient to their lord the king . . . [who] was so weakened that he had just barely enough [funds] to live, not even for his household, just for his own person and that of the queen. Things had reached such a point that there did not seem to be any way the lord king could recover his domains by any human agency." Quoted Contamine, "La 'France Anglaise'," pp. 18–19.

81 Even after the arrival of Joan of Arc and the victory at Patay, Charles, according to Waurin, knew "it would be impossible for him to resist the forces of the king of England and the duke of Burgundy joined together." Waurin, *Chronicles*, p. 202.

Philip could have responded by switching sides at this point, and almost certainly could have gotten from Charles, as the price of his transfer of loyalties, significant territorial contributions, public repentance for the murder of John the Fearless, and a voice in the French council.[82] The fact that a few years later he not only considered but actually did take that course indicates that such a decision was not out of the realm of possibility in 1430. But instead he shifted away from his position of *de facto* near-neutrality to resume active military cooperation with the English. As the government in London readied a royal expedition to carry Henry VI to France for his own coronation there, Burgundy even accepted the position of royal lieutenant for his English nephew.[83]

Some of his motives for staying true to the English alliance were same as they had been since 1423. He was still bound to Bedford by friendship, by mutual investment, and by his sister Anne. Charles's role in the murder of Philip's father, and most importantly[84] the potential stain on his honor that breaking his alliance with the Lancastrians might bring, still militated against switching sides. Moreover, during the years of the Truce of Chambéry the main bone of contention between the two dukes had lost its significance. Bedford had declined to support his brother's ambitions in the Low Countries, in effect allowing Philip to drive Gloucester out of Hainault (which he did April 1424) then to defeat and force the surrender of Jacqueline herself (finally in 1428). This, in combination with the pope's declaration that Jacqueline was legally married to Brabant rather than Gloucester, and various other developments, had made Burgundy's acquisition of Hainault, Holland, Zeeland, Friesland, Brabant, and Namur quite secure. And although recent events had weakened the English position in France, Henry VI's position was far from desperate. On the other hand, recent Lancastrian setbacks could be expected to raise English appreciation for Burgundian contributions, and prevent any future snubs *à la* Verneuil.

Indeed, already within days of the capture of the earl of Suffolk at Jargeau in June 1429, Bedford had again stepped up to make the strongest diplomatic step within his power to signal how much he valued the alliance and to increase the strength of Burgundy's commitment to Henry VI's position as king of France. As noted before, in 1424 the Lancastrian government of France had granted to Bedford a large apanage, including the rich county of Maine. Since Bedford and Anne remained childless, the prospect of his territories passing to a half-Burgundian heir were fading. But on 14 June 1429, Bedford sealed a new will that made Anne his heir to his French lands (now including the counties of Maine and Dreux,

---

82 Plancher, *Bourgogne*, p.j. no. LXX.

83 Vaughan, *Philip*, pp. 22–26; Monstrelet, *Chronicles*, vol. 1, p. 566 says the two dukes agreed "that each would exert his whole powers to resist their adversary Charles de Valois, and then solemnly renewed the alliances that existed between them."

84 See note 22.

the viscounty of Beaumont, and title to the duchy of Anjou)[85] not just with a life tenancy, but as recipient to full title over them.[86] That made it likely that the lands would ultimately go either to half-Burgundian heirs (Anne's children by Bedford or a later husband) or else, quite possibly, *actually become part of Burgundian territory*, since if Anne never had children, her heir would be her older brother Philip, or Philip's heirs. Assuming Henry VI's government honored the terms of Bedford's will,[87] there seemed to be only two likely ways that duke's extensive French lands might *not* come into the Burgundian family. The first would be if Bedford outlived Anne, but she was fourteen years younger than her husband and in good health, so this did not seem likely.[88] The second would be if Charles VII defeated the English and restored his power north of the Loire. Thus, Bedford's new will gave Burgundy a powerful incentive to ensure the survival of the Lancastrian domination of northern France. It was a clever move, and one that has not received the recognition it deserves from historians. (Operating on the same principle, in March of 1432 Bedford would arrange for the grant of the valuable county of Champagne, part of the royal domain, to Philip. Since it had at that time recently been largely overrun by the French, Philip would have to fight actively to gain the benefit of the grant, and he could not well expect to keep it if the Valois knocked the English out of France.)[89]

Five days after sealing his new will, Bedford urgently requested that Philip come to Paris for consultation. The Burgundian duke promptly did so. Bedford promised him financial recompense for the costs he incurred in resuming active campaigning against France. This packet of incentives aligned Burgundy's long-term and immediate interests with his inclinations. After several days of conferences the two dukes "promised one another . . . that each of them with all his power would employ himself in resisting against the enterprises of King Charles."[90] Philip then threw himself vigorously into the war against Charles VII, immediately providing

---

85 Stratford, *Bedford Inventories*, p. 12. Bedford also held a number of lesser but significant lordships, and the county of Harcourt, but most of these were actually part of Anne's dower-lands already and held by Bedford in her right.

86 Pocquet de Haut-Jusse, "Anne," pp. 311–12, 324–25.

87 This was not a given, since most of Bedford's French lands were technically held from Henry VI in tail male, which did not allow him to bequeath them to his wife, or for her to pass them on to Philip. Ibid. Note also that I have followed all the historians who have written on this subject in taking the will as meaning to transfer to Anne all Bedford's French lands, but it should be noted that it actually refers to his "conquestz immeubles," which could be taken to mean that the lands he had purchased (such as Dreux) would not be included. Still, the will itself indicates Bedford thought such an arrangement would stand, and it would have been hard for Henry VI's government to refuse to honor the will, since that would alienate Philip, whose support would be all the more needed after Bedford's death. This could have been done legally by declaring that the lands had reverted to the crown on Bedford's death, then making a new grant of them to Philip or someone of Philip's choosing.

88 This is nevertheless what did happen, as it turned out.

89 The grant is noted by Armstrong, "Double monarchie," p. 353.

90 Stevenson, *Letters and Papers*, vol. 2, pt. 1, pp. 101–06; Waurin, *Chronicles*, p. 190.

troops to garrison Meaux-en-Brie, the expected target of the anticipated French offensive. Bedford, in Rouen, set about collecting a field force, including reinforcements from England. Once Bedford's army had assembled, he pursued the army Charles VII had led into Champagne and challenged the French to battle. When combat seemed imminent, 700–800 Burgundian soldiers formed up with English and Lancastrian-French men-at-arms in the main battle line, and the banner of Lancastrian France was again carried by the Burgundian captain the sire de l'Isle-Adam.[91] This time he was not sent away. No battle ensued, but a little later, other Burgundian men-at-arms played a key role in repulsing the French attack on Paris. Then Duke Philip brought the largest force he could muster, 4,000 combatants, to the capital. At Bedford's urging, he undertook to defend Paris over the winter (1429–30), pending a joint spring offensive to recover the towns recently taken by Charles's army.[92] Despite French efforts to break him away from the English alliance, the widespread belief among his own people (including the large majority of his councilors) that it was time to make a separate peace with Charles, and his own preference for a campaign in a different direction, Philip, at Bedford's direction, led his men to Compiègne to try to recover it from the French.[93] When these troops captured Joan of Arc in a skirmish, Philip promptly delivered her to Bedford.

The degree of Philip's active support for English efforts in this period is all the more noteworthy because at the same time he was facing several other problems that made demands on his attention and his resources. Throughout 1430 he faced a rebellion in a substantial part of Flanders. In the summer and fall of 1430 he also had to deal with a separate war between the principality of Liège and his recently acquired duchy of Namur. In June 1430 one of his principal vassals suffered defeat at the hands of the French in a disastrous ambush in the Dauphiné; in December, another substantial Burgundian force was defeated in a battle near Bar-sur-Seine. In July 1431 Frederick, duke of Austria (induced by Charles VII) added to Philip's difficulties with a declaration of war against Burgundy. Practically no military action followed from this development, but it did add to the overall pressure on Philip to make terms with the French.[94]

Looking back from 1435, when Philip sealed the Treaty of Arras with Charles VII and switched sides, historians have tended to see the Anglo-Burgundian alliance as on a declining course from 1429, or even from 1423, and to depict its ultimate failure practically inevitable.[95] Yet, the truth is that Philip's actions in

91 Waurin, *Chronicles*, p. 200.

92 Ibid., *Chronicles*, pp. 210–11.

93 Ibid., *Chronicles*, 206; Stevenson, *Letters and Papers*, vol. 2 part 1, pp. 166–7; note also Armstrong, "Double monarchie," p. 372.

94 Vaughan, *Philip*, pp. 57–67

95 E.g. Vaughan, *Philip*, pp. 20; 27: "As early as 1423 the peace of Arras was in sight; ten years later, it was imminent." Historians almost invariably seem to treat the negotiations between Philip and partisans of the dauphin as steps towards Philip's switching sides, never considering that they could have led instead to him bringing more partisans to the Lancastrian side – as he did with the duke of Lorraine in 1422. Plancher, *Bourgogne*, p.j. no. XVII, p. xx.

this crisis for the Lancastrian government show he still had a strong preference for Bedford and Henry VI over Charles VII.[96] His resumption of active participation in the Hundred Years War was admittedly quite short-lived, since he signed a new and indeed more extensive truce with Charles in late 1431. But he did so because he was constrained to, not because he wanted to. The main reason was the failure of Henry VI's governments to provide him with adequate support, financial or military. During the siege of Compiègne the money that was to come from Henry to help pay for Philip's soldiers fell two months in arrears. The English contingent of his forces, under the earl of Huntingdon, also failed to receive its wages and had to retire from the operation for lack of funds. Philip thus found himself compelled to rely on his own resources to protect his lands. This he was not strong enough to do effectively against the resurgent forces of Charles VII, so that his territories were being ravaged and his revenues, in consequence, were drying up.[97]

He nonetheless expressed willingness to stay in the fight, provided that help from Henry would be forthcoming. When that aid proved not to be available, he made a truce with the French.[98] Even then, however, it is remarkable that he managed to get the French to allow a proviso in the agreement that would allow him to send 500 men-at-arms to Bedford's support without it counting as a violation of the truce.[99] This cannot have been something to which the French were happy to agree, so he must have insisted on it, which shows he was by no means endeavoring to minimize his ties to Henry VI in preparation for turning his coat. Essentially the same offers he ultimately did accept in 1435 were already on the table as far back as August 1429,[100] but Philip did not accept them then. Instead, he actually stepped up his commitment to the Anglo-Burgundian alliance until he could no longer do so without disastrous consequences for his own interests, and even then continued at least passive adherence to the Treaty of Troyes.

96 Cf. the contrasting opinion of Mark Warner, "Dual Monarchy," p. 123, who thinks Burgundy stayed with the English alliance in 1424–1427 only "because if he had not his discarded allies would have exploited the situation in Hainault and Holland to destroy him."

97 Plancher, *Bourgogne*, p.j. no. LCCV, pp. lcxxxv-lcxxxvi; Stevenson, *Letters and Papers*, vol. 2, pt. 1, pp. 156–181. His ambassadors made this point to Henry VI's council in November 1430 and April 1431, but it was not mere "flights of diplomatic hyperbole"; the same logic was reflected in an internal memorandum to Philip by one of his principal counselors in 1436, noting that in war many garrisons were needed, that maintaining them required great expenses, that "wherever war is waged and the countryside is destroyed and plundered by friend and foe alike and the populace is restless, little or no money can be raised," and that the ducal domains "which are mortgaged, sold or saddled with debts," could not raise the needed funds. Vaughan, *Philip*, pp. 102–3.

98 It was such a truce, not (as Vaughan, *Philip*, 26, suggests) a separate peace that the Burgundian ambassadors to England hinted at as the result if the promised English financial support was not delivered. See Plancher, p.j. no. LXXV, p. lxxxvj. For the truce and its revision, ibid., p.j. nos. LXXIX and CX.

99 Plancher, *Bourgogne*, p.j. no. XCI, p. cviij.

100 Ibid., pièce justificative, no. LXX.

The End of the Alliance

Since Burgundy had a broad truce with Charles between 1431 and 1435, the French were not really in a position to push him out of alliance with the English in that period. They also did not make any major new offers to lure him over to their side in those years. This is perhaps why historians have tended to think he was already ready or nearly ready to switch sides well before he did. But that leaves unanswered the question of why he did not indeed make a separate peace in 1431, instead of a truce that specifically allowed him to provide 500 men-at-arms to assist Bedford. What was different in 1435?

The main answer seems to be the breakdown of the personal relationship between the two dukes, beginning with the death of Anne of Burgundy, duchess of Bedford, in November 1432.[101] Philip, before 1432, was already being pressed by his subjects and counselors to reconcile with Charles VII. But his personal and familial bonds with Anne and Bedford worked against that, especially after Bedford made Anne (and through Anne, Philip) the heir of his extensive French lands in 1429. Anne's death took both of those bonding factors out of the equation.[102] It also had an important indirect effect on the undermining of the alliance. Quite soon after Anne's passing, Bedford married Jacquetta of Luxembourg, daughter of the count of St. Pol. Bedford's reasons were probably more personal than political,[103] but Philip took the marriage as a slap in the face. It was a matter on which he should have been consulted for two reasons, even aside from being the brother of Bedford's late wife: first, because he was the leading peer of Lancastrian France, and Bedford's marriage had major political and diplomatic implications; and second, because Jacquetta was the daughter of one of Philip's vassals and should not have been allowed to marry anyone without his consent. If he had been asked permission, moreover, he probably would not have given it. Jacquetta's uncle Louis of Luxembourg, the bishop of Thérouanne and chancellor of Lancastrian France, had been a thorn in Philip's side at least since 1427. Under Luxembourg's chancellorship, Philip had faced increasing difficulties with the

---

101 Doubtless the turn of the military tide against England also contributed to his decision, but it cannot have been a large factor, since that tide was flowing more against the Lancastrians in 1429–1430, when he *increased* his commitment to them, than between 1432 and the final Anglo-Burgundian breach in 1435.

102 I agree with Armstrong, "Double monarchie," that "the alliance would perhaps have survived" had Anne not died young. "Double monarchie," p. 369.

103 Williams thinks that Bedford's "main motive for remarrying was the belief that union with the House of Luxembourg would be the best means of strengthening the English position in France" at a time when "it was painfully clear to Bedford that the Anglo-Burgundian alliance was in the process of dissolution." *Bedford*, p. 223. But the House of Luxembourg, of whom Louis was by far the most important member at this stage, was already firmly bound to the success of the Lancastrian dynasty, on which his own position depended, and while there were certainly signs of stress in the Anglo-Burgundian alliance, it was at this stage far from obvious that it was doomed. Thus, if acting for reasons of state Bedford would have sought a marriage that would help rebuild the bond with Burgundy, not one that would threaten to break it.

Parlement (the high court) in Paris, which repeatedly interfered with his control over his French territories.[104] After Anne's death, Bedford's health went downhill, and he relied increasingly on the chancellor. Following his marriage to Jacquetta, the regent seemed to a contemporary Parisian to be leaving the governance of the realm to Louis.[105] One major reason for Philip to maintain the alliance was to retain a strong voice and privileged position in the political matters decided in Paris. If the man who undermined him there was not only to be given increasing authority, but even to be brought into Bedford's close family by marriage, what did that say about the future of Burgundy's relations with the Lancastrian crown?

Just weeks after Bedford's second wedding, his uncle, Cardinal Beaufort, made a strong effort to soothe Burgundy's hurt feelings and restore his relationship with Bedford. The two dukes came to St. Omer for a conference, but their relationship had so soured that a point of diplomatic etiquette that would surely have been resolved or at least sidestepped if Anne had still been alive was allowed to prevent any progress towards reconciliation. Bedford, as regent of France, thought that Duke Philip should recognize his superiority by coming to see him; Philip, who viewed himself as the regent's peer within his own territories (including St. Omer) thought both should meet at the appointed place. Neither budged; there was no meeting. Before this, Bedford had always stepped up to repair the Anglo-Burgundian alliance when it frayed, but not this time.[106]

Nonetheless, Philip still felt that he needed English support against the dauphinists, because he could not trust the latter to be faithful to their promises.[107] Then in late June 1433, just a month after the non-meeting at St. Omer, a palace coup eliminated Charles VII's chief minister, Philip's enemy George de la Trémoille. Among those most influential with the king thereafter were Philip's brothers-in-law Arthur de Richemont and Charles of Bourbon.[108] With this change, but not before, the transfer of allegiance of Duke Philip from the Lancastrian to the Valois dynasty, and the complete rupture of the Anglo-Burgundian alliance, became close

104 Armstrong, "Double monarchie," pp. 359–362, 370–71 (quotation 362). Already in 1430–1431 Philip had also already begun to meet rebuffs in his economic diplomacy with England; the matters under discussion were not under Bedford's control, but Philip seems also to have felt that the Anglo-Burgundian alliance was not helping him as it should on that front either. John H. Munro, "An Economic Aspect of the Collapse of the Anglo-Burgundian Alliance, 1428–1442," *The English Historical Review*, 85 (1970), pp. 232, 237–38.

105 Williams, *Bedford*, p. 223; Bourgeois de Paris, *Journal* (ed. Tuetey), pp. 295, 296 ("laissoit du tout regenter"). By early 1434, in Paris "there was no news of the regent, and no one ruled but the bishop of Thérouanne, chancellor of France," who was hated by the people because of his opposition to peace. Ibid., p. 298.

106 Williams, *Bedford*, p. 224.

107 Du Fresne de Beaucourt, *Histoire*, pp. 415–16, 456–7; Warner, "Dual Monarchy," pp. 112–14.

108 Mark Warner considers that "it was this change of regime which would be the principal factor in the reconciliation between Charles VII and Duke Philip." "Dual Monarchy," p. 114; see also M. G. A. Vale, *Charles VII* (Berkeley, CA, 1974), p. 72.

to inevitable. The Treaty of Arras, negotiated while Bedford lay dying in Rouen, followed in 1435.[109]

Conclusion

What, then, are the lessons for modern strategists in the variable success and the ultimate failure of the Anglo-Burgundian alliance during the Hundred Years War? Based on the analysis in this chapter, one can highlight three. First, it is crucial to appreciate the motivations of allies, which is the first step towards ensuring that the coalition's grand strategy can meet their needs sufficiently to maintain the alignment of interests of the partners. Allies' concerns must be appreciated in culturally sensitive ways, and with due regard to personal relationships, rather than being reduced to general considerations of *Realpolitik*. What matters most to them might not be what we would expect, or what would be our main interest if we were in their shoes. Second, that mutual action can strengthen the bonds between allies, and that this binding together is as much about being willing to trust a partner and to *receive* aid gracefully as it is about being willing to *give* aid. Third, that since the strength of an alliance will inevitably, during extended wars, have its ups and downs, and since shared success can strengthen alliances both by its own effects and by promising ultimate victory and its rewards, when a favorable military situation coincides with a period of strong attachment between allies, that is the time to vigorously press combined operations.

109 On the negotiations leading to the treaty, see Dickinson, *Congress of Arras*.

9

MEDIEVAL STRATEGY AND THE ECONOMICS OF CONQUEST[1]

It is not so long ago that many historians were unwilling to admit that "strategy" was a concept applicable to the Middle Ages.[2] Today, however, it would be difficult to find any specialist willing to support that position. It is now generally recognized that medieval rulers, who were both political and military leaders, and surrounded by advisers who moved back and forth easily and frequently between service in arms and service in government, naturally had a good understanding of how military force could be used to pursue political objectives, which is the fundamental question of strategy.

Since the essence of sound strategic planning is making rational decisions linking available means to desired outcomes, it should be worthwhile to study it using approaches borrowed from the discipline of economics, which at its core is also about rational decision-making, balancing costs and benefits of various courses of action. To be sure, we as historians know that not all actions are in fact based on rational calculation. Hatred, romantic love, immoderate greed, or sickness can push even a normally sensible person into making irrational decisions. The random chances of inheritance, in a system characterized by dynastic rule, can put a state under the rule of someone who is rash, foolish, bullheaded, downright stupid, or even insane. Strategy, moreover, is often set not by individuals but by compromise among groups whose members have partly conflicting interests, so that the outcome is not quite utility-maximizing for any party to the compromise. And perfectly rational decision making, as understood in economics, requires perfect information, which economists commonly assume, while historians – and

1 This article is based on a paper presented at the Universidad de Extremadura for a conference on Strategy in the Middle Ages, held in 2010. In addition to the organizers and participants in the conference, I would like to thank Don Kagay, Lawrence V. Mott, Guillaume Sarrat De Tramezaigues, Damian Smith, and Daniel P. Franke for assistance with this article, and to thank the United States Military Academy for financial support for attending the conference, and for the sabbatical leave during which this article was completed. The version published here includes minor corrections and a few additions to the previously published version.
2 For examples of this view dating from the 1940s to the 1980s, see the quotations and citations in John Gillingham, "Richard I and the Science of Warfare in the Middle Ages," reprinted in Matthew Strickland, *Anglo-Norman Warfare* (Woodbridge: Boydell, 1992), 195; Clifford J. Rogers, "Edward III and the Dialectics of Strategy," reprinted in Clifford J. Rogers, *Essays on Medieval Military History* (Farnham: Ashgate Variorum, 2010), I, p. 83.

DOI: 10.4324/9781003399971-12

A Note on Money[3]

Medieval money took three basic forms. The first two, which overlapped, were silver coins and money of account. The third was gold, whether in the form of coins, or of ingots or sacks of dust standardized by weight.

In Latin countries, from the time of Charlemagne money of account was based on the system that one libra (livre, libra, pound, etc.), abbreviated l. = 1.5 marks = 20 solidi (sous, sueldos, shillings, etc.), abbreviated s. = 240 denarii (deniers, dineros, pence, etc.), abbreviated d. Pennies were usually actual silver-alloy coins, but marks and pounds were not; they were simply quantities of pennies, either counted or weighed. Because the proportion of silver in coins varied from country to country, and in some cases from year to year due to debasement of the coinage, the value of a penny (and therefore of a pound) could vary greatly by place and time. For example, during the thirteenth century, 1 English pound or "pound sterling" (£) was commonly reckoned at 4 pounds of Tours (*livres tournois*, l.t.), but in 1299 one pound sterling could be exchanged for as much as 11 l.t. in devalued French coinage.

For the later part of the period dealt with in this article, the main gold coin minted in Latin Christendom was the florin. The standard florin was minted in Florence, starting in 1252, and contained 3.5 grams of pure gold. The French gold franc was originally minted at 3.82 grams and valued at one livre tournois; the écu, introduced later, was similar but slightly more valuable. In Aragon and Sicily, the *onza* or *uncia* was the main monetary unit, divided into 30 *tareni* each of 20 *grani*; up to 1284 an *onza* was one Sicilian ounce (21 grams) of gold, but thereafter it was a roughly equivalent value of silver coins. The term "bezant" was used both for Byzantine gold coins somewhat more valuable than a florin, and also for the equivalent gold *dinars* issued by Muslim states.

The following equivalencies for the year 1284, though subject to substantial variation over time, give a sense of the relative value of the monetary quantities noted in this article: 1.25 pound sterling (£) = 2 gold *onzas* = 4 livres parisis [l.p.] = 5 livres tournois (l.t.) or libras of Valencia (l. val.) or of Barcelona (l. barc.) = 6.66 gold bezants = 6.66 *doblas* = 10 Florentine florins = 25 shillings sterling (s. st.) = 60 *tareni* = 80 sous of Paris (s.p.) = 100 sous Tournois (s.t.) or sous/sueldos of Valencia (s. val.) = 300 pence sterling (d. st.) = 1,200 *grani* = 1,200 deniers tournois (d.t.) or dineros of Valencia (d. val.).

3 The exchange rates in this note are taken from Peter Spufford, *Handbook of Medieval Exchange* (London: Royal Historical Society, 1986).

particularly military historians – and medieval military historians perhaps more than any other sort – know that strategic decisions always have to be made inside the "fog of war," where information about enemy intentions and capabilities is usually very far from perfect.[4] Nonetheless, it remains true, I think, that when we read of a military commander making an apparently irrational strategic decision, it more likely to be one that merely *seems* irrational due to our lack of adequate sources or our lack of understanding of the commander's thinking.[5] In any case, if we look at the long run rather than at particular cases, we can assume that the general patterns of action in medieval warfare do reflect sensible calculations by decision-makers, and that changes in the patterns of the conduct of war over the course of the Middle Ages, or differences from one region to another, reflect differences in the underlying realities that structured the calculations of cost and benefit on which those decisions were based.

In this article I will focus on just a few such differences and their implications for strategic patterns. One is the presence or absence of slavery in a given place and time; another is variations in levels of urbanization and of encastellation. In order to keep the topic within manageable bounds, I will focus only on offensive rather than defensive strategy.

The place to begin our historical analysis is the same as the question that should come at the start of developing a strategic plan, according to no less an authority than Carl von Clausewitz, the brilliant military theorist whose *On War* should be carefully studied by all those seeking to understand strategy, medieval or otherwise. "The first of all strategic questions," he wrote, is for generals and statesmen "to establish . . . the kind of war on which they are embarking; neither mistaking it for, nor trying to turn it into, something that is alien to its nature." This dictum has so often been quoted out of context that its meaning has been obscured, but what Clausewitz meant here was that the strategist had to decide which of the two basic types of war he intended to wage: a war aimed at *forcing* the enemy to *surrender*, or one aimed at *persuading* him to *make concessions* in negotiations. Although Clausewitz did not use the terms, this distinction in his work is the main source of later theorists' and historians' differentiation between "total war" and

4 I.e. the "general unreliability of all information [that] presents a special problem in war." Carl von Clausewitz, *On War*, ed. and tr. Michael Howard and Peter Paret (Princeton: Princeton U.P., 1984), 140; cf. 101, 108, 117–18. Eugenia Kiesling has pointed out that Clausewitz does not actually use the phrase "fog of war," and argued that the importance of the "fog" metaphor to Clausewitz's thought has been exaggerated. Nonetheless, however, she agrees on the centrality of "uncertainty" to Clausewitz's argument. *"On War* without the Fog," *Military Review* (September–October 2001): 85–87.

5 A good example is Henry V's march to Calais in 1415, which has been described as "unaccountable" and "foolhardy to a degree of madness," but which was actually made for sound strategic reasons. Clifford J. Rogers, "Henry V's Military Strategy in 1415," in *The Hundred Years War: A Wider Focus*, ed. L. J. Andrew Villalon and Donald J. Kagay (Leiden: Brill, 2005): 399–427; but cf. Jan Honig, "Reappraising Late Medieval Strategy: The Example of the 1415 Agincourt Campaign," *War in History* 19 (2012).

"limited war."[6] In the modern world, where the international political system is mostly structured by undivided sovereignties and unambiguous territorial divisions, the distinction between these two basic objectives is more clear-cut than it was in the Middle Ages. In the modern world, conventional wars are usually ended either by a negotiated peace – a compromise generally resulting from some sort of military stalemate, for example in the War of Spanish Succession, the Korean War, or the Iran-Iraq War – or else by an unconditional surrender or dictated peace, often involving a change of government and resulting from the thorough military defeat of the losing side, as in World Wars One and Two or the Second Persian Gulf War of 2003. In the Middle Ages there were also outright conquests – for example of Lombardy by Charles the Great; of England by William the Conqueror; of Valencia and Majorca by Jaume the Conqueror; and of Normandy and Gascony by Charles VII of France. Such conquests were facts on the ground, and hardly required a negotiation with the defeated adversary.[7] There were also classic negotiated compromise peaces of various sorts. One type of compromise peace deserves special mention here, because it was very common in the Middle Ages but is relatively rare in the modern era. This type of peace involved the "losing" side making some sort of submission to the winning side and accepting a position of dependency – for example, the ruler of a formerly independent state might submit to his invading neighbor, accept his overlordship by doing homage, and offer a small annual tribute and/or military service in the future – but retaining most of his independence, his revenues, and his practical power within his territory. Such an outcome was not, however, a zero-sum result. The winner did gain what the loser lost, which *is* a zero-sum effect, but *in addition* the loser usually had some gain to offset his loss, in that he could thereafter expect his new overlord to extend to him the benefits of good lordship: favors, gifts, and above all protection from or assistance against third parties. Providing those benefits did cost the lord some resources, so the net transfer of utility from the defeated to the victor equaled the difference between what the benefits of good lordship were to the new vassal or tributary and what it cost him to obtain those benefits. That would on the face of it seem again to be a zero-sum proposition, but if the new lord's situation was such that he could provide important benefits to his vassal at low cost to himself, it would not be a zero-sum arrangement. For example, a powerful and warlike king could provide an important benefit to his dependents – protection by

---

6 *On War*, 88–89, 91–92, 69, 517, and Book Eight, chapters 7–9 (pp. 611–637), which analyze war plans according to the basic two kinds of war – limited-aim and total-aim – with the former subdivided into offensive and defensive variants.

7 All these cases did in the end involve negotiations and formal submissions of one sort or another. E.g. in 1066, William agreed to receive the submission of the surviving English magnates and swore to be a good lord to them before being crowned king. But this can hardly be said to be a compromise settlement. William the Conqueror fully achieved what he intended to, and indeed probably more (he can hardly have expected to have quite so many landed Englishmen killed during the conquest, or hence quite so much land available for him to distribute to his men); Harold had failed and was dead.

deterrence – at very little cost to himself. Thus, when a formerly independent lord became subject to another, there was inevitably some re-division of the wealth and power "pie," but the pie itself could also be made bigger.[8]

That is one reason why such outcomes were common in the Middle Ages, and why they could take forms that from a modern perspective seem at first glance rather bizarre. For example, the battlefield defeat of Pope Leo IX at Civitate in 1053 led – whether directly, as Geoffrey of Malaterra claims, or only indirectly and after a delay, as seems more likely – to the *winner* of the conflict, Robert, becoming the *vassal*, rather than the lord, of the papacy. This involved little sacrifice on Robert's part, since his subordination to the pope was largely nominal, but in return he gained from his new "lord" an important gift (one that the latter could give at little cost to himself, and yet that was very valuable to the recipient): the legitimization of Robert's subsequent offensives into Apulia, Calabria, and Sicily.[9]

Of course, one way to secure a compromise submission was to mount a credible threat of complete conquest. It was also possible to conquer an area – to compel its inhabitants to accept the aggressor as their actual, direct ruler, rather than as a nominal overlord – by employing the same methods that were more typically employed to extort political concessions, nominal submissions, and/or cash *parias* (payments of tribute as protection money): namely, for the invader to have his men "burn, pillage, and ravage all those lands which he had invaded, and to do all they could to instill terror in the inhabitants," as Robert Guiscard's men did when they first entered Calabria.[10] Still, though opportunities might be seized or decisions reconsidered based on the flow of events, one of the first strategic decisions to be made was how to proceed against the enemy: whether to aim at compulsion and conquest, or at punishment, pressure, and persuasion.

Each of those basic aims implied different modes of action, different costs, and different benefits, which brings us back to economics. Although there were exceptions, as already noted, generally persuasion was best accomplished by raiding and devastation, while a campaign of conquest typically implied the conduct of

8 On the other hand, the damage and expense of war, especially protracted war, could make the pie smaller.

9 Geoffrey of Malaterra, *The Deeds of Count Roger of Calabria and of His Brother Duke Robert Guiscard*, tr. Kenneth Baxter Wolf (Ann Arbor: U. of Michigan Press, 2005), 62; G. A. Loud, *The Age of Robert Guiscard: Southern Italy and the Norman* Conquest (Harlow: Longman, 2000), 120, 129–30.

10 William of Apulia, *The Deeds of Robert Guiscard*, trans. G. A. Loud (Leeds Medieval History Texts in Translation, at https://web.archive.org/web/20071007034752/www.leeds.ac.uk/history/weblearning/MedievalHistoryTextCentre/william%20ap%202.doc), 23; for the Latin see *La Geste de Robert Guiscard*, ed. and tr. Marguerite Mathieu (Palermo: Bruno Lavagnini, 1961), 150. Similarly, William the Conqueror's method of conquering Maine was described by his own chaplain as follows: "He sowed terror in the land by his frequent and lengthy invasions; he devastated vineyards, fields and estates; he seized neighbouring strongpoints and where advisable put garrisons in them; in short he incessantly inflicted innumerable calamities upon the land." William of Poitiers, *Gesta Guillelmi*, quoted in John Gillingham, "William the Bastard at War," in Strickland, *Anglo-Norman Warfare*, 150.

multiple extended sieges.[11] What were the benefits that could be expected to be gained from each of the two basic approaches?

Conquest by definition gave possession of the land and power over the people who inhabited it: one might say it gave the conqueror the *wealth* of the conquered area, as well as access to its future *income*. As Jaume the Conqueror put it when he decided to conquer Valencia rather than accepting an annual tribute from its Muslim ruler, he could thereby have both the hen now and the chicks in the future.[12] Raiding, on the other hand, involved seizure of a fraction of the region's wealth and its people – as many as the invader could get his hands on – and if successful might secure future tribute that amounted to a significant share of the population's income.[13]

11 As Stephen Morillo would note, this is based on an assumption of territorial powers competing over disputed land, and would not apply well to conquests by non-sedentary peoples (e.g. steppe nomads), and would also be somewhat less applicable to internal civil wars. See Stephen Morillo, "Battle Seeking: The Contexts and Limits of Vegetian Strategy," *Journal of Medieval Military History* 1 (2002), 21–41, esp. p. 29. For the details of what "devastation" or "ravaging" means in the context of medieval warfare, see Clifford Rogers, *The Middle Ages*, in the "Soldiers' Lives through History" series (Westport, Conn.: Greenwood, 2007), 85–90, and Clifford J. Rogers, "By Fire and Sword: Bellum Hostile and 'Civilians' in the Hundred Years War," in *Civilians in the Path of War*, ed. Mark Grimsley and Clifford J. Rogers (Lincoln: University of Nebraska Press, 2002), 33–57.

12 James [Jaume] I of Aragon, *The Book of Deeds of James I of Aragon*, tr. Damian Smith and Helena Buffery (London: Ashgate, 2003), cap. 243, p. 210.

13 John Gillingham has forcefully and correctly argued that before the tenth century across Europe, and in areas of cross-religious frontier fighting until the very end of the Middle Ages, slave-taking was of primary importance in the motivation for and conduct of raiding warfare. See his "Women, Children, and the Profits of War," in *Gender and Historiography. Studies in the Earlier Middle Ages in Honour of Pauline Stafford*, ed. Janet L. Nelson, Susan Reynolds and Susan M. Johns (London: Institute of Historical Research, 2012); "Christian Warriors and the Enslavement of Fellow Christians," in M. Aurell and C. Girbea (eds.), *Chevalerie et christianisme aux XIIe et XIIIe siècles* (Rennes: P.U. Rennes, 2011); "Crusading Warfare, Chivalry, and the Enslavement of Women and Children," in *The Medieval Way of War: Studies in Medieval Military History in Honor of Bernard S. Bachrach*, ed. Gregory I. Halfond (Aldershot: Ashgate, 2015); "A Strategy of Total War? Henry of Livonia and the Conquest of Estonia (1208–1227)," *Journal of Medieval Military History* 15 (2017). Slave-taking was also a central component of Anglo-Saxon warfare until the Conquest, and of Celtic war-making in Britain and Ireland into the thirteenth century. See David Wyatt, *Slaves and Warriors in Medieval Britain and Ireland, 800–1200* (Leiden: Brill, 2009). Slave-taking raids could also on occasion be used as means of conquest; for example, after two seaborne raids on the Tunisian island of Djerba carried off thousands of captives, the survivors secured permission from their overlord to surrender the land to the king of Aragon. Ramon Muntaner, *The Chronicle of Muntaner Translated from the Catalan*, trans. Lady Henrietta Margaret Goodenough (London: Hakluyt Society, 1920–1921) Reprinted online, with different pagination: (Cambridge, Ontario: In Parentheses Publications, 2000), at www.yorku.ca/inpar/muntaner_goodenough.pdf, 244–45. The practice of killing fencible men, brutalizing the remaining population, and carrying off as slaves their market-worthy women and children, or "andrapodizing," was also central to ancient warfare, as Kathy L. Gaca has explained in important publications including "The Andrapodizing of War Captives in Greek Historical Memory," *Transactions of the American Philological Association* 140 (2010): 117–61, and "Girls, Women, and the Significance of Sexual Violence in Ancient Warfare," in *Sexual Violence in Conflict Zones: From the Ancient World to the Era of Human Rights*, ed. Elizabeth D. Heineman (Philadelphia: University of Pennsylvania Press, 2011), 73–88.

Figure 9.1 Profits of War. Raiders could gain rich profits by seizing material goods, live-stock, and human prisoners. Of those three categories, the last usually brought by far the greatest payoff. Despite the artist's choice in this thirteenth-century illumination (depicting one of Saladin's raids into the Kingdom of Jerusalem), in areas where slave-taking was common the most prized and most frequently taken prisoners were girls and young women. [British Library, Yates-Thompson MS 12, fo. 161 r.]

But normally an overlord or tribute-taker could expect to get from a territory only a fraction of the revenues collected from its population by those who ruled it directly. In the case of Valencia, for example, the *paria* offered by its Muslim ruler to Jaume I was only about a tenth of what the city later generated in revenues for the crown of Aragon.[14] So conquest had a bigger payoff, but when economic actors make rational decisions, they are less concerned to maximize *revenues* than to maximize *profits*. The costs must be calculated, as well as the gains, and conquest was usually very costly to carry through to completion. When he first considered conquering Granada, Alfonso VI of Castile is reported to have decided instead to accept an annual tribute after reflecting as follows: "How should I aspire

14 Manuel de Bofarull y de Sartorio, *Rentas de la antiqua corona de Aragon*, vol. 39 of *Colección de documentos inéditos del archivo general de la corona de Aragon* (Bacelona: Imprenta del Archivo, 1871), 89; James I, *Book of Deeds*, 230.

to take it? . . . By force? No, my men will perish, my money will disappear and *my losses will be greater than any benefit* I could hope to derive should the city fall into my hands."[15] [Emphasis added.]

The mention of "hope" in the last phrase implies another consideration: that of "expected utility." The success of an attempted conquest was never certain, so it was not enough to simply weigh the gross benefit of success (B) against the direct cost of the effort (C). Instead a rational actor had to estimate probability of success (P), and weigh the anticipated gain (B*P) against the anticipated cost, including both the cost of the effort (C) and the risk of the positive harm that would follow from a defeat (H): (C+((1-P)*H). Thus, for example, if success were only 50% likely, the gross benefit of success (B) would have to exceed *double* the gross cost (C) by more than the amount of positive harm that would ensue from a defeat (H) to make the anticipated net value of the effort positive, and so to make the action economically rational:

Action is rational if BP > C + (1-P)*H.
If P = 0.5:
0.5B > C+ 0.5H
B > 2C + H

On the other hand, if probability of success was near 100%, then the estimated gross benefit of success would only have to slightly outweigh the estimated cost of effort to make the action rational.

The practical upshot of this is that a ruler would generally be unwise to undertake a siege unless success seemed quite likely. Major sieges had both a higher probability of failure and a higher expected cost of execution than did raids, which is one reason that the anticipated costs of attempted conquest often outweighed the expected gains, leading medieval strategists to pursue lesser strategic aims.

True, an invader could sometimes conquer a large area relatively cheaply if the target of occupation was not well fortified or its defenders obligingly came forth from their fortifications to be defeated in open battle, as for example in the conquest of England in 1066, or on a smaller scale Alfonso III of Aragon's brutal conquest of Minorca in 1287–88. The latter was hugely profitable. The Aragonese fleet of forty galleys was reinforced with 500 armored cavalry and probably 3,000 almogavars (light infantry) to reach a total of around 10,000 combatants. The force gathered in Barcelona around 1 November 1287, reached Majorca on 15 December, sailed for Minorca after Christmas, ran into a storm, and gradually collected off Minorca in early January 1288. Once enough forces were assembled, the Aragonese landed. The Minorcans, assisted by foreign mercenaries, met them in battle and after a hard fight were utterly defeated. The *almojarife* (the ruler of the

15 'Abd Allah b. Buluggin, *The Tibyan: Memoirs of 'Abd Allah b. Buluggin, Last Zirid Amir of Granada*, tr. Amin T. Tibi (Leiden: Brill, 1986), p. 90.

island) retreated into the castle of Mahon with a few survivors, and negotiated to surrender the castle, along with the town of Ciutadella and all other fortified places on the island, in exchange for being allowed to sail to Africa with his household. Alfonso took possession of Mahon on 17 January. The surviving inhabitants of the island (the women and children and a few men) were enslaved and sold at auction: 40,000 of them, according to the soldier and chronicler Ramon Muntaner. Their lands were distributed to Catalan settlers.[16]

The benefits of the conquest in loot alone – that is, the Muslim inhabitants of the island, their goods, horses, arms, cattle, and grain – reportedly amounted to 100,000 *livres tournois* (l.t.).[17] The total cost of the expedition, on the other hand, was probably a third or a quarter less than that amount.[18] Thus, it was quite profitable on that basis alone. The same was true, but even more so, for some Catalan raiding operations that did not involve actual conquest. For example, Roger de Llúria's pillaging expedition against the Byzantine Aegean in 1292, which

16 For the Minorca expedition and its aftermath, see Muntaner, *Chronicle*, pp. 341–46; Antonio de Bofarull, ed. and tr., *Crónica Catalana de Ramon Muntaner* (Barcelona, Jaime Jepús, 1860), 321–26; Cosme Parpal y Marques, *La Conquesta de Menorca en 1287 por Alfonso III de Aragón. Estudio histórico-crítico, con un apéndice de documentos* (Barcelona: La Casa Provincial de Caridad, 1901). Jaime Sastre Moll, "Breves notas sobre el saque de Menorca tras la conquista de Alfonso III (1287)," *Meloussa* 2 (1991): 49–58, is also valuable, though I do not find persuasive the argument that island was conquered practically without resistance.

17 Antonio Ortega Villoslada, *El reino de Mallorca y el mundo Atlántico, 1230–1349* (Sta. Christina, Oleiros, Spain: Netbiblo, 2008), 50–51, 51n; Biliothèque nationale de France [BnF], MS Lat. 10152, fo. 49v. The figure of 100,000 l.t. is credible, since it is less than ten times the proceeds from the sale of the inhabitants of Kerkenna (without their goods), and the latter is a much smaller island – about 1/11th the size – and arid. See p. 250.

18 This is based on the estimate that the fleet really carried 3,000 almogavars rather than the 30,000 claimed by Muntaner (in the manuscript as well as the printed edition, though the scribe who prepared the clean copy could have confused iij millia for xxx millia), and counting the costs based on three full months (November, December, January). For the number of the almogavars: Muntaner, *Chronicle*, 341–2; *Crònica* (Ms. 1803 de la Biblioteca Nacional, España), fo. 93v; but in support of the lower number, see Parpal, *Conquesta*, 17, and Muntaner, *Chronicle*, pp. 219, 262, 315, 384, 394, 422, 427, 515; but also on the other hand note also 333, 375. Wages for the almogavars are reckoned at the normal rate of 12 tareni per month (Lawrence Mott, personal communication), though they may have received the lower rate of 8 tareni until actually embarked (probably in mid-December): cf. Giuseppe La Manti, ed., *Codice diplomatico dei re aragonesi di Sicilia (1282–1355)*, vol. 1 (Palermo: Società siciliana per la storia patria, 1918), 624, for almogavar wages, and ibid., 497, with Mott, *Sea Power*, 236, 233, for numbers. I have accepted Muntaner's claim of 700 cavalry (as given in the manuscript *Crònica*, fo. 93v, rather than the 500 given in the printed version) on armored horses (rather than the usual mix of heavy horse and lighter, less expensive *janeti*) and calculated their wages at the normal fleet rate of 2 *onzas* per month, though again they may have received a lower rate until embarkation. Mott, *Sea Power*, 171, 293 n.108. The cost for the galleys and their crews (including 120 oarsmen and 30 crossbowmen each) is figured at 75 gold *onzas* per month. Ibid., 184. The total cost for the expedition then calculates out to about 30,000 *onzas*, with the *onza* at about 2.2 to 2.5 l.t. (Spufford, *Handbook*, 65, 74; 62). If there actually were 30,000 *almogavars* under pay, the cost of the expedition would be over 110,000 *onzas* and the expedition would not have been immediately profitable, though it would still have been so when the value of the land acquired is taken into account.

included the sacking of Monemvasia on Lesbos, earned the admiral five times the cost of the fleet.[19] But in the case of Minorca, we should add to the benefit column the value of the island itself. From Alfonso's perspective, this took two forms. First, he was able to give a substantial quantity of wealth, in the form of real property, to his followers. Second, he thereafter gained the profits of lordship over the island, which were not insignificant. Jaume the Conqueror had formerly received from it roughly 1,500 *livres tournois* worth of wheat, sheep, and cattle in annual tribute, and Jaume of Majorca reckoned that prior to its conquest by Alfonso he received from it 1,000 l.t. per annum in clear revenues, in addition to loans, special aids, and other intermittent benefits.[20]

Such rapid and profitable conquests were the exception rather than the rule, however, precisely because defenders had another option: a Vegetian strategy of holding their fortresses and attacking the invaders' logistics, which often promised better chances of success *and* the likelihood that if the strategy did fail, it would be a much better form of failure than the disaster likely to follow from a major battlefield defeat.[21] The reason for both of these advantages was precisely that by staying within their fortifications the defenders could force a would-be conqueror to engage in extended sieges, and the latter were so costly that they could easily collapse when the besieger ran out of the funds or supplies needed to sustain them. Moreover, even if the besieger did have sufficient means to capture a place by blockade, the large daily expense of keeping an army together gave the aggressor a strong incentive to cut short the duration of the siege by concessions to the defenders.

19 Muntaner, *Chronicle*, 244. This is not wild exaggeration: account records show that just the mastic (aromatic gum or resin) seized on Chios was worth more than the annual cost of fleet operations. Mott, *Sea Power*, 255. This immediately followed a raid on the Tunisian island of Djerba that resulted in the capture of 10,000 Muslims, and "made so much gain that the expenses of the galleys and the cost of fitting them out were cleared." Muntaner, *Chronicle*, 244. The fact that Muntaner is thinking in these terms is very significant.

20 Tribute: James, *Book of Deeds*, 132. For generalized prices for these goods in l.t. (in southern France in 1327), see M. Jusselin, "Comment la France se préparait à la guerre de cent ans," *Bibliothèque de l'école des chartes*, 73 (1912), doc. 1a, p. 221. Note that this tribute seems to have stopped flowing after 1285: see David Abulafia, *Mediterranean Emporium: The Catalan Kingdom of Majorca* (Cambridge: Cambridge U.P., 1994), 67. Jaume of Majorca's revenue: Ortega, *El reino de Mallorca*, 50–51, 51n; BnF MS Lat. 10152, fo. 49. Note the implication that the value of the people and goods in a given territory could exceed the purchase value of the lordship (based on the 10:1 purchase:revenue ratio discussed later) by around 10:1.

21 See John Gillingham's articles, "Richard I and the Science of Warfare in the Middle Ages" and "William the Bastard at War," both reprinted in Strickland, *Anglo-Norman Warfare*, for particularly cogent analysis of the Vegetian defense, and various articles in *The Journal of Medieval Military History* for a scholarly conversation on the subject of Vegetian strategy: in vol. 1 (2003), Clifford J. Rogers, "The Vegetian 'Science of Warfare' in the Middle Ages," and Stephen Morillo, "Battle Seeking: The Contexts and Limits of Vegetian Strategy"; in vol. 2 (2004), John Gillingham "'Up with Orthodoxy!' In Defense of Vegetian Warfare"; in vol. 8 (2010), L. J. Andrew Villalon, "Battle-Seeking, Battle-Avoiding or Perhaps Just Battle-Willing? Applying the Gillingham Paradigm to Enrique II of Castile."

Just how expensive was it to conduct a major siege? To understand the answer to that question, we have to consider both the *gross* cost of the siege and its *net* cost – that is, what it cost up front, and the extent to which that cost could be offset by the gains that came with the conquest. Let us start by examining the early fourteenth century, when the sources for addressing the question become particularly full, since we have theoretical, anecdotal, and archival material on the subject available. As a generalization that is significant though quite imprecise, we have the well-known observation of Pierre Dubois, writing at the start of the fourteenth century, that "a castle can hardly be taken in a year, and even if it does fall, it costs more from the king's purse . . . than the conquest is worth . . . Because of [this] . . . leaders are apt to come to agreements which are unfavorable to the stronger party."[22] Here we have a contemporary statement reflecting the two points I just made; Dubois takes it as uncertain that a siege of a castle, even a royal siege, will succeed, and suggests that if the fortress does surrender, it is likely to do so on favorable terms, which the king will be willing to grant in order to avoid the risk of failure and the high cost even of success, if the siege continues too long. DuBois's comparative "more . . . than the conquest is worth" is important in itself, as we shall see, but more precision would be more enlightening. Granted that sieges were likely to be unprofitable from an economic standpoint, it still makes a big difference if their economic costs only slightly outweighed their benefits, or if the typical conquest by siege cost several times the value it could be expected to return. The more sieges cost relative to the economic benefits the capture of the besieged places could provide, the less fiscally practical a strategy of conquest would be, and the stronger the non-economic motives justifying the operation would have to be, a point we will return to later.

Contemporary sources suggest that to capture a place by siege did not merely cost (as DuBois said) *more* than the conquest was worth, but rather *much* more. For example, the chronicler Edmond de Dynter tells us that the siege of Faquemont in 1329 cost four times what the lordship was worth.[23] That was actually quite a good bargain compared to the siege of Exeter in 1136. The town had paid a royal *ferm* (a fixed sum paid in lieu of the variable profits of lordship) of under £40 (40 pounds sterling) around 1108. Those revenues were hugely disproportionate with what it cost King Stephen to recapture the castle from the rebel Baldwin de Redvers: three months and, reportedly, £10,000 "in various expenses."[24] To

---

22 Pierre DuBois, *Summaria brevis et compendiosa doctrina felicis expedicionis et abreviacionis guerrarum ac litum regni Francorum*, ed. Hellmut Kämpf (Leipzig: Teubner, 1936), 3; translation from J. F. Verbruggen, *The Art of Warfare in Western Europe During the Middle Ages. From the Eighth Century to 1340*, tr. S. Willard and R. W. Southern. Second English Edition (Woodbridge: The Boydell Press, 1997), 306.

23 Sergio Boffa, *Warfare in Medieval Brabant, 1356–1406* (Woodbridge: Boydell, 2004), 173.

24 Ferm: "Austin Canons: Priory of Holy Trinity or Christchurch, Aldgate," *A History of the County of London: Volume 1: London within the Bars, Westminster and Southwark* (1909), pp. 465–475, at fn. 7. Cost: *Gesta Stephani*, in *Chronicles of the Reigns of Stephen, Henry II, and Richard I*, vol. 3, ed. Richard Howlett (London: Rolls Series, 1886), 25. The number seems high, but even if the actual amount were only one third as much, the point would still stand.

put that latter amount in perspective, consider that one generation earlier Robert Curthose had pawned the entire duchy of Normandy for one third less.[25]

Rather than relying on chroniclers' estimates, however, let us see what we can work out regarding a particular case for which we have relatively good documentation. The siege of Calais in 1346–47 cost Edward III upwards of £80,000 just in soldiers' and sailors' wages – that is, about twice the annual revenues of his crown during the first few years of his reign.[26] That is a staggeringly large sum of money in absolute terms, but what we want to know is how it compares to what the thriving port city was "worth" to the king, economically speaking. At the time of the siege the purchase price of a lordship was normally around ten times the revenues it produced. The Middle English "Rules for Purchasing Land" conclude that "in ten yere if thou wyse be/Thou schall ageyn thi money se": that is, with good management, lands could provide net annual revenues of a tenth their purchase price.[27] Similarly, perpetual rents, to be taken from secure revenues such as customs income, were typically priced in the early fourteenth

25 Charles Wendell David, *Robert Curthose, Duke of Normandy* (Cambridge, Mass.: Harvard University Press, 1920), 91–92, 95–96.

26 For an approximate idea of Edward's revenues, see James H. Ramsay, *A History of the Revenues of the Kings of England, 1066–1399* (Oxford: Clarendon Press, 1925), Table I, facing p. 2:292, but for caveats cf. Anthony Steel, *The Receipt of the Exchequer, 1377–1485* (Cambridge: Cambridge U.P., 1954), xxii-xxxiii. Cost: War wages for troops in Normandy, in France and at Calais (i.e. in the Crécy-Calais army, excluding forces in Gascony and elsewhere) from 4 June 1346 to 12 October 1347 amounted to £127,201 2s. 9½ d. Francis Grose, *Military Antiquities Respecting a History of the English Army*, 2d. ed. (London: Stockdale, 1812), 278; note also George Wrottesley, *Crecy and Calais from the Public Records* (London: Harrison and Sons, 1898), 205n. To isolate the cost of the siege itself, we must deduct from this total (1) the sum of wages paid to the army and the fleet from the start of the accounting period up to the opening of the siege and (2) the wages paid to the garrison of Calais between the end of the siege and the end of the account. To ensure that our estimate of the cost of the siege is conservative, we can make assumptions that will produce figures for those deductions that are likely to be higher than the actual figure. Following that principle, I have calculated the daily wage cost for the army as if it were at its full landing strength (using the numbers from the appendix of *War Cruel and Sharp*) from the start of the account through 2 September; this works out a bit under £32,000. A garrison averaging 2,000 men from the end of the siege through the end of the account would cost somewhat over £3,000. Wages for the full fleet of 738 ships (each with 1 master and 1 constable at 6 d./day) and 9,355 mariners (at 3 d.) for the period from 4 June through a week after the landing at La Hougue, then for 200 English ships and their crews through Edward's departure from Caen on 31 July, works out to around £6,400. Fleet numbers: Wrottesley, *Crecy and Calais*, 204. Wages: A. E. Prince, "The Army and Navy," in *The English Government at Work, 1327–1336* (New York: Medieval Academy of America, 1940), 385. This leaves well over £85,000, from which a relatively small sum would have to be deducted for fleet operations after 31 July and not directly related to the siege.

27 "The Rules for Purchasing Land," in *Codex Ashmole 61: A Compilation of Popular Middle English Verse*, ed. George Suffelton (Kalamazoo: Medieval Institute Publications, 2008), 71. Examples of properties evaluated as worth ten times their annual rents include J. Viard and A. Vallée, eds., *Registre du trésor des chartes* (Paris: Archives Nationales, 1979–84), III, part 2, nos. 5986 and 7333. See also Gregory Clark "The Cost of Capital and Medieval Agricultural Technique," *Explorations in Economic History* 24 (1988), pp. 269–76.

century at around ten times their annual value.[28] The revenues of Calais were later (in 1364) put at farm for 500 marks (just over £333) per annum,[29] suggesting that if Edward III had been able to purchase its lordship instead of besieging it, he might have expected to pay around £3,333. That means the siege had probably cost Edward something on the order of *twenty-five times* the "market value" of the lordship of the town.[30] Indeed, what he had expended to capture the port was the value of an entire principality, rather than a single town – the Dauphiné was acquired for the future Jean II for significantly *less* not long thereafter, and, had it been put up for auction, the lordship of Ireland would likely not have brought so much.[31]

The case is rather different with Edward's siege of Berwick in 1333, which was begun in early March by a very small army, and maintained from May until late July by a larger one. We unfortunately do not have the full pay records for this operation, but the surviving documents allow us to estimate the cost of the siege

28 Ibid. K. B. McFarlane notes that in fifteenth-century England, loans, land, and fixed rent charges typically brought only about a 5% rate of return. *The Nobility of Later Medieval England* (Oxford: Clarendon, 1973), 57. This was actually on the high side of the rates French towns paid on annuities they offered, which typically ranged from around 3–5%. Michael Wolfe, "Siege Warfare and the *Bonnes Villes* of France during the Hundred Years War," in Ivy A. Corfis and Michael Wolfe, eds., *The Medieval City under Siege* (Woodbridge: Boydell, 1995), 61. It may well be that interest rates fell in tandem with (and partly because of) the decline in returns on land investments, and the rise in surplus capital available, after the Black Death.

29 Dorothy Greaves, "Calais under Edward III," in George Unwin, ed., *Finance and Trade under Edward III* (Manchester: Manchester U.P., 1918), 324. This covered tallages, assizes, harbor fees, etc., but not customs revenues or mint profits.

30 Applying the same logic allows us to add another example. The ferm of the town of Bedford was £40, making the royal lordship of the town worth perhaps £400, but the siege of 1224 cost upwards of triple that amount. Cost: Emilie Amt, "Besieging Bedford: Military Logistics in 1224," *Journal of Medieval Military History* 1 (2002), 115–118. Ferm: Thomas Madox, *Firma Burgi, Or an Historical Essay Concerning the Cities, Towns and Boroughs of England* (London: William Bower, 1726), 131n.

31 The Dauphiné cost Philip VI and Jean II a total of about 400,000 *écus*, or around £68,000. Clifford Rogers, "The Anglo-French Peace Negotiations of 1354–1360 Reconsidered," in *The Age of Edward III*, ed. James Bothwell (York: York Medieval Press, 2001), 205 n. 40. Net royal revenues from Ireland were £2,120 in 1292–3; £1,424 in 1324. Harriss, *King, Parliament and Public Finance*, 524 and 524n. As another point of reference, the counties of Roussillon and Cerdagne were surrendered in bond for 200,000 florins (around 300,000 l.t. or roughtly £35,000) in 1462. Philippe Contamine et al., *Histoire militaire de la France. 1: Des origines à 1715* (Paris: Quadrige/ PUF, 1997), 210. That is practically the same as the 260,300 francs that Philip the Good of Burgundy's accountants calculated that his failed siege of Compiègne in 1431 cost. Plancher, *Histoire générale de Bourgogne*, vol. 4, p.j. lxxv. It is also comparable to the 200,000 francs (roughly 220,000 l.t.) reportedly spent by the government of Charles VI on the approximately thirteen-week siege of Senlis in 1418 before the costs of the siege became insupportable and the operation was called off. *Journal d'un bourgeois de Paris, 1405–1449*, ed. Alexandre Tuetey (Paris: SHF, 1881), 84–6; Peter Spufford, *Handbook of Medieval Exchange* (London: Royal Historical Society, 1986), 179, 191–93.

(mainly in war wages) at around £13,500, or somewhat more.[32] That was still a large sum of money, but reasonably proportionate to the royal revenues of around £1,000 per annum later generated by the port, which was by far the most economically significant town of Scotland.[33]

On the other side of the spectrum, there were sieges like Edward's siege of Tournai in 1340, which though it lasted only about two months was fantastically expensive. Wage costs alone amounted to around £40,000, on top of which there were huge expenditures in the form of subsidy payments to Continental allies.[34] Just how much of the roughly £376,000 in subsidies Edward's allies were promised from 1338 through 1340 should be included in the cost of the siege of Tournai

32 To make this estimate proved a challenging task, and space does not allow more than a summary explanation here. I have included as costs of the siege £3,720 in royal gifts that seem to represent block grants for the pay of Balliol's advance force. These sums were paid directly from tax receipts before the latter were handed over to the Wardrobe. (*Calendar of Close Rolls, 1333–1337*, 7–22.) The Wardrobe then for the rest of the year paid £5,629 5s. 7d. in war wages. Another £4,940 12s. 6d. was spent in "gifts made by the king to earls, barons, knights, esquires and others." Since gifts including horse restoration in 1329–31 averaged around £900 per annum, we can attribute around £4,000 of that sum to second-quarter payments for the men in Balliol's force and to *restauratio equorum* payments (for replacement of lost warhorses) related to the siege. Adding (arbitrarily) £150 of the £354 5s. 7d. ob. spent for various "works done at Berwick," including wages for the workers, we get within a pound of £13,500. The National Archives (Kew), Enrolled Accounts, Wardrobe and Household (E361/2); for the comparison to 1329–31 see J. H. Johnson, "The King's Wardrobe and Household," in *The English Government at Work*, 248–49. Note that I have not included the royal household's expenses; fees and robes of royal household members, including bannerets and knights; payments to messengers; etc., since these would have been nearly equally great if there had been no siege. I have also not added an estimate for the retinue of Henry of Lancaster, though he does seem to have been part of Balliol's force and is not among those listed as receiving grants in the *Close Rolls*.

33 Revenues: Even in 1333–34, after the city was cut off from the Scottish economy and still recovering from a long siege, Edward III was able to get over £650 in receipts for royal rents, the wine tax, fish, etc., in less than ten months. The National Archives, Kew, Pipe Roll 9 Edward III (E 372/180), m. 54d. Customs revenue from Berwick late in Edward III's reign seem to have been on the order of £700, similar to what it had been under Scottish rule. However, most of these revenues would have been from export of English wool and other goods, most of which would likely have come to the Crown through another port had Berwick remained Scottish. Harris, *King, Parliament and Public Finance*, 527; James Campbell, "England, Scotland and the Hundred Years' War," repr. in Clifford J. Rogers, *The Wars of Edward III: Sources and Interpretations* (Woodbridge: Boydell, 1999), 206; R. A. Nicholson, *Scotland: The Later Middle Ages* (Edinburgh: Oliver & Boyd, 1974), 108.

34 This figure is based on 1,400 men-at-arms from Brabant serving 45 days and 4,400 men-at-arms from Hainault, Guelders, Juliers, and Flanders serving 55 days, all at 15 florins/month (= 18 d. st./ day), for a total of £22,875, plus 15,000 Continental infantry at 3d./day for 55 days, plus an English force of around 1,300 men-at-arms (who received double the usual pay on this campaign), 1,000 mounted archers (at 6 d.) and 1,000 foot archers (at 3 d.), for 55 days, for a total cost of slightly over £10,000. Note that this does not include the wages paid while the army was gathering and marching to Tournai. For the numbers and durations, Rogers, *War Cruel and Sharp*, 204, 204 n. 28. For pay rates, ibid., p. 133, notes 32, 33.

is incalculable, but the sum should probably be reckoned as at least £150,000,[35] making the siege in total actually far more expensive even than the siege of Calais. Indeed, by the 10:1 pricing ratio used earlier, the siege of Tournai probably cost more than the purchase value of the lordship of the entire duchy of Aquitaine, even with its extent of 1324 (including the Agenais).[36] But the gross return on this immense investment of cash was essentially zero, because Edward ran out of money during its course, forcing him to break off the siege and agree to a truce with France.[37]

The point here is not that Edward III, in besieging Calais and Tournai, had made a pair of colossal errors of judgment, nor do I mean to suggest that he was not generally a rational actor. Tournai and Calais were not besieged primarily for the sake of acquiring those towns. Rather, the sieges were intended principally as gambits in Edward's battle-seeking strategy: he hoped that the battle-shy French would attack his army in order to rescue the besieged places, leading to a war-winning decisive battle.[38] And the value of Calais, once it was captured, was primarily strategic rather than economic: the military worth of having a secure port of disembarkation for troops and supplies when attacking France, and of gaining a base from which naval domination of the Channel could be greatly enhanced, was incalculable. But it does nonetheless make a large difference to the overall pattern of warfare if sieges *can or cannot* be justified in *purely economic terms*. If they can – as, perhaps, in the case of Berwick – then in one way or another the process of conquest can pay for itself. If they cannot – as in the case of Calais – then conquest must be fueled by resources provided by the conquerors. But the costs of a siege like that of Calais or Tournai were very large in relation to the income of a medieval king, even a very wealthy medieval king. Hence, the cost of extensive conquests in well-fortified and wisely defended regions simply could not be borne by any medieval state *unless* it was possible, in Cato the Elder's words, to make "war feed war."

In High and Late Medieval northern Europe, it seems fairly clear that sieges like the siege of Berwick, which were at least close to being economically worthwhile,

---

35 The majority (by a substantial margin) of all the military efforts the allies made on Edward III's behalf were comprised within the Tournai campaign. See Rogers, *War Cruel and Sharp*, chs. 6–9; for the amounts of the subsidies promised, see ibid., pp. 186 (Flemings), 147 (others).

36 Harriss, *King, Parliament and Public Finance*, 523, for clear revenues of £13,000 from Aquitaine in 1324.

37 Rogers, *War Cruel and Sharp*, 209–16. Perhaps even more fruitless was the 64,314 l.t. spent by Duke John the Fearless of Burgundy just to gather an army and ready materiel (including 200,000 crossbow quarrels, 20,702 pounds of gunpowder, 20,000 caltrops, 25 trebuchets, 120 cannon, literally an entire forest's worth of timber, etc.) for a siege of Calais in 1406 that never even actually started, much less succeeded. Bertrand Schnerb, "Un projet d'expédition contre Calais (1406)," in *Les champs relationnels en Europe du Nord et du Nord-Ouest des origines à la fin du Premier Empire: 1ère Colloque Historique de Calais*, ed. Stéphanie Curveiller et al. (Calais: N.P., 1994), 179–192.

38 Rogers, *War Cruel and Sharp*, chs. 9, 12.

were the exception rather than the rule. But we also know that in some medieval regions and periods, large-scale conquests were relatively common. In some cases this was because the area under attack was not well fortified (e.g. England in 1015–17 and 1066) or because the local population was not averse to being (re)conquered (e.g. Anatolia in 1097 or in 1304). But there were other cases in which neither of those circumstances prevailed, suggesting that in those places and times war was better able to feed war than usual, which in turn suggests that sieges cost less, or brought greater rewards, or both.

What, then, determined the cost of a siege? Particularly from the thirteenth century onwards, there were only three principal factors: the duration of the siege, and the size and composition of the army. Wages exceeded by far other incidental costs, such as hides and ropes purchased for siege engines, arrows and crossbow bolts, ransoms for soldiers captured while foraging or skirmishing, or even payments to compensate soldiers for warhorses lost in service.[39] For the eight-week siege of Bedford in 1224, for example, we have records of £58 8s. 5½ d. spent for ropes, £22 18 s. 6½ d. for roughly 20,000 crossbow bolts, £7 1s. 6d. for a few hundred pickaxes, and about £25 for various other materials, compared to £1,326 10s. ½ d. for wages.[40]

In the mid-fifteenth century, gunpowder artillery developed to the point that, for about a century (until new methods of fortification were created to counter the cannons' effect), sieges could on average be completed much more rapidly, which made them cheaper at just the time when governments were greatly increasing their per-capita revenues (and moreover when interest rates for borrowed money were falling), factors that together ushered in an era in which sweeping conquests were more practicable than they had been for a long time, as in the French reconquest of Normandy, conquest of Gascony, and occupation of Brittany, or the Spanish conquest of Granada, or the Ottoman conquests in the Balkans, or Charles VIII's conquest of Naples in 1494–95.[41] But before then, the duration of many sieges was determined principally by the amount of food stored within the walls, and – although this is a subject which has not been systematically studied, and should

39 It was in the fourteenth century that it became normal for all medieval soldiers to receive pay for the full duration of their service. Before then there was still some basic proportionality between the costs of a siege and its duration, but it was less precise, because some soldiers would typically owe some term of unpaid service, and others would be paid by their immediate lords rather than by the sovereign who was conducting the war.

40 Note that this does not take into account the large value of unpaid feudal military service devoted to the siege, which if factored would make the disproportion between human and material costs much greater still. Amt, "Besieging Bedford," 115–118.

41 Clifford J. Rogers, "The Medieval Legacy," *Early Modern Military History*, ed. Geoff Mortimer (London: Palgrave, 2004), 20–21; idem, "The Artillery and Artillery Fortress Revolutions Revisited," in Nicolas Prouteau, Emmanuel de Crouy-Chanel and Nicolas Faucherre, eds., *Artillerie et Fortification, 1200–1600* (Rennes: Presses Universitaires de Rennes, 2011): 75–80 (and chapter four of this Variorum volume).

be – the average length of successful sieges *seems* to have been fairly constant across the geographic and chronological range of the medieval period.

The other factor in the cost equation, the size and composition of the army, is highly variable. As a general rule, the size of army needed to sustain a siege was basically regulated by two factors. The first is the manpower needed to encircle the place with sufficient strength at each point to allow for the defeat of a sally by the garrison. The second is the force required to hold off or defeat a relief army. The first of these two elements is conditioned by the size and combat effectiveness of the force inside the place and by the physical size of the place being besieged; the second is determined by the available strength of the lord and the friends of the defenders. To have much chance of success, the besieging army would normally have to be at least as large as the larger of those two requirements, and preferably large enough to fulfill them both at once: otherwise, the besiegers might have to break off the siege to meet a relief army, in which case the defenders could resupply themselves basically as often as they could raise a relief force. The latter factor means that a fortified town of a given size and strength could generally be captured much more cheaply if it was the capital of its own city-state or independent small principality than it could be if it was a border city of a large, strong state.[42] (It also, incidentally, means that sieges can generally be undertaken much more cheaply after the defender has suffered a battlefield defeat than before, which helps explain why battle-seeking strategies were often favored by would-be conquerors.) Hence, it was easier to make war pay for war when targeting a group of small states, such as the *taifas* (Muslim-ruled independent principalities) of Iberia after the collapse of the Cordoban caliphate, than when targeting the provinces of a large state, which could muster a large relief army and so required the aggressor to muster a large besieging force, as for example with Edward III's sieges of Calais and Tournai.

It is, no doubt, a fairly obvious conclusion that it is more difficult to make conquests against a strong opponent than a weak one. The additional point being made here, however, is that against a strong opponent conquests are not only more difficult, but also, for any given scale of conquest, more expensive and so less economical. That is, they are less likely to allow for "self-funding" – or indeed, profit-generating – warfare of the sort that sustained the conquests of Charlemagne, for example.[43]

The geometric aspect of the cost of a siege is worth a bit more exploration here. The key insight to be derived is that from an economic standpoint, sieges of

---

42 Which is one reason a conquering power could generate a "bigger pie" by conquest, as noted on p. 232. The conquest of an independent city could (by raising the anticipated cost anyone else would have to pay to take it again and so, often, deterring the attempt) provide the city with a level of increased security greater than the costs it had to pay to its new overlord, at little or no cost to that overlord.

43 See Timothy Reuter, "Plunder and Tribute in the Carolingian Empire," *Transactions of the Royal Historical Society*, 5th ser., 35 (1985), pp. 75–94; Michael McCormick, "New Light on the 'Dark Ages': How the Slave Trade Fueled the Carolingian Economy," *Past & Present* 177 (2002): 43–46.

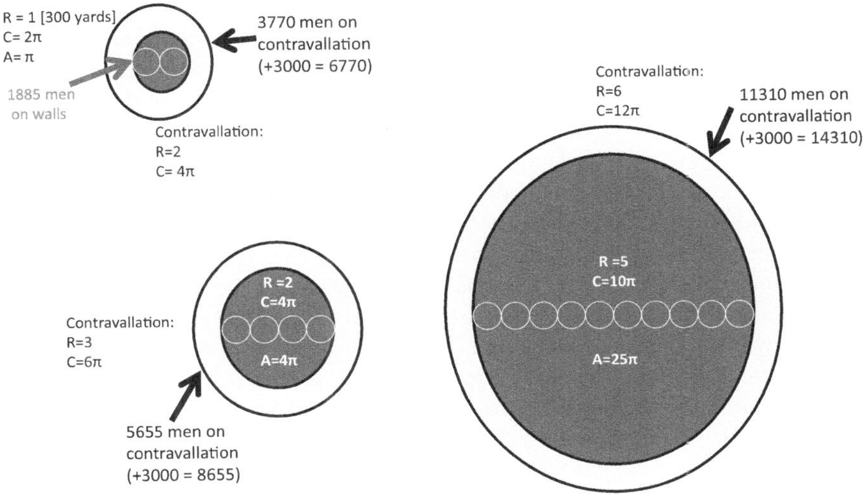

Figure 9.2 Geometric Relations and Siege Ratios

towns were more profitable than sieges of castles, and the larger the towns, the more profitable – not merely the more rewarding – their capture was likely to be. This is because, in a nutshell, the value of a place was roughly proportionate to its area – i.e. to the *square* of its radius – whereas the cost of capturing it increased by less than even a linear relation to its radius. In other words, when you double the radius of the place besieged, you quadruple its area, and so its value, but only double the size of its perimeter, and less than double the size of the perimeter needed to encircle it.[44] [See Figure 9.2.]

First, consider a circular-plan town with a radius 300 yards, assume that one man is required to defend each yard of perimeter, and set the size of the potential relief force at 3,000 men. In this case, the number of men needed to defend the perimeter of the town would be 1,885, while the attacking army would need 3,770 men to man encircling siege lines positioned 300 yards beyond the walls, plus 3,000 men for the counter-relief force, for a total of 6,770 men. In the second case illustrated on Figure 9.1 – a town with double the radius and quadruple the area – the requirement would be 5,655 men for the lines plus 3,000 for the counter-relief force, so a total of 8,655, or 28% more men – but the value of the place, proportional to its area, is quadrupled. In the third case, a city with a radius of 1,500 yards, it would take 11,310 men to man the siege lines, plus 3,000 for the counter-relief force, so 14,310 men. In other words, the manpower needed for the siege has

---

44 The following analysis was inspired by similar reasoning applied to a different point in John A. Lynn, "The *Trace Italienne* and the Growth of Armies: The French Case," reprinted in Clifford J. Rogers, ed., *The Military Revolution Debate* (Boulder: Westview, 1995), 174–79.

roughly doubled compared to the first case, but the value of the city, as suggested by its area (and therefore its population) has increased twenty-five times!

There are, to be sure, a number of oversimplifications in this analysis as I've just set it out – for example, the strength needed to defend the perimeter of the besiegers' lines would be affected by the number of armed men inside the city as well as the length of the lines, and would also be increased by the need to be able to concentrate sufficient men at any given point of the line in time to prevent a breakthrough, which would require more reserves for long lines than short ones. And of course most towns are not perfect circles in shape, and so the area enclosed is likely to grow somewhat less than with the square of the radius. On the other hand, an army large enough to besiege a city would in general practice not need to supplement its strength with a separate counter-relief force unless the defenders could threaten the deployment of a comparably sized army. Hence, in the third example, the size of the besieging army and therefore the cost of the siege could have been kept lower. Still, any reasonable assumptions to account for these secondary modifying factors would not change the basic conclusion: that the strength needed to besiege a town rises in a *much-less-than-linear* relation with its increases in population.

The essential point that this geometric approach helps to explain – that small places often require almost as many men to besiege as do large ones – is borne out by historical realities. For example, Edward III probably had an average of around 10,500 men in his siege-lines for the siege of Calais in 1346–47, and around the same, 10,000 or so, for the siege of Reims in 1359–60, even though latter city was perhaps three times as large and populous as the former.[45] Henry V employed over 12,000 men for the siege of Harfleur (with a population on the order of 6,000–8,000) in 1415, but significantly fewer – perhaps not much more than 7,000 – for the 1418–19 siege of Rouen, which was by a good margin the largest French city successfully besieged by the English in the entire Hundred Years War. The larger army was necessary at Harfleur because Henry had not yet won the battle of Agincourt and needed to be able to defeat an anticipated large French relief army. After Agincourt, that was much less of a threat, and so a much smaller army sufficed for a much greater siege.[46] Conversely, in 1346 Duke John

---

45 Reims had a population of around 20,000 in 1300. Michael [C.] Jones, "War and Fourteenth-Century France," in Anne Curry and Michael Hughes, eds., *Arms, Armies and Fortifications in the Hundred Years War* (Woodbidge: Boydell, 1994), 117. Calais: Rogers, *War Cruel and Sharp*, 276n.

46 For the siege of Harfleur and the strength of the English army there see Anne Curry, *Agincourt: A New History* (Stroud: Tempus, 2005), 70–71 and ch. 4. For Rouen, see eadem, "Rouen, Siege of," in Clifford J. Rogers, ed., *The Oxford Encyclopedia of Medieval Warfare and Military Technology* (New York: 2010); eadem, "Henry V's conquest of Normandy 1417–1419: the Siege of Rouen in Context," *Guerra y Diplomacia en la Europa Occidental 1280–1480. XXXI Semana de estudios Medievales. Estella 19–23 de julio 2004* (Pamplona: Gobernia de Navarra, 2005), pp. 237–54; and Richard Ager Newhall, *The English Conquest of Normandy* (New Haven: Yale U.P., 1924), 105–122. The perimeter of Rouen's walls was around five miles; thus Henry's army would have been barely adequate to man a line of contravallation around it, though the city's water defenses (which protected besiegers from sallies as well as defenders from assaults) would have facilitated Henry's task.

of Normandy, anticipating a battle with an English relief army, is estimated to have employed up to 15,000–20,000 men for the siege of the insignificant little town of Aiguillon, which could have been cut off from all land approaches with siege lines less than a thousand yards long.[47]

On the "benefit" side of the cost-benefit analysis, I have so far taken it as a simplifying assumption that the value of a town increases in proportion to its area and its population. Two refinements should be made to this generalization, however, which further strengthen the conclusion that it was more economical to conquer large towns than small ones. First, population actually tend to rise *more* rapidly than area, since the larger the population and area of a city, the more valuable the land at its center, and the more densely packed the inhabitants will tend to be, with more construction of two-story buildings and fewer gardens. Second, the wealth of a town will rise in a somewhat more-than-linear relation to population, since the inhabitants of large towns have to be per capita wealthier than the inhabitants of small towns in order to survive. This too is largely explained by the geometry of circles: the larger a town's population, the more square miles of cropland it needs to use to feed itself, and the greater the farthest distance that some of the grain has to travel to supply the city. Since it is the travel distance and hence shipping costs of the *farthest*-traveling grain that determines the marginal cost of grain delivered to the market, it is also that farthest-traveling grain that sets the price of grain in the city, and therefore the income needed for subsistence, and therefore the base level of wages. This was a very major factor in the Middle Ages because of the high cost of transportation, especially land transportation, of bulk goods.[48]

Thus, a large city with nine times the area and triple the perimeter of a small town is likely to have more than nine times as many people, with higher per-capita income and wealth, and might easily have, say, ten or twelve times rather than nine times the quantity of money and other valuable goods inside its walls, so that ten or twelve times the value of plunder could be expected if the city were taken by assault and sacked. True, most cities that fell to extended sieges, as already noted, were not taken by assault, but rather surrendered by negotiation, which typically meant that the inhabitants were allowed to keep their goods and the rights they had enjoyed under their previous lord. In that case, the typical pattern of medieval city development, in which growing cities usually acquired charters allowing them to pay *fixed* sums for at least some of the income streams their lords derived from them, might make the actual revenues which could be derived from a large city rise *less* than proportionally with its population, wealth, and economic activity. Nonetheless, the greater wealth of a large town might still be available to be tapped by a conqueror in other ways. For example, at the end of the siege of 1418–19, as part of its surrender terms, Rouen agreed to pay an indemnity of £50,000, in addition to building a new ducal palace for Henry V. This was a huge

---

47 Jonathan Sumption, *The Hundred Years War, v. 1: Trial by Battle* (London: Faber and Faber, 1990), 485–87.

48 See John Landers, *The Field and the Forge* (Oxford: O.U.P., 2005).

sum, and more, by a good margin, than the total wage cost of the operation. Henry in addition gained the property of any townsmen who refused to do him homage.[49]

Here a special consideration should be taken into account: whether or not the context of the siege would allow for the enslavement of the population of the town when it fell. This was usually done, in High Medieval and Late Medieval Europe, only in the case of war across religious lines.[50] The points made above about the *relative* profitability of besieging small vs. large towns would remain basically valid either way, but it makes a very large difference to the *absolute* profitability of the siege. This made the logic of siege and conquest significantly different in the case of the Iberian *reconquista* or the Norman conquest of Muslim Sicily than in the case of a war between Normandy and Maine, for example.

The key thing to appreciate here is that the value of the human inhabitants of a town was likely to be greater than the value of the material goods that could be plundered. In Carolingian Europe, slaves seem to have typically fetched around a pound of silver, the value of four swords or 20 mature sheep or yearling oxen.[51] This may not seem a huge amount, but it was likely the equivalent of around two months' pay for a well-equipped soldier.[52] A comparable ratio appears again in the mid-fourteenth century, when the Catalonian Admiral Bernat de Cabrera got 27,084 florins for 1,447 prisoners captured in a naval battle: about 20 florins each, which works out to around 50 days' pay for a man-at-arms (or around

49 Newhall, *English Conquest*, 157–58; Christopher Allmand, *Henry V* (Berkeley: University of California Press, 1993), 126; Adolphe Cheruel, *Histoire du Rouen sur la domination anglaise* (Rouen: E. Le Grand, 1840), notes section, pp. 55–56. This was 37.5 times the indemnity paid by Louviers, after a siege of 16 days by Henry's army during its march to Rouen. Newhall, *English Conquest*, 100. It is worth noting that the records of both the costs and the profits of the conquest of Normandy are fairly extensive, and would merit an in-depth analysis.

50 William D. Phillips, Jr. *Slavery from Roman Times to the Early Transatlantic Trade* (Minneapolis: U. of Minnesota Press, 1985), 93, indicates that a major exception to this generalization is that the Normans, during their conquests in southern Italy in the twelfth century, often engaged in large-scale enslavements. This, however, seems to be incorrect. Cf. Timothy Smit, "Pagans and Infidels, Saracens and Sicilians: Identifying Muslims in the Eleventh-Century Chronicles of Norman Italy," *Haskins Society Journal* 21 (2009): 83.

51 Michael McCormick, *The Origins of the European Economy: Communications and Commerce, AD 300–900* (Cambridge: Cambridge U.P., 2002), 756–57; Carl I. Hammer, *A Large-Scale Slave Society of the Early Middle Ages. Slaves and Their Families in Early Medieval Bavaria* (Aldershot: Ashgate, 2002), 56, 80.

52 The Berkshire Domesday entry has each 5 hides of land sending a soldier for expeditionary service, with the soldier being provided 20 s. (i.e. one pound sterling) "for his food and pay for two months," which works out to 4 d. per day; an individual holding five hides would have been of the upper (thegnly) class. A hide of land was a variable measure, nominally enough to support one household (including servants) and a plow-team of eight oxen. Stephen Morillo, *Warfare under the Anglo-Norman Kings* (Woodbridge: Boydell, 1994), 66–67. The *lithsmen* (soldier-oarsmen) of Cnut in 1018 earned, on an annual basis, either the same wage or 2/3 as much (depending on the interpretation of the "mark" at the time). Richard Abels, "Household Men, Mercenaries and Vikings in Anglo-Saxon England," in John France, ed., *Mercenaries and Paid Men. The Mercenary Identity in the Middle Ages* (Leiden: Brill, 2008), 158.

Figure 9.3 The Sack. As this fifteenth-century illumination emphasizes, the sack of a city taken by storm could produce a jackpot of wealth for the besieging soldiers. [British Library, Manuscript Royal 14 D I, fo. 97 r.]

200 days' pay for an infantryman).[53] Slaves sold on Majorca between 1240 and 1259 averaged about the same value.[54] After the Black Death the shortage of labor raised the value of slaves, so that in the late fourteenth and early fifteenth

53 Pere III of Catalonia (Pedro IV of Aragon), *Chronicle*, tr. Mary Hillgarth, notes by J. N. Hillgarth (Toronto: Pontifical Institute of Mediaeval Studies, 1980), 478n. For 12 fl./month for men-at-arms in French fleets in 1327, and 12.4 fl./month for Gascon men-at-arms (and one quarter that for foot-men) in Castilian service at Algeciras in 1343, see Jusselin, "Comment France se préparait," 222; Nicolas Agrait, "Castilian Military Reform under the Reign of Alfonso XI (1312–50)," *Journal of Medieval Military History* 3 (2005): 104–5. For 15 fl./month in 1346–47 (in English service) and in 1369 and 1375 (esquires in the Dauphiné): Grose, *Military Antiquities,* 278; Contamine, *Guerre, état et société,* 628. For 10 fl./mo in Burgundy in 1360 (for esquires), ibid.

54 Averaging 100 s. Valencian (15 florins) in the 1240s, 200 s. in the 1250s (30 florins) and (with smaller numbers of sales) from 150 s. (22.5 florins) in the 1260s to around 280 s. (42 florins) in the 1290s. Ricard Soto y Company, "La situació dels andalusins (Musulmans i Batejats) a Mallorca després de la Conquesta Catalana de 1230," *Mélanges de la Casa de Velázquez* 30 (1994): 206. Captives who had the connections or resources to be redeemed (ransomed) instead of enslaved were on average more valuable, but not greatly so – except when large military operations drove down the price of common slaves through excess supply.

centuries prices of 45–60 florins were common.[55] Long-term trends both in slave prices and in soldiers' wages affected the ratio; also, there was a wide range of prices for individual slaves, determined by sex, appearance, age, skills, etc., and the value of slaves could vary greatly even in the short term due to fluctuations of supply and demand.[56] Successful military operations that involved the capture of many slaves tended to glut the market and drive down prices, sometimes quite severely.[57] Still, it seems from the data available that usually the median price of a slave exceeded two months' wages for an average armored soldier. Hence, even if it took a besieging army six months to capture a town, if the size of the besieging army was as large as the number of men of military age of the town, and each of the latter had two to three dependent women or children, the sale-value of the population would likely be at least equal to the wage cost of the siege, and quite possibly 50% or 100% greater, or even more. When Malaga fell after a siege of around three months in 1487, for instance, the value of the enslaved inhabitants was approximately triple the expense of capturing the place.[58] Any material goods within, plus the value of the city itself, would be additional profit.

Twelve hundred and fifty-four Muslim captives sold after the Aragonese seized the island of Kerkenna in 1286 brought an average of about 113 *tareni* (nearly 23 florins) each.[59] The value of these slaves, along with a few other Muslims who were ransomed, totaled almost fifty times the annual tribute the island had formerly paid to the Aragonese.[60] The cost of the operation was less than half the proceeds

---

55 Jeffrey Flynn-Paul, "Tartars in Spain: Renaissance Slavery in the Catalan City of Manresa, c. 1408," *Journal of Medieval History* 34 (2008): 350, 353; Nikos G. Moschanos, "Der Sklavenmarkt im östlichen Mittelmeerraum in der Palaiologanzeit," *Südost-Forschungen* 65–66 (2006): 40–41, nn. 83–84 (the *hyperpa* being worth about half a florin).

56 Kate Fleet, *European and Islamic Trade in the Early Ottoman State: The Merchants of Genoa and Turkey* (Cambridge: Cambridge U.P., 1999), 42–50, and Soto, "Situació dels Andalusins," 95–201, provide good data and examples.

57 Fleet, *European and Islamic Trade*, 46, for examples, including that "at the end of the thirteenth century the price of a Turkish slave fell below that of a sheep as a result of the military successes of the Byzantine general Philanthropenos." According to Muntaner, during the siege of Collo in 1282 the Spanish captured so many Muslims that a slave cost only a single *dobla* (about 1.5 florins). *Chronicle*, 103; Spufford, *Handbook*, 160–61, 173, 139.

58 Barbara Holmgren Firoozye, *Warfare in Fifteenth-Century Castile* (Ph.D. Dissertation, UCLA, 1974), 40. Cf. also Caffarro, quoted n. 66.

59 Lawrence V. Mott, *Sea Power in the Medieval Mediterranean. The Catalan-Aragonese Fleet in the War of the Sicilian Vespers* (Gainsville: University Press of Florida, 2003), 129 (4,727 ounces, at 5 *tareni* to the florin per Spufford, *Handbook*, 65, for the year 1288). The total was equal to the wages of 1,250 fleet crossbowmen for over six months, or the same number of rowers for over fourteen months, or the same number of knights with armored horses for somewhat less than two months at high pay (2 *onzas*), or over three and a half months at low pay (one *onza*). Ibid., 157–58, 170–71, 175.

60 Mott, *Sea Power*, 130.

of the slave auctions alone.[61] Hence, the *net* profit from the human captives was much greater than the value of their tribute revenue stream into perpetuity, even at a 10% rate of return on capital.

Another example: Jaume the Conqueror in 1235 captured the small fortress of Montcada outside Valencia after a siege of around eight days, and thereby gained "one thousand one hundred forty-seven Saracens . . . and many fine goods, and pearls, necklaces, and gold and silver bracelets" the value of which together "easily amounted to one hundred thousand *bezants* [33,333 florins]."[62] This was a huge sum, roughly equivalent to a year's income for 800 *caballeros*.[63] The most interesting thing about this case, however, is that the king tells us that he later sold 100 of the Muslims for 17,000 bezants (5,100 florins), and "would have got thirty thousand" had he not been forced to sell in a hurry.[64] Jaume did get first choice of the prisoners, but even if his picks were on average each worth three times as much as the others, that would still leave the human plunder of the operation, even if sold at around half its full value, equal to around *three times as much* as the material plunder – the pearls, necklaces, gold, silver, and other "fine goods" seized.[65] The same king was later offered a tribute of 2,000 bezants per year in order to refrain from the conquest of Valencia. No wonder he declined, since if he profited from the capture of Valencia in proportion to the capture of Montcada,

61 Figuring the duration of the operation as one month, though it may have been more like two weeks (the fleet sailed for Kerkenna in late August and was auctioning the slaves in Sicily at some point during September). Note that the forces used for the operation – 18 galleys, 2 other ships, 2,000 almogavars, and 100 cavalry, for a total manpower of nearly 5,200 armed men – were determined by the size of the active fleet, and far exceeded what was needed to occupy the island. Calculated based on Mott, *Sea Power*, 233, 238, 129, 293n, 171.

62 James, *Book of Deeds*, 184.

63 Figuring the bezant at 3 s. Valencian (James, *Book of Deeds*, 391), and the annual income of a *caballero* at 373 s. (per note 69).

64 James, *Book of Deeds*, 184–85. If his were worth 17,000, the others would then (at 1/3 the average value of his) be worth (1/3)*(1047/100)*(17,000) = 59,330. That would make the sum of the value of his captives and theirs 76,300. The material plunder would then be (100,000–76,300) = 23,700. Hence the ratio of the value of the captives to the value of the material plunder would be 76,300:23,700, or 3.21:1.

65 Ibid. We can fairly assume that in areas less materially rich (for example Slavic areas), the ratio would be even more tilted towards human plunder. Where slavery did not exist, the value of human plunder (in ransoms) could still exceed the value of material plunder, though generally by a lower ratio. For example, in a study of records of plunder gained by French garrison soldiers in the fifteenth century, Philippe Contamine found that, overall, ransoms brought in about 2.5 times as much as the sale of captured armor, weapons, horses, livestock, and other goods. Philippe Contamine, "Rançons et butins dans la Normandie anglaise (1424–1444)," in idem, *La France aux XIVe et XVe siècles. Hommes, mentalités, guerre et paix* (London: Variorum Reprints, 1981), VIII, 261–62. When whole urban populations were captured, ransom was not really an option (since for each captured townsman, both the people who would be most likely to pay a ransom – his family members and friends – and their wealth would also be already in the hands of the captors) and was rarely practiced. As noted earlier, however, towns did sometimes agree to pay large indemnities as part of surrender negotiations, which could be viewed as, in large part, collective ransoms for the population.

he might immediately gain as much, gross, in one year as he otherwise would receive in tribute in two hundred years (if we conservatively take the population of Valencia, including refugees, to have been 20,000) or even five hundred years (if we accept the king's own statement that 50,000 men and women were in the city).[66] On this basis, war, even a war of long sieges, could feed war, and conquest could more than pay for itself.

Of course, even in the Reconquista sieges usually ended by negotiation, and therefore usually without the enslavement of the population. The Valencians, for example, were not enslaved, but rather allowed to depart their city freely. But even in such cases the *potential* that if the siege lasted to the bitter end the inhabitants would be enslaved (as at least 10,000 of the women and children within Almería's main town reportedly were in 1147, for example)[67] played into the dynamic of the conquest. When cities surrender – and especially when the inhabitants surrender and also agree to give up their real property and become refugees – they do so because of threats, and threats are judged not just by their severity but also by their credibility: that is, by the probability that they will actually be carried out. Threats that if carried out will be very costly for the threatener are likely to be perceived as empty. For example, the threat "If you don't surrender now, I'm going to keep up this siege until I capture your city and kill you all, even if it takes every penny I have or can borrow in order to do it," may lack credibility, whereas the threat "If you don't surrender now, I'm going to keep up this siege until I capture your city and enslave you all, even if it costs me a significant fraction of what I get by selling you all into slavery" is more believable, and hence more frightening and more effective.[68]

66 James, *Book of Deeds*, 230. Robert I. Burns, *Islam under the Crusaders: Colonial Survival in the Thirteenth-century Kingdom of Valencia* (Princeton: Princeton U.P, 1973), 76, estimates 15,000, but James's number is not fantastic, when we consider the likelihood that its population was swelled by refugees, and that (as noted previously) 1,147 Muslims were captured in the tower of Montcada, after which the population of outlying areas would have been driven to take refuge within the city, rather than in its satellite fortifications (as James explicitly intended: *Book of Deeds*, 181), and that at Almería also 50,000 were said to have been killed, enslaved, or ransomed (see next note).

67 Caffaro's chronicle is somewhat unclear, but I interpret it to mean that when the main town was taken by assault 20,000 Muslims were killed, and 10,000 women and children shipped off to slavery in Genoa, after which another 20,000 inhabitants who had taken refuge in the upper city (*suda*) ransomed themselves: "ciuitas tota capta est usque ad subdam. et illo die de Sarracenis uiginti milia interfecti fuerunt, et ab una parte ciuitas scilicet deripata .x. milia fuerunt, et in sudam .xx. milia; et inter mulieres et pueros decem milia Ianuam adduxerunt. Sarraceni uero infra quattuor dies sudam et personas reddiderunt, et miliaria marabotinorum triginta milia dederunt, ut personas euaderent." *Annali Genovesi di Caffaro e de'suoi continuatori*, vol. 1, ed. L.T. Belgrano (Rome: Fonti per la storia d'Italia, 1890), 84. Cf. Joseph O'Callaghan, *Reconquest and Crusade in Medieval Spain* (Philadelphia: University of Pennsylvania Press, 2003), 140, who interprets the number enslaved as 30,000.

68 Threats can be very effective even if not explicitly communicated. Situations like this can profitably be analyzed in terms of game theory, to which Thomas C. Schelling's brilliant *The Strategy of Conflict* (Cambridge: Harvard U.P., 1960) remains an excellent introduction.

In the case of Valencia, where the inhabitants did surrender and were expelled in 1238, King Jaume and his army profited by gaining all the territory of the city. In addition to the large tracts of land distributed to the great nobles and to the urban communities that had sent troops, Jaume provided 380 individual *caballeros* with new fiefs (*heredades*).[69] If we work on the principle that the capital value of a fief was ten times its annual income, then this suggests that the value of the real property acquired by the siege was at minimum enough to support around 3,800 *caballeros* for one year – or to support a mix of cavalry and cheaper infantry more than strong enough to conduct the siege. To make the same point more concretely: around this time a typical knightly income in the Crown of Aragon was 373 sous of Valencia,[70] suggesting the sale value of a proper knightly *heredad* would be some 3,730 sous. The knightly *heredades* distributed from the conquest of Valencia may have been only half the normal size, but if we conservatively assume that at least as much land collectively went to the nobles and towns as to the individual knights, that indicates a total land value of 1,417,400 sous or more.[71] That, in turn, would be about 3/4 the cost of supporting for the five and a half months of the siege an army of 13,000 men, including 1,000 *caballeros* with armored horses, 3,000 crossbowmen, and 9,000 *almogavars* or other equivalently paid infantry, which is a reasonable guess at the size of the force conducting the siege.[72] On top of this, the city itself was, after its repopulation by Christians, a cash-cow for the Aragonese monarchy, providing the royal fisc with annual revenues later estimated at nearly 80,000 sous of Valencia per year. In addition, control of the city ensured (or at least made possible) control of the surrounding region, with large additional revenues.[73] So in this instance again we have a case in which the siege was sufficiently rewarding to bear its own costs, provided that Jaume's soldiers were able and willing, in essence, to serve "on credit."

The basis of my analysis so far has been the assumption that conquest could fund itself when the value of what was conquered matched or exceeded the

---

69 James, *Book of Deeds*, 228–36.

70 Robert I. Burns, *The Crusader Kingdom of Valencia* (Cambridge, MA: Harvard U. P., 1967), 1:xii.

71 Half size: James, *Book of Deeds*, 233. James later observed that 8 Valencian *jovates* (a total of about 60 acres) was not enough to be a good *heredad* for a "man of worth." Ibid., 347–8, 392. The 1:1 ratio for knightly:other land disbursements probably leads to an underestimate of the total land disbursements, since (per the next note) the 380 knights who received fiefs accounted for less than half the knights in the army.

72 James, *Book of Deeds*, 221, gives the size of the army as 1,000 knights and 60,000 foot; the former is perfectly credible, but the latter appears exaggerated. The figure of an average strength of 12,000 infantry is simply an educated guess (cf. ibid., pp. 78 n. 49, 87, 60, 139, 151, 323). Even if there were twice that many almogavars, however, it would not greatly change the basic point: the value of the lands disbursed, as estimated earlier, would then be about half (rather than three quarters) the wage cost for five months of siege. Cost: calculated using the later-thirteenth-century Aragonese fleet wages of 2 onzas/month for cavalry with armored horses, 18 tareni for crossbowmen, and 12 tareni for almogavars (see note 17), and an exchange rate of 50 sous of Valencia to the onza. For the exchange see Spufford, *Handbook*, 61, 66, 68, 108, 160–61.

73 Bofarull, *Rentas*, 89.

rewards needed to get soldiers to participate in the conquest. The same basic logic works, *mutatis mutandis*, whether soldiers are receiving wages, as under Edward III, or are fighting in anticipation of future rewards. Although space does not allow for a full development of the point, I should note here that the whole logic changes fairly dramatically when military systems based on self-funding offensive military obligation, such as knight service owed for fiefs, are employed for conquest. In such cases, the *gross* cost of mounting short to medium-length expeditions can be, for the ruler, near zero. When you have short-term "use it or lose it" military forces available to you each year, which you have in effect paid for in advance by the granting of fiefs, it is economically most efficient to use them in operations that can be completed before the feudal service term is over, which favors a strategy based on raids rather than sieges, and so pursuit of compromise peace agreements such as nominal submissions or exaction of tribute, rather than true conquests.

Let me conclude by recapitulating a few main points that have emerged from this analysis. First, application of an economic approach, based on calculation of costs, benefits, and profits, can help clarify the broad patterns of medieval strategy, particularly with respect to the key decision of whether to aim to persuade the enemy to make concessions, or to aim to force him to surrender. Second, forcing an enemy to surrender usually meant besieging his fortifications, and sieges were so expensive, because of the high cost of sustaining many soldiers over long periods of time, that conquest of an area by a series of sieges might be impossible except by the injection of huge quantities of money and resources from external sources. However, third, under some circumstances conquests, even conquests involving protracted sieges, could be made to pay for themselves. Fourth, in part because of the geometric relations involved, sieges of large towns were, though more expensive in gross terms, nonetheless more profitable and economically viable than sieges of small towns, and much more economically viable than sieges of strong castles, which cost nearly as much as sieges of towns but brought far smaller economic benefits to offset the costs.

Fifth, because the cost of a long siege included the cost of sustaining a field army capable of holding off a relief force, it was cheaper and more economical to besiege a city that belonged to a small state than one that belonged to a large state, or to conduct a campaign of conquest in the wake of a major battlefield victory than without one. Sixth, where the enslavement of a defeated population was *possible*, even if it was rare, conquest was more feasible, because the *threat* of enslavement would tend to make sieges succeed more often, more quickly, and with less favorable terms granted to the defenders if they did not surrender quickly. These general conclusions would lead one to predict that conquest would be most feasible in areas that are less encastellated and less united – which is fairly intuitively obvious – but also in areas that are more urbanized, and where the conquerors and conquered are of different religions, which are two points that may be counter-intuitive to some. If we look at the Norman expansion into Sicily, or the most successful periods of the Iberian Reconquista, or the establishment

Figure 9.4 Bothwell Castle, Scotland. Because of their smaller extent and their nature, it was possible to site castles on difficult heights, or within small river bends. The quantity of resources it would take to give one city powerful fortifications could equally build a dozen strong castles, each of which could withstand a prolonged siege. Thus, conquest was especially difficult and costly in areas where castles were common. [Photograph by Clifford J. Rogers.]

of the Crusader States, we are looking at areas where all these factors come together: where the targeted area is urbanized, politically divided, relatively lacking in castles, and filled with people who can be enslaved. Other major conquests saw at least several of these factors, though not all of them, present: Anglo-Saxon England in 1066 was relatively urbanized, had almost no castles, included a large number of slaves among its inhabitants,[74] and was suffering from internal divisions (as the death of Earl Tostig, fighting on the Norwegian side at Stamford Bridge, testifies), which helps explain why Harold did not attempt a Vegetian defense. The Letts and Estonians whose conquest is described by Henry

74 Perhaps 30% of the entire population, according to Wyatt, *Slaves and Warriors*, 31. See also David A. E. Pelteret, *Slavery in Early Medieval England, from the Reign of Alfred until the Twelfth Century* (Woodbridge: Boydell Press, 1995), 70–74.

of Livonia, though certainly not urbanized, were also lacking in castles, were divided into small, weak polities, and (as pagans) were subject to enslavement.[75]

Although the analysis in this article has rested on and been illustrated by the warfare of Europe during the Middle Ages, most of the basic ideas developed here should also apply to other eras and other regions. The Persian, Macedonian, Roman, and Ottoman empires, for example, were all built up by conquests in fairly urbanized areas with relatively few castles or castle-like fortresses. All these imperial powers made use of the threat of mass enslavement to facilitate their conquests, and all sometimes engaged in that practice and gained large profits by doing so. I hope that other scholars will test my generalizations both by applying them in other contexts, and also with deeper consideration of case studies within the scope of medieval warfare.

75 Henricus Lettus, *The Chronicle of Henry of Livonia*, tr. James A. Brundage (New York: Columbia University Press, 2003); Gillingham, "A Strategy of Total War?" See also Sven Ekhdal, "The Treatment of Prisoners of War during the Fighting between the Teutonic Order and Lithuania," in Malcolm Barber, ed., *The Military Orders: Fighting for the Faith and Caring for the Sick* (Aldershot: Ashgate, 1994), 263–269. The same applies, though not as strongly, to the Slavic areas gradually conquered by the German Empire. There too we see the pattern where after the fall of a stronghold, "all of the adults were killed, while the youths and maidens were led off as slaves," and distributed among the soldiers of the conquering army. E.g. Widukind of Corvey, *Deeds of the Saxons*, ed. and tr. Bernard S. Bachrach and David S. Bachrach (Washington, D.C.: Catholic University of America Press, 2014), 50.

INDEX